Thermodynamics
of the Glassy State

Series in Condensed Matter Physics

Series Editor:
D R Vij
Department of Physics, Kurukshetra University, India

Other titles in the series include:

One and Two Dimensional Fluids: Properties of Smectic, Lamellar and Columnar Liquid Crystals
A Jakli, A Saupe

Theory of Superconductivity: From Weak to Strong Coupling
A S Alexandrov

The Magnetocaloric Effect and its Applications
A M Tishin, Y I Spichkin

Field Theories in Condensed Matter Physics
Sumathi Rao

Nonlinear Dynamics and Chaos in Semiconductors
K Aoki

Permanent Magnetism
R Skomski, J M D Coey

Modern Magnetooptics and Magnetooptical Materials
A K Zvezdin, V A Kotov

Series in Condensed Matter Physics

Thermodynamics of the Glassy State

Luca Leuzzi
INFM - National Research Council (CNR)
Italy

Theo M. Nieuwenhuizen
University of Amsterdam
the Netherlands

CRC Press
Taylor & Francis Group
Boca Raton London New York

CRC Press is an imprint of the
Taylor & Francis Group, an **informa** business

CRC Press
Taylor & Francis Group
6000 Broken Sound Parkway NW, Suite 300
Boca Raton, FL 33487-2742

First issued in paperback 2019

ISBN-13: 978-0-7503-0997-4 (hbk)
ISBN-13: 978-0-367-38841-6 (pbk)

Library of Congress Cataloging-in-Publication Data

Leuzzi, L. (Luca), 1972-
    Thermodynamics of the glassy state / L. Leuzzi, T.M. Niewenhuizen.
        p. cm. -- (Series in condensed matter physics)
    Includes bibliographical references and index.
    ISBN 978-0-7503-0997-4 (hardback : alk. paper)
    1. Spin glasses. 2. Glass. I. Nieuwenhuizen, Theo M. II. Title.

QC176.8.S68L48 2006
530.4'13--dc22   2007025699

Visit the Taylor & Francis Web site at
http://www.taylorandfrancis.com

and the CRC Press Web site at
http://www.crcpress.com

Preface

This book fills a hole in the literature on glassy systems.

In the last fifteen years great progress has been made on our theoretical understanding of structural glasses. This is also due to the use of many ideas and concepts that have been derived in the framework of spin glasses (i.e., amorphous magnets) that are a different kind of amorphous system.

Spin glasses have experienced a very rapid development starting thirty years ago; this was mainly due to the existence of solvable, but interesting models, with infinite range forces, that display rather complex behavior. The need for analyzing a solvable model in all its aspects has pushed theoreticians to forge analytical tools that have been useful in many other fields, among them structural glasses. Later, solvable models also for glassy systems were introduced and studied in great detail.

The injection of these new ideas, that partially formalized old arguments, led to a global rethinking of all the properties of the glassy state, starting from basic thermodynamics properties. However, this new point of view is only presented in original papers and in specialized monographs dedicated to more specific aspects of the glassy states.

This book presents a comprehensive account of the modern theory of glasses starting from the basic principles, i.e., thermodynamics, and from the experimental analysis of some among the most important consequences of thermodynamics (i.e., the Maxwell relations).

Immediately after the Introduction, the book underlines one of the most crucial properties of glasses at low temperature: the existence of two temperatures in these off-equilibrium systems. Thermodynamics must be modified in a deep and nontrivial way in order take care of this new and unexpected phenomenon, that is at the basis of the modified fluctuation dissipation relations that are appropriate for glassy systems.

These concepts are carefully investigated using solvable (or nearly solvable) models in which many different subtle properties can be studied in detail. In this way one can see in an explicit and immediate manner the physical origin of the aging phenomenon that is one of the hallmarks of glassy behavior.

At this point the reader is ready to tackle the approach based on the potential energy landscape, whose features are discussed in general and studied in simple models. Finally, more detailed theories of the glassy states are presented and analyzed where different microscopic mechanisms are discussed also for realistic or quasi-realistic models of glasses.

This book will certainly be extremely useful to anyone who approaches for the first time the study of glasses because it first describes general properties of the glassy states and later shows how these properties are present in specific models: in this way the reader is not lost in a multitude of different models that are used to derive general properties, as often happens in the literature. This book will also be useful to the experienced researcher, who sometimes may overlook the less technical consequences of his or her own work. It is always very stimulating to read a well-done reflection on the basic results in a developing field where the new conceptual points are discussed in a systematic way. I am sure that this book will remain a reference text in the field for a long time.

Giorgio Parisi
Rome, April 2007

Acknowledgements

The origin of this book goes back to 1996, when Giorgio Parisi, in a visit to Amsterdam, drew attention to the paradox concerning the Prigogine-Defay ratio in the traditional thermodynamic description of glass. We thank him for constructive interactions and support throughout the years. Special thanks are also due to Bernard de Jong, for encouragement and advice with the start of the book.

In the course of our research on the subject and the writing of this book, we have benefited from the interaction with many friends, colleagues and collaborators, of whom we mention, in alphabetical order, Armen Allahverdyan, Luca Angelani, Gerardo Aquino, Andrea Baldassarri, Emanuela Bianchi, Desiré Bollé, Carlo Buontempo, Andrea Cavagna, Fabio Cecconi, Claudio Conti, Andrea Crisanti, Leticia Cugliandolo, Silvio Franz, Adan Garriga, Alessandra Gissi, Claude Godrèche, Eric Hennes, John Hertz, Jorge Kurchan, Emilia La Nave, Jean-Marc Luck, Luca Paretti, Maddalena Piazzo, Claudia Pombo, Andrea Puglisi, Felix Ritort, Giancarlo Ruocco, David Saakian, Francesco Sciortino, David Sherrington, Leendert Suttorp, Wim van Saarloos, Gerard Wegdam, Emanuela Zaccarelli.

We also thank the students of the course "Theories and Phenomenology of Structural Glass," held at the Department of Physics of the University of Rome "Sapienza," in the academic years 2005-06, 2006-07 for their critical observations and suggestions.

VIII

To Jurriaan,
Isabela
and Aurora

Acronyms

AG	Adam-Gibbs
CRR	Cooperative rearranging region
DB	Disordered backgammon
FDR	Fluctuation-dissipation ratio
FDT	Fluctuation-dissipation theorem
FEL	Free energy landscape
HNC	Hyper-netted chain
HO	Harmonic oscillators
HOSS	Harmonic oscillators-spherical spins
IID	Identically independently distributed
IS	Inherent structure
LJ	Lennard-Jones
LJBM	Lennard-Jones binary mixture
LW	Lewis-Wahnström
MC	Monte Carlo
MCT	Mode coupling theory
NM	Narayanaswany-Moynihan
OTP	Orthoterphenyl
PEL	Potential energy landscape
PES	Potential energy surface
PVC	Polyvinylchloride
REM	Random energy model
RFOT	Random first order transition
RKKY	Rudermann-Kittel-Kasuya-Yosida
ROM	Random orthogonal model
SCE	Small cage expansion
SPC/E	Simple point charge extended
SSBM	Soft spheres binary mixture
TAP	Thouless-Anderson-Palmer
TTI	Time translation invariant
VF	Vogel-Fulcher

Symbols

α:	slow processes carrying the structural relaxation in the glass
α_T:	coefficient of thermal expansion
α:	localization parameter in the density functional approach to the random first order transition theory
α:	logarithm of the total number of PEL basins
β:	processes occurring on short timescales with respect to glass relaxation times
β:	inverse temperature in units of the Boltzmann constant k_B
χ:	susceptibility, integrated response
G:	Gibbs free enthalpy
G:	response function
G:	shear modulus
K:	fragility index
k_B:	Boltzmann constant
R:	gas constant
S_c:	configurational entropy
T_0:	Vogel-Fulcher temperature
T_d:	dynamic glass transition temperature, crossover temperature
T_e:	effective temperature
T_f:	fictive temperature
T_f:	final temperature (in Kovacs and PEL equilibrium matching protocols)
T_g:	glass transition temperature, glass temperature
T_{is}:	effective temperature in the framework of IS
T_K:	Kauzmann temperature
T_{mc}:	mode-coupling temperature
τ_{eq}:	relaxation time to equilibrium
τ_n:	nucleation time
τ_{obs}, τ_{exp}:	observation or experimental time
t_{erg}:	time after which ergodicity is recovered

Contents

Introduction

Pars Syriae, quae Phoenice vocatur, finitima Iudaeae intra montis Carmeli radices paludem habet, quae vocatur Candebia. ex ea creditur nasci Belus amnis quinque milium passuum spatio in mare perfluens iuxta Ptolemaidem coloniam. lentus hic cursu, insaluber potu, sed caerimoniis sacer, limosus, vado profundus, non nisi refuso mari harenas fatetur; fluctibus enim volutatae nitescunt detritis sordibus.

tunc et marino creduntur adstringi morsu, non prius utiles. quingentorum est passuum non amplius litoris spatium, idque tantum multa per saecula gignendo fuit vitro. fama est adpulsa nave mercatorum nitri, cum sparsi per litus epulas pararent nec esset cortinis attollendis lapidum occasio, glaebas nitri e nave subdidisse, quibus accensis, permixta harena litoris, tralucentes novi liquores fluxisse rivos, et hanc fuisse originem vitri.

Plinius, Historia Naturalis, book XXXVI, 190-191[1]

Thermodynamics is a theory dealing with energy balance, the most fundamental aspect of the universe. It puts constraints on the behavior of physical systems but has by itself virtually no predictive power. It tends to verify whether processes are commensurate with or in violation of this energy balance. Thermodynamics does not deal with time, that being the domain of kinetics, or with rheological parameters such as viscosity. It indicates direc-

---

[1]That part of Syria which is known as Phoenicia and borders on Judea contains a swamp called Candebia amid the lower slopes of Mount Carmel. This is supposed to be the source of the River Belus, which after traversing a distance of 5 miles flows into the sea near the colony of Ptolemais. Its current is sluggish and its waters are unwholesome to drink, although they are regarded as holy for ritual purposes. The river is muddy and flows in a deep channel, revealing its sands only when the tide ebbs. For it is not until they have been tossed by the waves and cleansed of impurities that they glisten. Moreover, it is only at that moment, when they are thought to be affected by the sharp, astringent properties of the brine, that they become fit for use. The beach stretches for not more than half a mile, and yet for many centuries the production of glass depended on this area alone. There is a story that once a ship belonging to some traders in natural soda put in here and that they scattered along the shore to prepare a meal. Since, however, no stones suitable for supporting their cauldrons were forthcoming, they rested them on lumps of soda from their cargo. When these became heated and were completely mingled with the sand on the beach a strange translucent liquid flowed forth in streams; and this, it is said, was the origin of glass [Pliny, 1972].

tions, such as, for instance, the famous dictum of Clausius concerning the flow of heat from high to low temperature, but it does not assert anything more specific, as, for example, the amount of flow or its velocity. It is able to state that 70-million-year-old glassy rocks represent a nonequilibrium state, but it does not explain what prevents this system from relaxing to its crystalline ground state.

At issue in this book is a question that has bothered and still bothers generations of physicists: can these limitations be eliminated and can a thermodynamic theory be developed for a notorious nonequilibrium system, the glassy state, in which time and viscosity are of prime importance? We shall argue how this is, indeed, possible in certain situations, by including one extra parameter, the effective temperature, we will analyze up to which extent this description accurately encodes the main out-of-equilibrium features of the glassy state and we will consider and discuss what lies beyond the boundaries of validity of this representation.

Before going into this, however, we will mention some of the classical notions and observations over which mankind has pondered since the dawn of civilization up until our present time.

Glass, a tool for mankind

Glass is a very present element in everyday life, so much so that it becomes a complicated and long exercise to imagine being without it: no windows, no spectacles, no bottles, no street lamps, no screens for computer, mobile phone and television, no windscreen in the car, no light bulbs, no watches, just to mention a few examples. Indeed, it is a necessary material. Yet, besides its practical use, glass, e.g., in the form of lenses, mirrors, flasks or pipes, has been fundamental in the development of scientific research and visual art, helping and guiding the human sight to look at things in different perspectives and on widely different length-scales, from the microscopic world of cells and bacteria to the open space of planets and stars.[2]

Where and when glass first appeared is not exactly known. Naturally occurring glasses, such as obsidian, were employed in the making of arrowheads, blades and even early mirrors since the neolithic age. Obsidian was produced during volcanic eruptions by the sudden cooling down of silica-rich magma. Its manipulation was very advantageous because it could be easily fractured and the cut edge of the blades could be made very sharp (the cut edge of obsidian is theoretically one molecule thick!).

The place of birth of man-made glass is usually set in the Middle East, possibly in more than one region. Clear evidence has been discovered in Egypt,[3]

---

[2]For a fascinating account of the role of glass in science development, see the book of Macfarlane & Martin [2002].

[3]Glass factories have been discovered in Amarna, dating back to the reign of Akhenaton (1353/51 BC-1336/34 BC), the sun god, and in Qantir-Piramesses, in the Eastern Nile delta,

and in Mesopotamia, but also in the Caucasus. The date is more uncertain: while some archaeologists date the first glass crafts back to 3500 BC, some other studies, based on the detection of glazing traces on ceramics artifacts, go back to 8000 BC. The ancient Roman historian Pliny the Elder (AD 23/24-79) reports in his *Historia Naturalis* how Phoenician merchants incidentally (re)discovered glassmaking (likely ca. 2000 BC). Certainly, glass was used and diffused around 1500 BC, since many discovered glass objects date to that period. From that age on, the knowledge of glass manufacturing began to spread off in Eastern Asia and Europe.

The development of the making of glass (as of any other manufacture) is often nonlinear and not always the incidental discovery of the glass brought a civilization to work out techniques and build glass objects for practical uses in a continuous way. In some cases (in China and Japan, for instance) the knowledge of glassmaking was acquired at some point but later forgotten and learned again or imported in later times.

Next to the *core formed* and the *wound* techniques, *glassblowing* appears around 100 BC, and it will be a key point in the further introduction of new, lighter and transparent glass objects and the general expansion in the use of glass. Such a technique required a very fluid glass former and, therefore, temperatures much higher than in other procedures. This implied good knowledge in the construction of furnaces and ovens.

Under Romans, the diffusion of glass around the Mediterranean Sea and in Central and North Europe grew immensely. Roman glassmakers were able to produce recipients at low cost, affordable for almost everyone. They made, as well, precious refined pieces, symbols of richness and a high social status. Also the making of transparent glass, in the form of lamps, ink pots, and beverage containers was very much worked out. The main incentive to manufacture the latter was to valorize wine, allowing the appreciation of its color, besides its flavor. Windows, as well as lenses, instead, were not diffused under the Roman empire, even though glassworkers had the expertise to make them (actually some instances of glass windows have been discovered in Pompeii and other places in the Italian peninsula). The making of windows spread after the fall of the Western Roman Empire, mainly in Northern Europe, especially in churches. Stained glass windows have been recovered in the french city of Tours (5th century AD) and in England (7th century). A new impulse was given to this kind of production by Benedictine monks (from the foundation of their order in AD 1066) who considered the use of glass in churches as a way to glorify the Lord and, in the north, by Gothic architecture, probably the best examples being preserved in Notre Dame's cathedral in Chartres (AD

---

operational during the time of Ramses the Great in the 13th century BC. The ancient Egyptians manufactured glass jewelry with great care. An instance is the manufacture of Tutankhamen's vulture necklace and collar, a dedicated and intricate craftsmanship, encrusted with blue and red glass. The eyes in his gold mask are made of colored glass inlays, making them similar, in appearance, to lapis lazuli.

1220).

After the fall of the Roman empire the most advanced region of the world for manufacturing glass remained the Middle East, in particular, the territory corresponding to present-day Syria, Egypt, Iraq and Iran, that was initially under the rule of the Sassanid Empire and then, from the 7th century on, under the influence of the Arabs and Islam. The inherited Roman techniques were further refined and the extensive trade of glass objects of Islamic making spread this knowledge very far away to Russia, Eastern Africa and China. Arab glassmakers positively interacted with scientists, providing flasks for chemistry and particular instruments for the studies of optics. These were globes of transparent glass, filled with water, that were able to refract and decompose light and to magnify objects. Ancient Arabs, however, did not invent lenses, even though the technology existed to produce them and, moreover, they would have been very useful in the scientific research developed at that time. The apex of the use of glass was reached between the 13th and the 14th century, after which, in correspondence with the Mongol invasions of the Middle East, glass production collapsed: in AD 1400 the Mongol invader emperor Tamerlane ordered the destruction of all laboratories in Damascus and the deportation of all glassmakers, symbolically sanctioning the end of the brilliant glass manufacture of the Middle East.

The making and use of glass recovered and grew, instead, in Europe, between AD 1100 and AD 1700, to include, besides windows, also beverage glasses, lenses, spectacles, prisms and mirrors. The Republic of Venice was, since the 14th century, the main center for glass manufacturing. Its island of Murano jealously guarded all the most modern secrets, collected in the course of the centuries thanks to the continuous exchange of the merchants with the Middle East and further developed to yield new qualities such as, e.g., crystal glass and multicolored glass. Very skilled glassmakers were active, as well, in Bohemia, Anvers and the Netherlands and other centers in France, Germany and, eventually, in England where the manufacture of glass was industrialized (88 factories were counted in 1696). There, at the end of the 17th century, George Ravenscroft invented *lead glass*: a combination of silica[4] with potash and lead oxide. The effect of potash is to lower the melting temperature of the silica, that is otherwise around 2100 degrees Kelvin. The lead oxide, instead, increases, depending on its concentration, the refractive index of the glass, and thus its luster, with respect to, e.g., the glass made in Venice. The realization of lead glass brought, in the 18th century, the construction of long-range telescopes. It is nowadays still widely used, also because of its property of shielding X-ray radiation.

In other regions of the Eurasian continent, glass diffused as well, although with a different impact on the lives of common people and on the development of technical and cultural innovation. Nevertheless, archaeological discoveries

---

[4]Silica is the common name for silicon dioxide, whose chemical formula is SiO_2. Silica-based glasses are the most used ones and are often referred to as "window" glasses.

witness the use and manufacturing of glass crafts by the major civilizations of Eastern Asia: India, China and Japan. In India, glass objects were in use already around 1400 BC (discovery of Paiyampalli, Tamil Nadu) [Sen & Chaudhary, 1985], mostly consisting of beads and decorative objects. There is evidence that, in the first five centuries AD, they commonly circulated and that the knowledge of glassblowing was acquired. In ancient India, however, the applications did not evolve far beyond: glass was mainly employed to imitate other objects (and for creating false gems), it was considered a surrogate material and did not have a social nor a religious role. Even in alchemy and medicine, the *ravadanis* (alchemical-medical chemists) preferred containers made of compressed earth or clay to glass receptacles [Subbrarayappa, 1999].

In China, the development of glass manufacture was at a level far lower than the one, actually quite advanced, reached in ceramics, metallurgy, printing and weaving. Glassblowing arrived in China about 500 years after its spread in the Middle East but it caused no substantial improvement in production. Glass never substituted Chinese traditional porcelain in the making of receptacles, or greaseproof paper in the fabrication of windows, to give a few typical examples.

In Japan, the role played by glass until the 19th century was almost marginal, apart from certain periods. The most ancient glass artifacts discovered belong to the Yayoi era (ca. 300 BC-AD 300). With the advent of Buddhism (AD 538) the making of glass shrines propagated all over the country. Later, during the Nara era (710-794), the glassblowing technique was acquired by Japanese artisans and laboratories arose in the temples for manufacturing religious ornamental objects. Then a decline came: from the Heian era (794-1185) to the arrival of Jesuit missionaries and Western merchants at the end of the 16th century the glass industry practically disappeared. Portuguese and Dutch navigators brought to Japan lead glass and *crown glass*[5] that were adopted and commonly used, but only in the Meiji restoration (1868-1912) did a real industry of glass objects eventually start [Blair, 1973].

In the rest of the world there is no strong evidence of the presence of man-made glass, but natural rock glasses such as obsidian were extensively employed. In Mesoamerica, for instance, obsidian was used to make flakes for religious offerings and household rituals, in butchery and hunting at the daily life level and, eventually, in war, in the form of swords, projectiles, axes, spears and arrowheads. Its use was common in any period and civilization, by the Mayas (classic period ca. AD 250-900), the Toltecs (ca. AD 900-1100) and the Aztecs that confronted the *conquistador* Cortes with obsidian weapons in 1521.

---

[5]The so-called crown glass is made by twirling glass blobs, or "crowns," at the end of a punty and cutting the glass disk obtained this way in rhomboidal, square or circular plugs. It was introduced at the beginning of the 14th century in the region of Rouen (France) and remained one of the most common processes for making windows until the 19th century.

Contemporary glass

"Window glass," the archetypal glass, is the most familiar and has revolutionized architecture with each new technical innovation: from crown glass windows to the optical good quality sheets of unlimited dimensions made of *float glass* (manufactured by pouring out the glass on molten tin beds, a procedure invented in 1959 by Pilkington Brothers). Other types are just as familiar: optical glasses with varying refractive indices used in lenses for reading, telescopes and cameras, container materials such as bottles and drinking glasses, or television and computer screens. Less well known forms exist such as the preservation of food by bringing it in a glassy state. Currently, communication in our daily lives heavily depends on the astounding technology embedded in silica-based fiber optics with impurity densities reaching the level of a few parts per billion range (milligrams per 1000 kilogram) and with a capability of transmitting a signal over 200 km without amplification.

Window glass, existing in a myriad of types, is silica-based. Its typical chemical composition is about 70% SiO_2, 15% CaO and 15% Na_2O, the last two elements being added to lower the melting point of pure silicon dioxide from about 2100 to 1600-1800 degrees Kelvin. Other oxide-based glasses are manufactured such as borate, phosphate, and germanate glasses. They all have as prime requirement an oxygen coordination of the glass-forming cation of two, three or four, as postulated, more or less *ad hoc*, by Zachariasen [1932].[6]

Numerous oxygen-free glassy compounds also exist. Glasses containing elements of sulfur, selenium or tellurium, i.e., chalcogenide-based glasses, are actively researched for novel properties, as well as halide glasses, the principal one being beryllium fluoride (BeF_2), that is topologically equivalent to cristobalite[7] in its crystalline form. The latter displays an attenuation one thousand times smaller than the one of silica-based optical waveguides, in the optical length-wave range around 1500 nm, making it suitable to transmit a signal without amplification around the whole globe. Unfortunately, the absence of proper drawing technology has prevented its use so far.

Further vitreous materials are the metallic glasses, containing no anions, that find numerous applications, especially in magnetism, because of the lack of grain boundaries. Finally, carbon-based polymer glasses constitute an important part of our daily life going by the name of nylon, polyvinyl chloride (PVC) bottles or wraps.

---

[6]That low coordination, commonly the tetrahedral, fourfold one, is important, can be inferred from the occurrence of *glassy ice*, that, because of hydrogen bonding, occurs as a tetrahedrally coordinated solid with oxygen taking the place of the cation, and that, in one of its crystalline forms, is topologically equivalent to cristobalite, the high temperature SiO_2 polymorph.

[7]Cristobalite is a polymorph that is stable only at very high temperature, above 1650 K, but can exist as a metastable crystal also at lower temperature. The stable crystal phase is, in this case, quartz.

Some inkling of the large number of glassy systems can be found in the books of Donth [2001]; de Jong [2002]; Rao [2002]; Mysen & Richet [2005].

The nonequilibrium nature of glass

From the point of view of physics, all glasses represent an excited state, and may, in due course, relax to the crystalline ground state. Crystallization involves two steps: (i) the *nucleation* of microscopic bubbles of crystal in the liquid phase and (ii) their *growth*, rapidly transforming the whole material into a solid whose structure is an ordered pattern. Nucleation processes tend to be extremely slow in glassy systems on experimental times.

Such observation time may be very long, 70-million-year-old silica-rich volcanic glasses not being uncommon. It is the nature of this structural relaxation which we shall consider in this book, formally encoding it in a generalized thermodynamic framework for glassy systems.

As a typical example consider window glass. Each window glass everywhere in the world is far from equilibrium, a cubic micron of such glass neither being a crystal nor an ordinary undercooled liquid. It is, in some sense, an undercooled liquid that in the glass-forming process has fallen out of its own metastable equilibrium. As mentioned before, the glassy state is inherently a nonequilibrium state: a substance that is glassy in daily life, on a timescale of years, may behave like a liquid on a geological timescale. A thousand-year time lapse movie of a window glass would show a sequence of events akin to the popping of a soap film.

Any liquid cooled down to low enough temperature sufficiently fast will become glassy, i.e., it will lack time to evolve into a long-range ordered crystalline array. Two types of mechanisms conceivably play a role in this quenching process: (i) fast interactions, which happen on a short timescale, characterize relaxation processes that are rapid enough to remain in thermal equilibrium at every step of the cooling (β relaxation processes), and (ii) slow mechanisms, mainly reconstructive transformations involving many molecules, that practically "carry" the structural, off-equilibrium, relaxation (called α processes). In principle, the larger the variety of molecules is, the easier the glass formation becomes. The α processes start lagging behind the thermal state of the system during cooling already in the molten state. Their relaxation time, then, exceeds the time needed to reach equilibrium and they progressively get out of phase with the forcing thermal field, becoming increasingly more decoupled from it.

The viscosity

Relaxation times in glassy systems scale with viscosity, which indicates the resistance to flow of a system and is a measure of its internal friction. The International System unit of viscosity is Pa s = kg/(m s). An older unit is Poise, 1 Poise = 0.1 Pa s. Water of 20 degrees Celsius has viscosity 1 centiPoise. The viscosity of other substances of common use are reported in the following table.

Hydrogen at 20 °C	0.008 6 cP
CO_2 gas at 0 °C	0.015 cP
Air at 18 °C	0.018 2 cP
Water at 20 °C	1.002 cP
Mercury at 20 °C	1.554 cP
Olive oil at 20 °C	84.0 cP
Pancake syrup at 20 °C	2,500 cP
Maple syrup at 25 °C	3,200 cP
Honey at 25 °C	2,000-3,000 cP
Chocolate syrup at 20 °C	25,000 cP
Peanut butter at 20 °C	250,000 cP
Tar or pitch at 20 °C	3×10^{10} cP
Soda glass at 575 °C	1×10^{15} cP
Earth upper mantle	3 to 10×10^{23} cP
Earth lower mantle	2 to 3×10^{25} cP

An undercooled liquid is called glass, when it has a viscosity of 10^{13} Poise, one thousand-million-million times as large. Viscosity has, in general, a strong temperature dependence varying, e.g., in silica-based glassy systems, over fifteen orders of magnitude as shown in Fig. 1.

As a parameter, viscosity fixes the different manufacturing processes in a glass tank and industrial terms like strain point, anneal point, softening and working point are all defined for a specific viscosity. Despite its ubiquitous operational use in industry, it is perhaps the least understood of all glass properties. In silica-based systems the viscosity is dominated by the silica concentration in the system, with high silica percentage having an enormously higher viscosity. This manifests itself in nature in very fluid lava for low-silica-containing fluids, or else as explosive volcanism for high-silica-containing ones, where the high viscosity prevents softer energy release.

The point at which the viscosity, or relaxation time, are so large that equilibrium no longer exists between the thermal state of the glass-forming system and the surrounding heat bath, is called the glass transition temperature, commonly occurring at about two thirds of the melting temperature in silica-based glasses. This transition temperature, with its measurable heat effect, discriminates between a glass and an undercooled liquid.

There are two typical phenomenological behaviors of the viscosity as a function of the temperature, as temperature decreases towards the glass transition, that have been identified so far. The first one is the so-called Arrhenius relaxation law, according to which viscosity (as well as relaxation time) grows exponentially at a low temperature, as $\exp(A/T)$, where A is the activation energy for viscous flow. The second one is the Vogel-Fulcher, or Vogel-Fulcher-Tammann-Hesse, law, expressed as $\exp[B/(T - T_0)]$, that diverges even faster than the Arrhenius one, as it can be expressed defining a temperature-dependent activation energy $A = BT/(T - T_0)$, even diverging

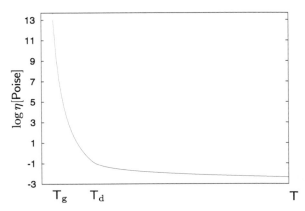

FIGURE 1

The most significant property of a glass-forming liquid is the large increase of the viscosity, η. In a relatively narrow temperature interval it increases from, say, some centiPoise, the value for water at room temperature, to 10^{13} Poise, where the liquid vitrifies. In ordinary life one may think of sugar-syrup. When heated, it flows easily, when put into the fridge, it becomes solid. We show a pictorial representation of the viscosity behavior (in logarithmic scale). Below the temperature indicated by "T_d" the enhanced increase occurs up to the glass temperature "T_g," under which no viscosity can be measured because the material is in the (amorphous) solid state.

at the low but finite temperature T_0. Both laws indicate a very large increase in viscosity or, equivalently, in relaxation time, preventing the material from reaching thermal equilibrium, that, instead, gets stuck in the glassy state. During the last decennia two categories of glasses have been distinguished according to the above-mentioned temperature dependence around the glass transition: the strong glasses and the fragile glasses.[8] The distinction is based on the flow behavior of glasses in the molten state. The materials belonging to the Arrhenius family are designated as strong. They display a very high viscosity above the melting point. For instance, SiO_2 has a viscosity of $2.4\,10^3$ Pa·s about 300 degrees above its melting point (ca. 2100 K). The materials whose viscosity follows the Vogel-Fulcher law are designated as fragile.

The putative flow of window glass

Because of the time frame over which it happens, it is difficult to envision flow of very highly viscous fluids. A case in point is the supposed flow of

[8]We immediately stress that neither the "strong" nor "fragile" property of glass refers to resistance to crashes or heating, rather to the difficulty for macroscopically rearranging its amorphous packing (into another, equivalent, amorphous packing), following an external perturbation. We will come back to the difference between "strong" and "fragile" in Chapters 1 and 3.

windows in cathedrals, the observation being that those glass panes tend to be thicker at the bottom relative to the top, thus being an indicator of flow over time due to gravitational pull. However, a number of instances have been reported where the inverse was the case, the thinner side down, clearly questioning this view. By invoking old flat glass manufacturing technology, either cutting glass cylinders open or making crown glass, it has been argued that flat glass thickness would vary in antiquated technology and that artisans would systematically put the thick side down. The jury is, of course, still out on this one. Firstly, the flow observation in ancient windows presumes that they have never been restored or cleaned and have remained in situ since the 12th century. Secondly, some fragments may flow and others may not, depending on the varying silica concentration used by different manufacturers. This concentration, as we have already mentioned, completely dominates the flow behavior of any silica-based glass. The problem will be addressed more in detail in the next chapter but we anticipate that estimates for flow timescales at environment temperature have been computed, exceeding the age of the universe.

Crizzling: the terminator of medieval glass

Silica (SiO_2) is an acidic oxide, which means that it is only soluble in a basic solution and below a pH of about eight it is virtually insoluble. It is, therefore, not surprising that all strong acids (hydrochloric, sulfuric, nitric, ...) except for hydrofluoric acid, are kept in silica-based glass bottles. Resistance to acid leaching or weathering is strongly linked to the amount of silica in the glass and the type of modifier added to it.

For instance, medieval glass is under a continuous threat not due to flow but due to a phenomenon called crizzling or crisseling, causing every large museum to have a conservator for medieval glass and making transport of ancient glass artifacts for an exhibition in other museums a delicate matter. Crizzling is a chemical instability in glass caused by an imbalance of the chemical components of the glass former (the "batch"), typically silica, soda and lime, when they are not properly mixed and not homogeneously dispersed during melting. This instability of the glass, including the archetypal soda-lime window glass, lowers its defenses against attacks by atmospheric moisture (the glasses absorbing moisture from the air are said hygroscopic). Among the most hygroscopic glasses, one has potassium, rubidium or cesium silicates, and, to a minor extent, sodium silicate. The mixture of sodium with calcium, or lithium, can decrease this absorption.[9]

Particularly harmful is an excess of alkali or a deficiency of stabilizer (usually lime). It has been observed that ancient glass is often wet and, when dried, turns wet again. Alkalis usually present in the aqueous film covering the glass, generate a local basic environment in which silica is very soluble,

---

[9]The latter combination is a manifestation of the mixed alkali effect.

thus producing a network of micro-cracks on the surface. In due time, these cracks grow and may suddenly destroy the glass. This picture is confirmed by the fact that in archaeology relatively little glass is recovered. Crizzling can be slowed or even halted by treating the glass with soluble lithium silicates, but the dissolved silica cannot be replaced: the degeneration of the material cannot be reversed.

The failure of equilibrium thermodynamics for glasses

The term *thermodynamics* was originally coined to describe processes dealing with the flow of heat. Flow processes are notoriously nonequilibrium processes and it is, therefore, surprising that the word thermodynamics implicitly and tacitly got the predicated equilibrium attached to it. Actually, in the older literature, one may still encounter the concise word "thermostatics" for what we now call "equilibrium thermodynamics."

In the 1950s, 1960s, 1970s, equilibrium thermodynamics was believed to describe the glassy state. This may be underlined from the following quote from [Gibbs & Di Marzio, 1958] on page 373: "In any event, we can categorically state that a glass-forming material has equilibrium properties...." This, actually is only partly the case, e.g., vibrational modes are in equilibrium, but configurational collective modes are not. A different, somehow complementary, widespread opinion concerns the inapplicability of many thermodynamic approaches to glasses, depending on the identification of thermodynamics with a theory exclusively concerning equilibrium processes (see, e.g., the recent discussion of Kurchan [2005]). Though similar opinions are diffused in the scientific community, especially among senior scientists, this is clearly a faulty notion because thermodynamics, the dynamics of heat flow, is intrinsically not constrained to equilibrium, which is but an extremely limited case in the total set of thermal flow processes. The success of Josiah Willard Gibbs' (1839-1903) equilibrium thermodynamics has presumably led to this narrowing of the scope of thermodynamics.[10]

---

[10]We mention an anecdote to further illustrate this consensus. During a visit to the Newton Institute in 1997 one of us, Th. M. N., was introduced to the director of the Institute for Applied Physics. When hearing that the visitor's current research area was glasses, the director responded spontaneously by asserting that "Thermodynamics does not work for glasses, because there is no equilibrium." The correct statement of the Cambridge research director and other senior physics peers, from whom we heard the same reaction, should have been that "Equilibrium thermodynamics does not work for glasses, because there is not a good enough equilibrium," a non-surprising and non-embarrassing statement. It is reminiscent of Einstein's opinion "The second law will hold valid, I presume, as long as its premises are fulfilled," both being mere tautologies, that do not answer the central question at stake: within which borders can thermodynamics be formulated? A more recent variant of the same mistake concerns Hawking radiation by black holes. Though not always stated in theoretical considerations, it is a two temperature problem, involving the Hawking temperature of the black hole and the 3 K temperature of the cosmic microwave background. Clearly, this is a nonequilibrium situation, akin to glasses [Nieuwenhuizen, 1998b].

Phase transitions of crystalline materials have commonly been classified as being first order, as a consequence of discontinuities in various thermodynamic functions, or continuous, displaying a growing length-scale around the critical point.[11] In mean-field approximations, i.e., when thermodynamic fluctuations can be neglected, there may occur a continuous transition with discontinuous features. For glasses, a smeared discontinuity in thermodynamic variables such as heat capacity, thermal expansion and compressibility is observed in the vicinity of the glass transition temperature. These discontinuities tend to be damped and smeared out, looking somewhat similar to continuous phase transitions of the mean-field type. The analogy to mean-field type phase transition is not perfect, however, not only because of the smeared out nature of the transition, but also because of the smaller specific heat value recorded below the glass transition temperature.

In first order phase transitions, there holds the well known Clausius-Clapeyron relation between discontinuities of thermodynamic quantities and the slope of the transition line. For continuous phase transitions, there is a pair of similar relations, called Keesom-Ehrenfest relations [Keesom, 1933; Ehrenfest, 1933]. Let us just sketch the confusing situation regarding their application to glasses, as it has existed for decades in literature, postponing a formal discussion to the next chapter. It was investigated experimentally whether the jumps in properties of glass versus liquid still satisfied the two Keesom-Ehrenfest relations for crystalline materials. In a very well-known review, Angell [1995] discusses that one Keesom-Ehrenfest relation, involving a discontinuity in the compressibility is always violated, whereas the other, involving a discontinuity in the specific heat, is commonly but not always satisfied. It has become fashionable to combine these two relations by introducing the so-called Prigogine-Defay ratio. For equilibrium transitions this quantity should be equal to unity. Values below 1 were expected not to be possible, and for glasses typical values are supposed to range between 2 and 5 even though very careful experiments on glassy polystyrene led to a value around 1. The Keesom-Ehrenfest relations are characteristic for equilibrium thermodynamics and their failure in applications to glasses has prevented the construction of an equilibrium thermodynamics for glassy systems. Even so, still in 1981, Di Marzio belabors the issue in a paper entitled "Equilibrium theory of glasses" with a subsection "An equilibrium theory of glasses is absolutely necessary" [Di Marzio, 1981]. In view of the inherent and pronounced

---

In fact, it is a drawback of the success of Gibbsian statistical mechanics, that led nowadays to the too often observed consensus that thermodynamics is an "old fashioned subject" that has no impact on modern research and can be taught shortly or even treated as some side issue in undergraduate courses. This consensus, as we will see, is not well founded as the search for a thermodynamic theory of nonequilibrium systems is a lively field.

[11] A known example is the boiling of water, which exhibits critical opalescence near its critical point at 220 atmospheres and a temperature of 374.2 Celsius, when the correlation length of the liquid molecules becomes micron-sized, comparable with the wavelength of visible light.

nonequilibrium character of the glassy state, we consider such an approach untenable and intrinsically flawed. We shall present in this book a number of instances where a quasi equilibrium fails to describe the physics, such as in its assertion that the original Keesom-Ehrenfest relations are always satisfied for glassy systems, violating experimental observation.

The challenge, then, lies in developing a thermodynamic description valid for systems not close to equilibrium, with very large orders of magnitude variation in the time dependence of their flow properties, ranging from the picosecond regime to, for silic-rich glasses, the age of the solar system, covering a range of twenty five orders of magnitude. We may expect naively that each time window of a couple of orders of magnitude has its own characteristic dynamics, which is approximately independent of the ones above and below it, and that the existing huge nonlinearity could be segmented in quasi-linear fragments. We will see whether this expectation is met, and under which conditions, in the following chapters.

Our book is structured as follows. In the first chapter we will review most of the known basic properties of glasses and glass-forming liquids and we will set a unique notation for quantities and phenomena that will hold in the rest of the book. In the second chapter we progressively introduce a "two-temperature" thermodynamics, starting from experimental observations and theoretical intuitions in the last sixty years, that is a generalization of the equilibrium thermodynamics. We will introduce the idea of "effective temperature" and we will make a survey of its use and misuse in recent literature. Among other things, we will show that, contrary to common belief, the "mechanical" Keesom-Ehrenfest relation is automatically satisfied if properly interpreted in the effective temperature framework and we will show how the "calorimetric" Keesom-Ehrenfest relation, the one dealing with the heat capacity, is modified because of the generalization of a given Maxwell relation to reflect the lack of equilibrium. In Chapters 3, 4 and 5 we present and study some simplified models, holding all the characteristic of glass formers, yet being analytically approachable, with the aim of exemplifying and discussing the properties, possibilities and limits of the proposed thermodynamics for the glassy state. Chapter 6 is dedicated to the potential energy landscape approach, widely used in numerical simulations of computer glass models, where the concept of effective temperature has been thoroughly investigated. In the last chapter we dedicate some space both to well established theories that we often recall in the book, such as the mode-coupling theory for undercooled liquids, or the replica theory for mean-field glasses, both when quenched disordered interactions are explicitly introduced and when disorder is self-induced, and to recent theories, such as the avoided critical point theory and the random first order transition theory.

1

Theory and phenomenology of glasses: a short review

Who when he first saw the sand and ashes by a casual intenseness of heat melted into a metalline form, rugged with exercise and clouded with impurities, would have imagined that in this shapeless lump lay concealed so many conveniences of life as would, in time, constitute a great part of the happiness of the world..... This was the first artificier in glass employed, though without his knowledge or expectation. He was facilitating and prolonging the enjoyment of light, enlarging the avenues of science, and conferring the highest and most lasting pleasures; he was enabling the student to contemplate nature, and the beauty to behold herself.

Dr. Samuel Johnson (1750)[1]

In this first chapter, we present a general introduction to the items that are at the basis of the book. We will give a fundamental description of materials and phenomena, referring to different approaches in literature, both experimental and theoretical, trying to link them among themselves and with the further aim of setting a non-ambiguous notation valid throughout the text. We will mainly concentrate on the aspects that will be discussed in the following chapters in order to provide a sort of glossary that can be consulted at any moment, starting from the very definition of glass.

1.1 Processes, timescales and transitions in glass-formers

A glass can be viewed as a liquid in which a huge slowing down of the diffusive motion of the particles has destroyed its ability to flow on experimental timescales. The slowing down is expressed through the relaxation time, that

---

[1] From McGrath & Frost [1937].

is, generally speaking, the characteristic time at which the slowest measurable processes relax to equilibrium.

Cooling down from the liquid phase, the slow degrees of freedom of the glass former are no longer accessible and the viscosity of the undercooled melt grows several orders of magnitude in a relatively small temperature interval. As a result, in the cooling process, from some point on, the time effectively spent at a certain temperature is not enough to attain equilibrium: the system is said to have fallen out of equilibrium.

The preparation, indeed, plays a fundamental role to get a glass out of a liquid, thus avoiding the crystallization of the substance. Depending on the material, the ways of obtaining a glass are very diverse and consist not only in the cooling of a liquid but also include compression, intense grinding or irradiation of crystals with heavy particles, decompression of crystals that are stable at high pressure, chemical reactions, polymerization, evaporation of solvents, drying, deposition of chemical vapors, etc. Many kinds of materials present a glass phase at a given external condition if prepared in the proper way. We already named silica, halide and chalkline based glasses in the Introduction, as well as carbon-based polymer glasses, e.g., polyvinylchloride (PVC). Some others will appear in the following, such as germanate dioxide (GeO_2), orthoterphenyl (OTP), $K^+Ca^{2+}NO_3^-$ and open network liquids. For an exhaustive literature the reader can refer to [Angell *et al.*, 2000; Debendetti & Stillinger, 2001; de Jong, 2002; Rao, 2002; Mysen & Richet, 2005], limiting ourselves to the current century.

Without going into details on the huge specificity of glassy compounds, a good glass former can be defined as a system in which noncrystalline packing modes of the molecules are intrinsically at low energy and different modes are separated by high energy barriers. Apart from rare, often explicitly constructed exceptions, the crystal state is always at lower energy, but the probability of germinating a crystal instead of a glass during the vitrification process is negligible when cooling fast enough: the nucleation of the crystal phase is practically inhibited. In a nucleation event a small but critical number of unit cells of the stable crystal state combine on a given characteristic timescale, the *nucleation time* τ_{nuc}. In a good glass former, the number of molecules involved in the nucleation must be much larger than the number of molecules cooperating in the structural relaxation of the glass phase (composing what is called a cooperative rearranging region - CRR), yielding, in this way, a nucleation time much longer than the *structural relaxation time*, τ_{eq}. A large nucleation time means that the probability that a fluctuation takes place, allowing a critical number of unit cells to form a crystal, is low. In certain (computer) binary solutions [Kob & Andersen, 1994, 1995a,b; Hansen & Yip, 1995; Parisi, 1997b], at low temperature, the amorphous state is even thermodynamically preferred to any crystalline structure. The same appears to occur for the atactic vinyl polymers, whose lowest energy conformations cannot pack in a regular structure [Gibbs & Di Marzio, 1958].

Many processes are involved at the glass transition, or better *around* it, since the transition region depends on the way it is reached in an experiment, and the timescales of the processes play an essential role for the properties and the behavior of the glass former. The literature about glass is immense and manifold and so is the notation. For what concerns the characteristic times and the processes involved in different temperature domains, the list of names is rather long, sometimes redundant or even confusing. For the sake of clarity, we will initially define all the terms used in this book, trying to be as precise, schematic and faithful as possible.

Imagine following a liquid glass former during a cooling procedure, starting from a high temperature (look at Fig. 1.1 as a guide, starting from the left side). Already in the warm liquid, different processes occur on different timescales. At a given temperature, that we will simply denote as T_{cage}, the thermal movement of particles is slow enough for the diffusion to be hindered by the formation of *cages*. A cage is, in this case, a dynamic concept relative to each particle in the liquid, whose motion is constrained to occur next to other particles around it, with which it collides like in a three dimensional (floating) pinball machine. This is different than the purely collisional motion taking place in the warm liquid. We will refer to the timescale of such an inside-cage process as the *rattling time*. Cooling further, a first bifurcation of timescales takes place between the relative fast rattling time and the relaxation time for the system. The relaxation time is, here, the characteristic timescale of the process of diffusion from the cage, that becomes longer and longer as the temperature decreases. This process is named in many different ways in literature. Staying close to the notation of Donth [2001], we choose the name $\alpha\beta$. The reason will become clear in a short while.

In summary, the relaxation time in this temperature region is the characteristic time needed to have one long distance diffusion process of a particle while it is rattling with a high frequency among its neighbor particles forming a cage around it.

1.1.1 *Dynamical* glass transition

Cooling further, in the so called *crossover region* (always refer to Fig. 1.1), a second bifurcation of timescales takes place between processes involving a global rearrangement of the system, thanks to the large cooperativeness of the particles (we will call them α processes), and processes that involve only a limited number of molecules in a local, microscopically small, rearrangement, thus not contributing to the structural relaxation of the glass former. The latter are usually called β processes.[2] We reserve the label α for the slow-

---

[2]In many amorphous materials, besides the relatively fast "rattling-in-the-cage" processes, sometimes called β_{fast}, other thermalized processes take place on longer timescales. Though slower, these β processes evolve on timescales of some order of magnitude shorter than those of the α processes. An example are the the Johari-Goldstein β processes relaxing according to an Arrhenius law [Johari & Goldstein, 1971]. See also Sec. 6.1.3.

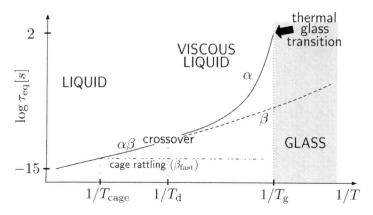

FIGURE 1.1

Arrhenius diagram for the relaxation time τ of a glass former (in seconds) vs. the inverse temperature. From left (high temperature) to right (low temperature) a first bifurcation of the characteristic times of the warm liquid occurs at temperature T_{cage}. Then, at T_{d}, a second bifurcation occurs, more important for the onset of the glass formation, in the dynamic crossover region where the *dynamic glass transition* takes place. Eventually, at T_{g}, the material freezes and becomes a glass, since the structural relaxation time becomes longer than the experimental time. At room temperature the structural relaxation is still longer than this, possibly reaching geological timescales.

est processes, needing a huge cooperativeness to occur below the crossover region, stressing that, in general, different molecular mechanisms may be responsible for the lower temperature α processes and the higher temperature $\alpha\beta$ processes. In the crossover region, thus, $\alpha\beta$ processes bifurcate in α processes - whose characteristic timescale is the *structural relaxation time* - and β processes with a much shorter characteristic time: this way, we are in the presence of a *separation of timescales*, that becomes more and more enhanced as temperature is lowered.

The α-β bifurcation, and the relative crossover, corresponds in mean-field theories for glasses to an *arrested state*, i.e., to a transition to a non-ergodic phase. We will come back to this in Chapter 7. Depending on the community of scholars, the temperature at which the crossover (or the mean-field transition) takes place is named either as T_{c} (crossover or critical) [Donth, 2001; Angell, 1995; Bouchaud *et al.*, 1998], as T_A (arrest) [Kirkpatrick & Wolynes, 1987b; Kirkpatrick *et al.*, 1989], as T_{d} (dynamic) [Mézard & Parisi, 1999a, 2000] or as T_{mc} (mode coupling) [Götze, 1984, 1991; Götze & Sjögren, 1992]. We will adopt T_{d} and we will refer to the crossover as *dynamical transition*. We stress that this is just a matter of convention.

What happens at the crossover? In terms of the free energy landscape (FEL) description of the phase space, where metastable states are represented by local minima and stable states by global minima, a two level structure appears: some minima of the free energy are separated by very small barriers and between them β processes take place; groups of those minima are contained in larger basins separated by barriers requiring a much bigger free energy variation to be crossed. To make the system go from a configuration in one of these basins to another configuration in another basin, i.e., to have an α process, a longer time is needed. Indeed, the typical crossing time τ_{eq} is related to the free energy barrier ΔF separating the valley where the system is currently located from the rest of the landscape, $\tau_{eq} \sim \exp \beta \Delta F$. The timescale on which these processes are occurring is, however, at T_d and below (but well above the temperature of the formation of the solid glass) still very short in comparison with the *observation time* (elsewhere called the *experimental time, τ_{exp}*; we will invariably use both terms). To be qualitative, we can say that for silica-based glass-formers this time is much less than $O(10^2)$ seconds. Below T_d the system is, thus, still at thermodynamic equilibrium. The phase is disordered but the number of minima of the free energy increases and some local minima become deeper. The dynamics of the slowest processes ($\alpha\beta$ for $T > T_d$, α for $T < T_d$) displays a huge slowing down, but the temperature is nevertheless high enough for the system to attain equilibrium on experimental timescales.

1.1.2 *Thermal* glass transition

Decreasing further the temperature, an increase in depth occurs of the global and local minima of the thermodynamic potential, corresponding to different stable and metastable states. The barriers between them become higher and higher until some states become unreachable during the experimental time.

And here we come to the actual *glass transition* from a liquid to a solid amorphous phase, or *thermal*, or else *calorimetric*, glass transition, to better differentiate it from the above-mentioned dynamic glass transition. To avoid any kind of confusion due to a sometimes not too mindful conventional notation, however, we immediately stress that this "transition" is not a true thermodynamic phase transition. On the contrary, its origin is strictly kinetic: it is actually another crossover, that takes place when the structural relaxation time of the cold liquid glass former becomes longer than the observation time. The temperature at which this happens is called the *glass temperature T_g* [Kauzmann, 1948]. Its nonuniversal nature cannot be expressed better than by stressing that its value depends on the cooling rate and, more generally, on the preparation protocol.

It is usually observed experimentally that the relaxation evolves from a Debye exponential at $T > T_d$ to a two step process at lower temperature, that is more and more enhanced as $T \rightarrow T_g$. In terms of time correlation functions (e.g., of density), this means that they first decay rather quickly to a plateau and, on a longer timescale, start to decay again towards equilibrium, see Fig. 1.2 for a pictorial example holding for $T < T_d$. In the frequency domain this is expressed by the structure of the loss part of the dielectric susceptibility (or

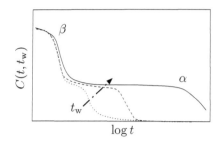 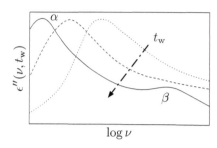

FIGURE 1.2

Typical time correlation function for undercooled liquids. A first decay occurs on the timescale of β relaxation and the decay to equilibrium on the timescale of α processes. At low enough temperature, where time translational invariance is broken, C depends on the value of the waiting time, the plateau length increasing towards longer t as t_w grows.

FIGURE 1.3

Double peak structure of the imaginary part of dielectric susceptibility as a function of the frequency. The peak shifts towards lower frequency as t_w increases, reflecting the increase in relaxation time. As the system is deeply in the aging regime a secondary peak appears, corresponding to Johari-Goldstein β processes.

of any other activity chosen to probe the material in a response experiment). Its structure changes from a single-peaked spectrum to a shape with two peaks, the one at lower frequency being the α peak (cf. Fig. 1.3). This two step/two peaks pattern strongly depends on the time spent since the initial preparation of the sample, the so-called waiting time t_w, before the measurement is performed.

At T_g, the heat capacity decreases in a clear way going from liquid to glass (see Fig. 1.4) and, on reheating, an abrupt but different change shows up as well,[3] as it is shown in Fig. 1.5. Moreover, discontinuities of this kind also occur in the compressibility and the thermal expansivity. This looks similar to a continuous mean-field phase transition, although the analogy is far from perfect, because of the smeared nature of the discontinuities and because the smaller specific heat value occurs below the glass transition, rather than above, as would normally occur in liquid-crystal phase transitions (cf. Fig. 1.4). Though the freezing of modes allows for an analogy with mean-field phase transitions, as will be discussed in more detail in the next chapter, we are not in the presence of a real thermodynamic phase transition.

Nevertheless, the glass state below the glass temperature T_g is often referred to as the "thermodynamic" state of a vitrified substance. It is true that, for not too long observation times and/or well below T_g, parameters practically do not show any time dependence and the amorphous solid seems, therefore,

---

[3]Some generic behavior in the cooling-heating process is analyzed in Sec. 2.6.

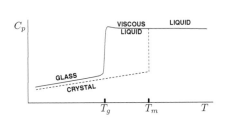

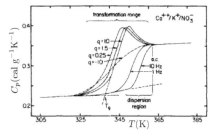

FIGURE 1.4

Typical pseudo-discontinuity at the thermal glass transition in specific heat at constant pressure, typical of the vast majority of glassy materials. The abrupt change occurs around the glass temperature T_g where, in the supercooled glass former (the viscous liquid), the characteristic times of relaxation of the slowest α processes become longer than the observation time: $\tau_{eq} \gg \tau_{exp}$. As a comparison, the melting temperature T_m is also shown, at which the true phase transition to the crystal state takes place if the freezing into an amorphous solid phase can be avoided, e.g., by very slow cooling.

FIGURE 1.5

C_p behavior around T_g, for heating and cooling experiments in liquid $Ca^{++}K^+NO_3^-$ solutions [Moynihan *et al.*, 1976; Torell, 1982; Angell & Torell, 1983] by differential scanning calorimetry. In heating an overshoot takes place, larger as the rate q (in K/min) increases (cf. Sec. 2.6). The glass transition is identified in the "transformation range." Data obtained by two AC measurements (at 1 Hz and 10 Hz) are also plotted: the jump occurs at higher temperature showing that the operative identification of T_g depends on the technique used. Reprinted figure with permission from [Angell & Torell, 1983]. Copyright (1983) by the American Institute of Physics.

to be in a properly defined thermodynamic state. However, even in this case, the glass and the liquid phase cannot be connected by any path in the time independent parameter space, nor can an adiabatically slow state change *ever* connect the liquid phase to the glass phase below T_g. Time will always play a fundamental role in the formation and description of the glass, the fundamental reason simply being that it is not an equilibrium state. Glassy substances that look like a solid on experimental timescales, of seconds or years, may look like a liquid on geological timescales. If every 500 years a snapshot would be taken of a window glass, the movie composed of them would look much like that of a soap bubble (supposing temperature remains stable enough over the 500 years). All of this would be impossible for a crystalline state. The non-static nature of glass is seen the best near the glass transition. To mention one aspect: after bringing a substance slightly into its glass phase by cooling down a glass former fast enough, the simple act of stopping the cooling and waiting long enough, will bring it back in its thermodynamic equilibrium state: the liquid.

The glass temperature T_g rather marks the transition from *ergodic* to (practically) *non-ergodic* behavior. At equilibrium the system configurations are distributed with the Boltzmann-Gibbs distribution. Changing the state of the system, this is ergodic if the configurations that the system can sample during its evolution do not depend on the initial condition, i.e., if it forgets about the change. The whole configuration space will be visited according to the Boltzmann-Gibbs distribution at that temperature and, as a consequence, the ensemble average is equal to the time average. Below T_g, the system degrees of freedom leading the structural relaxation, i.e., the diffusion processes that make the material flow, are frozen. This implies, e.g., that in a given time t_w (the time waited after the quench to the glass state, see also Sec. 1.3), only a limited region of the configuration space, connected to the initial configuration, can be visited in the evolution of the system. The system cannot explore all energetically accessible parts of the configuration space in this time. As a consequence, ensemble average and time average are no longer equivalent, at least on the time windows considered for the glass formation. [4]

We may be allowed to call this separation of explorable subregions of the configurational space *ergodicity breaking* but of the *weak* kind. By weak we specify that the system undergoing the glass transition does not fall into one unique metastable state with a practically infinite lifetime as, for instance, the diamond state. That would be ordinary or "strong" ergodicity breaking. A precise definition of weak ergodicity breaking has been devised in the framework of trap and spin-glass models [Bouchaud, 1994; Cugliandolo & Kurchan, 1995]: a system is in a weak broken ergodicity phase if the time needed to explore an infinite system is infinite. In a long but finite time t_w spent under T_g, the glass is able to go arbitrarily far away from the metastable state in which it initially vitrified, and nevertheless it is unable to visit the whole configuration space. In general, there exists a time t_{erg}, though, beyond which the system can be considered ergodic. The point is that, for most of the glass compounds, t_{erg} is much longer than a human lifetime and it can be considered as infinite, making the glass practically, if not theoretically, non-ergodic. We will reformulate this phenomenon in the next chapter (Sec. 2.8) in terms of correlation functions.

The property displayed by the glass, of being able to go around many, not extremely deep, metastable states, even if stuck out of equilibrium, does not imply that a stable state (crystal) does not exist. The fact that the material does not relax towards it but stays out of equilibrium, spending a huge amount

---

[4]In a magnetic system with Ising symmetry, a phase transition occurs to a phase where the magnetization is either "up" or "down." Broken ergodicity means here that the transition between these two states takes an enormous time, so it does not occur in practice. In glasses a similar phenomenon occurs, though it is much softer, and a process which would take a "too long time" in one experiment, may still be achievable in another setup.

of time exploring the phase space depends on the random initial conditions in which the sample is prepared. The resulting very slow relaxation is called either *annealing*, if it is externally driven, or *aging*, when it is a spontaneous phenomenon. All of Sec. 1.3 is dedicated to the aging phenomenon. The properties of the glass, therefore, strongly depend on its *age* and history, i.e., on the *waiting time* (t_w) elapsed since the preparation of the sample till the beginning of the measurements, on what happened to the system during t_w and on the preparation itself (see Figs. 1.4 and 1.5).

In general, the location of the empirical glass transition temperature T_g depends on the cooling rate, the pressure and the composition, but also on the experimental conventions adopted for its operative determination. Indeed, a convention has to be established to fix the proper cooling and heating rates since the glass transition is kinetic in origin. Moreover, the choice of a given susceptibility to refer to for response measurements is made by convention, since the behavior of the loss parts of different activities (dielectric, shear, thermal or compression) versus mobility is not unique. Finally, a further convention identifies which point of the smeared vitrification step, in specific heat, one has to adopt, because no true cusp, strict discontinuity or divergence, occurs in the curves.

In practice, e.g., in silica-based glasses, the empirical glass temperature T_g is usually determined making use of a very slow cooling rate ($\dot{T} = 10K/\text{min}$) for which in a rather small transition interval, of $O(10K)$, the viscosity increases up to 15 orders of magnitude until it reaches 10^{13} Poise, and the equilibration time is of the order of 10^2 s. It is related to the slowest possible experiment one can realistically carry out.

Around T_g, any observation time sets the timescale between the relatively fast β processes and the practically quenched α processes. In other words, for $T \sim T_g$, the relaxation time of α processes becomes exceedingly long with respect to common experiments, so long that they will never reach equilibrium on laboratory timescales (for $T \ll T_g$ not even on geological timescales [Zanotto, 1998]). Many more local minima of the free energy landscape appear and the deepest local minima, corresponding to metastable states, become separated on the time scale of the experiment. The system has a very slow aging dynamics, proceeding by *activated processes* rather than thermal fluctuations (see later, Sec. 1.4).

1.2 Strong and fragile glass formers: laws for structural relaxation

The relaxation time of a glass-forming liquid depends on temperature and, to a certain approximation, is related to viscosity in a proportional way, following the mechanical Maxwell equation

$$\eta \sim G_g \tau_{\text{eq}} \tag{1.1}$$

where G_g is the infinite frequency shear modulus ($\sim 10^9$-10^{12} Pa).

Certain materials exhibit an Arrhenius behavior of viscosity for temperature above T_g:

$$\eta(T) = \eta_0 \exp \frac{A}{T} \tag{1.2}$$

where η_0 is the limiting high temperature viscosity, of the order of 1 cP for a large number of substances. These show a high resistance to structural changes, usually small jumps of specific heat (with the exception of cases where hydrogen bonds play a major role), their vibrational spectra and radial distribution functions show little reorganization in a wide range of temperature and the potential energy hypersurface (or *landscape*) has few minima and high barriers. They are called *strong* liquids. Examples of strong glass formers are silica, germanate dioxide (GeO_2) and open network liquids such as boron trioxide (B_2O_3).

In other materials, instead, the viscosity temperature dependence presents a large deviation from the Arrhenius law and the viscosity pattern is phenomenologically reproduced by the so called Vogel-Fulcher (VF) law:[5]

$$\eta(T) = \eta_0 \exp \frac{B}{T - T_0} \tag{1.3}$$

where B and T_0 are fit parameters, whose possible physical meaning will be discussed in the following section in connection with configurational properties of glass formers. These materials are referred to as *fragile* liquids. In fragile glass formers the *microscopic* amorphous structure at T_g can be easily made collapsing and, with little thermal excitation, it is able to reorganize itself in structures with different particle orientations and coordination states.

The term fragile (or strong) does not refer to a particular brittleness of the material because of a crush or a fall. Indeed, one usually refers to fragile and strong glass-forming *liquids*, as well. The word rather qualifies the easiness (respective difficulty) of the system to change from a glassy state to another glassy state energetically degenerate. In strong glasses large cooperative rearrangements responsible for this glass-to-glass transformation are more rare than in fragile ones. In terms of the free energy landscape in which the glassy system is evolving at a given temperature, the fragile glass presents many more degenerate minima, separated by sensibly smaller (though large) barriers than the strong glass, see Fig. 1.6. It is the shape of the configurational space that matters in this classification. On the contrary, the structure of

---

[5]A more proper name would be Vogel-Fulcher-Tammann-Hesse law [Vogel, 1921; Fulcher, 1925; Tammann & Hesse, 1926]. In this book, however, we will use the shorter and most used inscription VF.

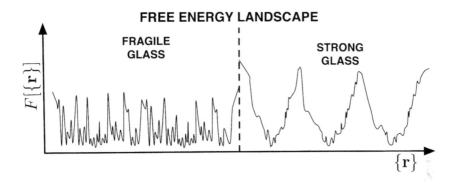

FIGURE 1.6

One dimensional pictures of the free energy landscape F in fragile and strong glass formers. The horizontal axis represents the one dimensional projection of the configurational coordinates of the degrees of freedom $\{r\}$. The crystal global minimum is omitted.

intermolecular forces and the geometrical properties of the amorphous material in real space determining the resistance to cracks do not play any specific direct role in the strong/fragile classification.

In fragile glasses, usually big jumps of specific heat occur and the increase in number of apart valleys of the free energy landscape very much increase in order to account for the large configurational entropic contribution arising in correspondence with the difference in specific heat. As we said already, relatively low free energy barriers, in comparison with those occurring in strong glass landscapes, are present between minima. We will study in Sec. 1.6 how a more quantitative classification can be made for the degree of fragility of a glass former.

Some examples of fragile glass formers are $K^+Ca^{2+}NO_3^-$, $K^+Bi^{3+}Cl^-$, OTP, toluene and chlorobenzene. In general these are liquids characterized by simple nondirectional Coulomb interactions or, else, van der Waals interactions (e.g., for the aromatic substance OTP [Hodge & O'Reilly, 1999; Moynihan & Angell, 2000]). The most fragile substances known are polymeric [Struik, 1978].[6]

The parameter T_0 entering Eq. (1.3) depends on the material and, in practice, even on the range of temperatures in which the fit is performed. At this level, it is not a physical parameter. In general, in quantitative analysis of the relaxation time dependence on temperature, one has to be very precise

---

[6]In specific model cases (binary mixtures) the glassy state can even be lower in energy than the crystalline one and is thermodynamically stable with respect to any crystal configuration, see, e.g., [Kob & Andersen, 1994; Parisi, 1997b] for Lennard-Jones and soft spheres binary mixtures, respectively, or Secs. 2.8, 7.3 and Appendices 6.A.1, 6.A.2.

in identifying physically robust parameters. The region under investigation has to be carefully chosen in order to find a meaningful interpretation for fit parameters and a number of reasonable conventions has to be adopted. In particular, T_0 is often identified with the "Kauzmann transition" temperature that will be discussed in section 1.4.2.

Also a generalized VF law is used in fitting experimental data for the viscosity pattern of glass-forming liquids:

$$\eta(T) = \eta_0 \exp \left(\frac{B}{T - T_0} \right)^\gamma \qquad (1.4)$$

The exponent γ is usually set equal to 1, and an argument for setting $\gamma = 1$ was originally given by Adam & Gibbs [1965]. An alternative explanation for this choice is provided in the framework of the so-called random first order transition (RFOT) theory of Kirkpatrick *et al.* [1989] to which Sec. 7.5 is dedicated. However, these studies do not exclude exponents $\gamma \neq 1$, always compatible with data, merely affecting the width of the fitting interval. On the contrary, analytic approaches [Kirkpatrick & Wolynes, 1987b; Parisi, 1995] yield $\gamma = 2$ in three dimensions. Though, usually, the estimated value of the exponent is $\gamma \geq 1$, glass formers also exist for which Eq. (1.4) fits the viscosity data with $\gamma < 1$. The same broadening can be implemented for strong glass formers, for which a generalized Arrhenius relaxation law, i.e., Eq. (1.4) with $T_0 = 0$, can be used to properly fit the data of undercooled liquids, see, e.g., Bässler [1987]. In one of the exactly solvable glass models that we will treat in Chapter 3, the exponent γ will simply be a model parameter, so that this standard picture can be investigated and qualitatively different dynamic regimes can be associated to different values of the exponent.

In Chapter 7 we will discuss more intrinsic theoretical explanations for the above-mentioned behaviors of viscosity and relaxation time.

1.3 Aging

The *aging* phenomenon is the property of a slowly relaxing system to depend on its history when subjected to a certain class of measurements [Struik, 1978; McKenna, 1989; Barrat *et al.*, 1996]. This is strictly connected with the weak ergodicity breaking taking place in glassy, slow relaxing systems, evolving towards equilibrium structures on timescales longer than the characteristic timescales of the experiments. In other words, aging is a consequence of the fact that glassy systems are out of equilibrium even on macroscopic timescales. The aging regime is, nevertheless, not the most general out-of-equilibrium situation. In this dynamical regime a certain degree of universality in the nonequilibrium behavior can still be identified. In particular, in mean-field systems even if in the aging regime the dynamics of two time observables, such as correlation functions and susceptibilities, is not time translational invariant (TTI), one can see that a kind of *covariance* yet occurs [Bouchaud *et al.*, 1998]. Namely, after a transient, the dynamic evolution of a glassy system of age t_w is described by the same equations characterizing a system whose age is a fraction (or multiple) of t_w up to a rescaling in time ($(t - t_\mathrm{w})/t_\mathrm{w}$ like).[7]

Before reaching the aging regime, when the time t, elapsed since the beginning of the experiment, is much shorter than t_w, we have what is sometimes called a *stationary regime*. It is the earliest regime of the dynamics, in which the system evolves towards the highest metastable state available and, from the point of view of the measured observables, it seems like relaxing to equilibrium. The system, instead, is evolving "rapidly" towards a metastable state. Since $t - t_\mathrm{w} \ll t_\mathrm{w}$, it does not have enough time to leave this state and to start exploring the rest of the configuration space, so it behaves like if it was thermalizing to a ground state. Experiments on glycerol [Grigera & Israeloff, 1999] have shown that already for $t - t_\mathrm{w} \sim 10^{-5} t_\mathrm{w}$ this initial regime is overcome (see also Sec. 2.8).

In the opposite time limit, the aging regime is overcome when the observation time becomes so long that the whole configuration space of the system can be visited according to the equilibrium Boltzmann-Gibbs probability distribution. This very long timescale (geologically long, for what concerns glasses at room temperature) is what we call the ergodic time t_erg.

If we carry out a response experiment, that is if we weakly perturb the glassy system at a "waiting" time t_w with an external field and we measure the response (*activity*) of its conjugated variable at some later time t, we observe that aging takes place. In addition to what is displayed in the case of externally induced, irreversible response function, aging occurs, as well, in the time correlation functions of thermal fluctuations (though in qualitatively different ways). This makes the predictions of the fluctuation-dissipation theorem (FDT) [Kubo, 1985] not reliable any more. One speaks sometimes of "violation" of FDT, even though the theorem has no reason to hold since its hypotheses (based on the presence of local equilibrium) are not fulfilled. Yet, one is free to *define* a fluctuation-dissipation ratio (FDR) between the derivative of correlation function and the response function, as a measure of the distance of the system from equilibrium. In the aging regime taking place in slowly relaxing systems, such a ratio indicates some kind of *effective temperature* (different from the heat-bath temperature at which the experiment is carried out) at which the system evolves *as if it were at equilibrium*. We will more carefully analyze the meaning of this statement in Chapter 2. This possibly holds exclusively on a given time window, defined as the period of time during which the relaxation of fluctuation and response is practically independent from the history, i.e., they are roughly constant. The lower the heat-bath

---

[7]In literature sometimes t denotes the actual age of the system, i.e., $t > t_\mathrm{w}$. We use this convention in the book. Some other times t stands for the time elapsed since the experiment began (at t_w) and corresponds to what we call here $t - t_\mathrm{w}$.

temperature, the longer the time window ("time sector"). Whether and up to which extent this ratio can be considered as a temperature in the proper calorimetric sense, will be the subject of many of the following chapters.

1.3.1   Time sector separation

Let us call t_0 the characteristic time of microscopic fast mechanisms, usually of the order of picoseconds. If, as it seems from the analysis of the aging regime in glasses, there occur universal properties when it is both $t_0 \ll t_w \ll t_{erg}$ and $t_0 \ll t_w + t \ll t_{erg}$ we can assume a *separation of timescales* on which physical processes take place in the very slow relaxation dynamics. In general one can describe the time covariance mentioned above substituting the simple scaling $(t - t_w)/t_w$ with a generic form $h(t)/h(t_w)$ [Cugliandolo & Kurchan, 1994; Bouchaud *et al.*, 1998], where h is a generic function determining the dynamic evolution of out-of-equilibrium quantities inside a given time sector (i.e., for a given timescale, separated from the others). It is the called the *time sector function* and represents a sort of effective age of the out-of-equilibrium system. Indeed, if one expands $h(t) \simeq h(t_w) + (t - t_w)h'(t_w)$ for small $t - t_w$, a scaling $(t - t_w)/\mathcal{T}_w$ is formally recovered, where $\mathcal{T}_w \equiv h(t_w)/h'(t_w)$ is the *effective age*. Out of equilibrium, the time translational invariance does not hold anymore and the two time functions (correlation and response) can display a rich structure for $t, t_w \gg t_0$. If we make the assumption that different processes act on different, well-separated, timescales this would yield multiple scaling structures describing the evolution of two time observables on different time sectors.

   These multi-scaling forms can be described by different time sector functions $h(t)$. The dynamical equations in this approach [Cugliandolo & Kurchan, 1993, 1994; Bouchaud *et al.*, 1998, 1996] become invariant under any monotonous reparametrization $t \to h(t)$. The function $h(t)$ can, in principle, be determined only by matching with the short time solution ($t \sim t_0$) or by numerical simulations of dynamics with two times. Examples for time sector functions and effective ages are:

$$ h(t) \sim \left(\frac{t}{t_0} \right)^{\theta} \qquad \to \qquad \mathcal{T}_w \sim t_w \qquad (1.5) $$

for the spherical p-spin model [Cugliandolo & Kurchan, 1993, 1994] and the trap model [Bouchaud, 1992]. This is also the time sector function behavior that we will find in Chapter 3, for the harmonic oscillator class of models where $\theta = 1/2$ above zero temperature,[8] for the strong glass case or for the fragile glass case above the Kauzmann temperature (see Sec. 1.4.2). In the latter case the aging regime can be probed also below the Kauzmann temperature and there θ is model dependent: in particular, it is inversely proportional to

---

[8]At exactly zero temperature it is $h(t) \sim t/t_0(\log t/t_0)^2$.

the generalized VF exponent γ, cf. Eq. (1.4). For the backgammon model, analyzed in Chapter 4, it is $\theta = 1/2$.

Other, more general h-laws can be devised, as for instance

$$h(t) \sim \exp\left[\frac{1}{1-\mu}\left(\frac{t}{t_0}\right)^{1-\mu}\right] \qquad \rightarrow \qquad \mathcal{T}_{\mathrm{w}} \sim t_{\mathrm{w}}^{\mu} t_0^{1-\mu} \qquad (1.6)$$

introduced by Struik [1978] to account for aging experiments in polymer glasses or

$$h(t) \sim \exp\left[\log(t/t_0)\right]^{\nu} \qquad \rightarrow \qquad \mathcal{T}_{\mathrm{w}} \sim \frac{t_w}{\nu\left[\log(t_{\mathrm{w}}/t_0)\right]^{\nu-1}} \qquad (1.7)$$

(see [Vincent *et al.*, 1997]). The last two cases reduce to the first one when $\mu = 1$ or $\nu = 1$.

For the sake of completeness, we also report here the existence of different kinds of aging dynamics. One has *full aging* when for both $t_{\mathrm{w}}, t \gg t_0$ the system forgets the value of t_0. *Sub-aging* takes place when the effective aging time is smaller than the waiting time, for instance for $\mu < 1$ or $\nu > 1$ in the above examples [Rinn *et al.*, 2000]. Eventually, *super-aging* occurs when the full aging scaling is broken and the value of t_0 is important also for $t_{\mathrm{w}} \to \infty$. The reader can refer to the reviews of Vincent *et al.* [1997] and Bouchaud *et al.* [1998] where the aging covariance has been extensively investigated.

We also mention that, even if the time sector function is a valuable theoretical tool, it is, in practice, very difficult to distinguish between different forms of $h(t)$: in numerical simulations because of possible re-equilibration due to finite fields or to finite size effects, and in real experiments because of the rather hard task of separating the asymptotic behavior of two time observables from sub-asymptotic contributions.

1.4 Configurational entropy and the Kauzmann paradox

Upon cooling a glass-forming liquid, a transition occurs to a glass phase of *smaller* specific heat (caused by the fact that slow modes no longer contribute to it), see Fig. 1.4, and lower entropy and, thus, to a phase of *larger* free energy. Why then is the material in the glass phase?

This happens because the condensed amorphous system has lost entropy by having to select one out of the many equivalent metastable states available. The entropy contribution accounting for such a large number of practically equivalent selectable physical realizations is called *configurational entropy* or complexity.

1.4.1   Kauzmann paradox

In 1948 Kauzmann pointed out the paradox that in some glassy materials the difference between the liquid entropy and the crystal entropy (i.e., the entropy of the most organized state for the system), would extrapolate to zero at temperatures definitely above zero.[9] To circumvent this unphysical result, he proposed a scenario in which the glass was the only experimentally attainable form of supercooled liquid at these temperatures, therefore inhibiting any reasonable, meaningful extrapolation of data relative to the liquid arbitrarily below the glass temperature T_g. He called the temperature below which no distinction between (supercooled) liquid and glass could ever occur, not even for experiments far longer and accurate than the feasible ones, a "pseudo-critical" temperature [Kauzmann, 1948]. We will refer to it as T_K, the *Kauzmann temperature*. Some years later, Gibbs & Di Marzio [1958] proposed, instead, the occurrence of a true thermodynamic phase transition at the temperature T_K where the difference between the entropy of the undercooled liquid and the entropy of the vibrational modes of the crystal that could in principle be formed (the *residual* entropy), is supposed to vanish. This difference is usually referred to as *excess* entropy. Besides the vanishing of excess entropy, such a thermodynamic transition would be characterized by a rigorous discontinuity of the specific heat and by the mathematical divergence of the relaxation time.

If compared with the temperature T_0 at which the viscosity would diverge, according to a fit by means of the VF law, cf. Eq. (1.3), the ratio T_K/T_0 stays in a range between 0.9 and 1.1 for a huge variety of amorphous systems [Angell, 1997]. The supposed divergence of the viscosity at T_0, moreover, has been used as an argument in favor of the occurrence of a thermodynamic transition at that temperature, enforcing the identification $T_0 = T_K$. As already noticed by Kauzmann for boron trioxide, however, in some glasses such as window glass, germanate oxides and others mentioned in Sec. 1.1, the excess entropies usually do not extrapolate to negative values: no Kauzmann transition occurs at finite temperature. Indeed, in the latter case, $T_0 = 0$ and the relaxation time happens to follow the Arrhenius law, Eq. (1.2), instead of the Vogel-Fulcher one. These glass formers are those belonging to the strong glass group, while the VF law is a feature of fragile glass formers.

The excess entropy is somewhat connected with the entropy contribution induced by all the configurational states that the system at low temperature (but above the Kauzmann temperature) can visit, even though it still contains contributions from fast, equilibrated processes occurring also in the solid glass. Already around the dynamic critical temperature T_d the space of states changes qualitatively with respect to the high temperature phase, characterized by only one relevant equilibrium state at any timescale (metastable states

---

[9]Kauzmann also considered other thermodynamic variables, such as the internal energy. For the present discussion, they are less relevant.

can also be there, but they are separated by too small barriers to be of any relevance for hindering the dynamics of the system even at short experimental times). Different local and global minima appear in the thermodynamic potential corresponding to different metastable and stable states, respectively. The configurational entropy is the logarithm of their number.

1.4.2   Static phase transition and Kauzmann temperature

If the supposed identification between excess entropy and configurational entropy holds, the temperature T_K is, then, defined as the point of vanishing configurational entropy; this corresponds to a scenario in which the system is stuck forever in one state, an *ideal glass state*. Often, after the theory of Gibbs & Di Marzio [1958], one assumes that a thermodynamic phase transition occurs at this temperature, although this fact is hardly checkable in experiments. From an experimental point of view, T_K would be the glass transition temperature T_g in an idealized adiabatic cooling procedure, at which a real discontinuity of specific heat would appear and the viscosity would diverge. This prediction is impossible to test experimentally, since the relaxation time is too long to actually perform such an experiment. Nevertheless, it has a mean-field analogue in spin-glass models (i.e., models with *ad hoc* quenched disorder) whose frozen phase is organized in a two level hierarchy of metastable states [Kirkpatrick & Thirumalai, 1987b], very similar to the one conceived for the glass. Moreover, mean-field theories for more realistic glass models that assume the existence of a Kauzmann transition, provide interesting results for the glass former behavior at higher temperature in agreement with the standard knowledge [Mézard & Parisi, 1999b], and allow for an analysis of the glass state below the Kauzmann temperature.[10] Eventually, numerical simulations of widely investigated computer glass models, such as soft spheres and Lennard-Jones binary mixtures, show that they display a nonzero Kauzmann temperature [Coluzzi *et al.*, 1999, 2000a]. We will consider these models in Chapter 6 (see Appendix 6.A) and we will analyze the mean-field theories (both with and without quenched disorder) in Chapter 7.

We notice that, contrary to what happens in ordinary continuous phase transitions, at this candidate phase transition the divergence of the relaxation time would not be algebraic in temperature, but exponential, and no susceptibility diverges at the critical point.

1.4.3   "Classic" versus "modern" configurational entropy

Traditionally, the configurational entropy S_c was defined as the logarithm of the number of different configurations. Indeed, for any molecular gas, the kinetic motion can be easily integrated out, leaving the configurational

---

[10]The existence of T_K also will be implemented in one of the models with facilitated dynamics discussed in Chapter 3.

partition sum as the remaining, hard part. As an example, we may mention the Gibbs & Di Marzio [1958] polymer model, where the configurational sum involves summing over different configurations of polymers of arbitrary length.

In the modern view, that, among others, is reflected in the various models and theories to be discussed later on, configurational entropy is the part of the entropy arising from modes that have a characteristic time larger than the observation time. This distinction automatically arises in mean-field models, such as the p-spin model, in their dynamical regime, as initially devised by Kirkpatrick & Wolynes [1987a], and it was proved to be exportable to glass systems in the mean-field approximation [Monasson, 1995; Mézard & Parisi, 1996; Mézard & Parisi, 1999b; Nieuwenhuizen, 1998a].[11]

With this definition of configurational entropy in mind, we shall continue to discuss historical issues such as the Adam-Gibbs (AG) relation (Sec. 1.5), having stressed that the traditional approach can be tricky and lead to ambiguous, or wrong, prediction. Looking back at the historical situation, say for the Gibbs-Di Marzio model, one can only wonder why, e.g., polymers with just a few units, so small that they have a rapid dynamics, have ever been thought to have any bearing on glassy behavior. Indeed, only the part over large polymers, having an equilibration time at least comparable to the experimental time, should be relevant for slow behavior and, thus, counted in the configurational entropy. The Gibbs-Di Marzio evaluation of the "classical" configurational entropy for a compressible polymer melt [Gibbs, 1956; Gibbs & Di Marzio, 1958; Di Marzio & Gibbs, 1958], indeed, showed up to be inconsistent with the results of numerical simulation of polymer melts, even to the point that the numerical configurational entropy, as opposed to the Gibbs-Di Marzio one, does not vanish at any finite temperature [Wolfgardt *et al.*, 1996].

A related problem comes from the identification of the configurational entropy with the excess entropy $\Delta S = S_{\text{liq}} - S_{\text{cryst}}$, where the temperature behavior of the liquid entropy is extrapolated to temperatures below T_{g}. This relies mainly on two hypotheses: (i) the entropic contribution of the fast vibrational degrees of freedom is equal in the undercooled liquid and in the crystal, so that they cancel out in the difference and that only the contribution of well-separated states remains; (ii) in an undercooled liquid it is possible to separate the vibrational contribution from the configurational one.

The first hypothesis is violated, see, e.g., the experiments by Goldstein [1976]. Different contributions to the excess entropy, besides the configurational one, are indeed, identified, as the system changes from one configurational state (i.e., well-separated from the others, in a deep basin of the thermodynamic potential) to another one energetically degenerate: (a) a contribution due to the change in the degree of anharmonicity, (b) a contribution

---

[11]The distinction between equilibrium entropy, related to fast processes, and configurational entropy, related to slow processes, will be fully incorporated in our later discussion of solvable glassy systems, e.g., in Chapters 3, 4 and 5.

coming from the variation in the number of fast degrees of freedom, and (c) a last contribution related to the change in vibrational frequencies. Eventually one can conclude that the excess entropy, that actually is the only experimentally measurable "configurational" quantity, is only partially related to the configurational entropy, and does not coincide with it.

The second hypothesis corresponds to the assumption that the free energy landscape of the undercooled liquid is very similar to the one of the vitrified solid, even if there are no well-separated metastable states and the vibrational motion mixes with the diffusive motion. What actually happens is that molecules do vibrate but among ever-changing neighbors. In the glass-forming liquid above T_g there is only one global minimum of the free energy and this kind of entropy decomposition is an assumption with no topographic support.[12]

1.4.4 An intrinsically dynamic "state" function

The just-defined configurational entropy is intrinsically a dynamic quantity and depends on the experimental timescale. The total entropy of the system will receive contributions both from the entropy of fast (equilibrated) processes, accounting for all configurations visited by the system in the time it has at its disposal, and from the configurational entropy, accounting for all those configurations hidden because the system has fallen out of equilibrium. As the observation time increases - at fixed thermodynamic parameters - the latter will decrease, since larger regions of the phase space will be reached and will contribute to the entropy of fast processes. What happens in glasses, is that even for extremely long, say "geological" or inconceivable (it is the same for human) times, the configurational entropy still yield a finite contribution, connected to the particular structure of the FEL.

In fragile glasses, as the temperature decreases, S_c tends to zero at some temperature below any experimentally realizable T_g but strictly above zero and apparently independent from the observation time (as far as it is $t_{exp} \ll \tau_{eq}$). This is called the Kauzmann temperature and, when the system is cooled down across T_K, the conjecture is made that it ends up in one *stable* state and cannot move to another one even in an infinite time, $t_{erg} = \infty$. The only contributions to the total entropy will be, thus, given by the configurations belonging to the ergodic component reached by the system at T_K. In other words, in this framework, the system is thought to undergo a thermodynamic phase transition.[13]

---

[12]The potential energy landscape approach of Chapter 6 is, actually, based on this assumption. Even though not supported by direct observation, the analogy between undercooled liquid phase and relative glass phase is a source of many interesting analyses and predictions in numerical simulations of glass models.

[13]The temperature T_K was initially defined as the temperature at which ΔS extrapolated to zero [Kauzmann, 1948], whereas, now, we identify the Kauzmann point as the one at which $S_c(T_K) = 0$ and a phase transition occurs [Di Marzio & Gibbs, 1958]. Even though

The rare fluctuations making the system pass from one glassy metastable state to another one, are the *activated processes*. As the experimental timescale grows in some order of magnitude, the subset of the fastest processes "activated" in the shorter experiment, will simply consist of equilibrium processes. One can argue that the configurational entropy accounts for the slow degrees of freedom, whereas the equilibrium entropy takes into account the fast ones. The boundary between slow and fast and, therefore, between what contributes to equilibrium entropy and what contributes to configurational entropy, is, once again, established by our observation time.

If this picture seems too "dynamic," or slippery, one can otherwise start from mean-field theories where "meta"-stable states are actually eternal and nothing depends on their lifetime (it is infinite and that is it). In these theories, at T_d, *dynamic arrest* takes place [Götze, 1984; Kirkpatrick & Thirumalai, 1987a; Crisanti *et al.*, 1993; Kurchan *et al.*, 1993] and the system gets stuck in the basin of attraction of the metastable state to which its initial condition belongs. The arrest occurs because the barriers between the deepest metastable states are of the order of the size N of the system and, therefore, diverge in the thermodynamic limit. Even though global minima, corresponding to thermodynamic stable states, exist, in the mean-field approximation of model glasses, they cannot be reached by any dynamic evolution because the probability of being attracted by one of the metastable states at higher free energy is practically one. The activated processes are completely absent and we have true ergodicity breaking, unless small systems ($N \ll 10^{23}$) are considered.

In the thermodynamic limit of mean-field models, the configurational entropy is, then, defined as the logarithm of the number of ergodically separated states. That is, it counts the *alternative* possible histories of the system. Such a definition, though operationally requiring many different equivalent experiments, is however very clear from a theoretical point of view and does not depend on any timescale. The fast processes are those responsible for changes inside an ergodic sector (and contribute to the equilibrium entropy), whereas the slow processes are simply stuck. The configurational entropy can be counted with no reference to operational parameters and conventions, it being the entropy determined by the number of states existing in the range of $T \in [T_K < T < T_d]$.

1.5   Adam-Gibbs entropic theory

Below the dynamical crossover region, local thermal motion cannot maintain molecular mobility when the free volume available drastically decreases. This

---

it is often assumed that $\Delta S \neq S_c$ the two situations are conceptually different.

means that cooperation between molecules is necessary to continue evolving towards equilibrium. When the free volume reduces, the motion of a single molecule is strongly influenced by the motion of the others: the rearranging movement of one particle is only feasible if a whole cooperative rearranging region (CRR) of molecules contributes to the motion. In many materials CRRs are of the order of nanometers [Matsuoka, 1992].

Adam & Gibbs [1965] considered CRRs as independent subsystems that can be analyzed by statistical mechanics. At fixed temperature and pressure (we will, therefore, consider enthalpy H, instead of energy in the Gibbs measure), the partition function of all configurations of particles inside a subregion of size n will be given by

$$Z(n) = \sum_{E,V} g(E, V, n) \, e^{-\beta H(E,V)} \tag{1.8}$$

where $\beta = 1/(k_B T)$, $k_B = 1.3806505 \times 10^{-23}$ Joule/Kelvin is the Boltzmann constant and g is the density of configurations of n particles having energy E and volume V. Adam and Gibbs also defined another partition function exclusively counting those configurations of n particles actually allowing the subsystem to undergo a complete rearrangement: $Z_r(n)$. The frequency by which a rearrangement of the whole subregion occurs is, then, proportional to Z_r/Z. Considering the Gibbs free energy $G = -\log Z(n)/\beta$, and its analogue for the free energy of the rearranging configurations, G_r, at fixed temperature, $\mu = G/n$ is the chemical potential.

The probability that a subsystem with n particles makes a cooperative rearrangement at temperature T is, thus,

$$w(n, T) \propto e^{-\beta n \Delta \mu} \tag{1.9}$$

where $\Delta \mu = \mu_r - \mu$. The structural relaxation time of the whole system is proportional to the inverse of the average of the above probability.

Because of the assumed independence of CRRs the total number of particles can be written as

$$N = \sum_{n=1}^{N} n \, \mathcal{N}(n, T) \tag{1.10}$$

where $\mathcal{N}(n, T)$ is the number of subsystems of size n at temperature T. Notice, that small sizes also are considered in the sum, whereas it is unconceivable that the rearrangement of small regions can play any role in the global structural relaxation of the glass former. Hence, a *minimal size for cooperativeness*, $n^\star$, is introduced, and the average of Eq. (1.9) reads

$$\bar{w}(T) = \frac{1}{N} \sum_{n=n^\star}^{N} n \, \mathcal{N}(n, T) \, w(n, T) \tag{1.11}$$

$$\sim \frac{n^\star \mathcal{N}(n^\star, T) e^{-\beta n^\star \Delta \mu}}{N} \sum_{n=n^\star}^{N} \frac{n \, \mathcal{N}(n, T)}{n^\star \mathcal{N}(n^\star, T)} e^{-\beta(n-n^\star)\Delta \mu}$$

At this point, the extra assumption is done that the number of CRRs, $\mathcal{N}$, practically does not depend on its size n.[14]

At low temperature, $\beta\Delta\mu$ is large, implying $1 \gg e^{-\beta\Delta\mu} \gg e^{-2\beta\Delta\mu}$, so that we can keep the first term in the sum in Eq. (1.12), corresponding to CRRs of size $n = n^\star$. This way, the relaxation dynamics of the glass turns out to be dominated by the smallest admissible CRRs, i.e., the average rearrangement probability is

$$\bar{w}(T) \sim \frac{n^\star \mathcal{N}(n^\star, T)}{N} e^{-\beta n^\star \Delta\mu} \tag{1.12}$$

One has, eventually, to estimate such a minimal threshold size. It can be obtained by considering the configurational entropy S_c of the whole system and the number of subsystems containing $n^\star$ particles, $\mathcal{N}(n^\star) = N/n^\star$. The configurational entropy per CRR is $s_c^\star = S_c/\mathcal{N}(n^\star)$, and this leads to[15]

$$n^\star = \frac{N s_c^\star}{S_c} \tag{1.13}$$

so that the relaxation time can be expressed in terms of the configurational entropy as

$$\tau_{\text{eq}} \propto \exp\left\{\frac{C}{T S_c(T)}\right\} \tag{1.14}$$

This is the Adam-Gibbs (AG) relation connecting the dynamical relaxation through CRRs to the configurational entropy S_c. The parameter in Eq. (1.14), $C = N s_c^\star \Delta\mu / k_B$, is proportional to the variation of Gibbs free energy when computed on a sub-ensemble of rearrangeable configurations ("free energy" G_r) rather than on the whole ensemble of configurations (free energy G): it is, thus, an extensive quantity proportional to $G_r - G = n^\star \Delta\mu$.

In practice, the variation of S_c with temperature is often estimated by integrating c_P/T, where c_P is the specific heat at constant pressure. As we have abundantly stressed in Sec. 1.4 this is the excess entropy that contains contributions from fast processes too, thus, leading, in our view, to approximate results, whose reliability must be checked *a posteriori*.

Even though the configurational entropy in Eq. (1.14) is not well defined and strong, even unrealistic, assumptions have been made in the derivation of the AG relation (e.g., the assumption that the volume of the system is irrelevant for the number of feasible CRRs!), the AG prediction has been often confirmed by experiments and the entropy theory is still one of the best recognized theories in glass physics and chemistry. This is related to its

---

[14]We notice that this implies the quite counterintuitive effect that increasing the volume of the system at constant density the number of CRRs does not increase. We, however, stick here to the original derivation of the Adam-Gibbs relation, postponing to a later part of the book (Chapter 7, Sec. 7.5) a critical discussion of the theory.

[15]The number of configurations in a subsystem was taken as equal to 2 by Adam & Gibbs [1965], implying $s_c^\star = k_b \log 2$.

exponential shape, that well describes the very slow relaxation of glasses. We will reconsider the derivation of the AG relation when we present the random first order phase transition theory of Kirkpatrick *et al.* [1989] and its recent developments (see, e.g., [Bouchaud & Biroli, 2004; Lubchenko & Wolynes, 2007]) in Sec. 7.5.

1.5.1 Absence of flow in cathedral glasses

Several scholars have estimated the timescale for flow of glasses. This is easily found to be a geological timescale, or even larger than the age of the Earth. To mention one example, Zanotto [1998] discusses the problem and studies the relaxation times starting from the Maxwell equation Eq. (1.1), $\tau_{eq}(T) \sim \eta(T)/G_g(T)$.

Using a value for the "equilibrium viscosity" of the supercooled melt extrapolated to room temperature, Zanotto arrives at $\tau \sim 10^{32}$ year. But this should be seen as an upper estimate. Together with Gupta [Zanotto & Gupta, 1999] he writes "additional remarks," where he employs arguments from the Adam-Gibbs theory. The viscosity is then

$$\eta = \eta_0 e^{C/(S_c T)} \tag{1.15}$$

and the configurational entropy is

$$S_c(T_f) = \int_{T_0}^{T_f} dT \, \frac{\Delta c_p}{T} \tag{1.16}$$

Here T_f is the *fictive temperature*, introduced by Tool [1946] and formally defined by Narayanaswamy [1971] as discussed in Sec. 2.2. For the present example T_f can be considered as a parameter stating how far from equilibrium is the solid glass with respect to the equilibrium it would have attained if extrapolated from the liquid phase. It has to be compared with the heat-bath temperature T.

It was shown by Richert & Angell [1998] that the variation of specific heat, Δc_p, in the region above T_g is well approximated by B/T, where B is a constant, so that $S_c(T_f) \sim \Delta c_P(T)(T_f/T_0 - 1) \sim B(T_f - T_0)/(T_f T_0)$ and

$$\eta = \eta_0 \left\{ \frac{CT_K}{B} \frac{T_f}{T(T_f - T_0)} \right\} \tag{1.17}$$

For soda-lime-silica plate glass Scherer [1986] reports

$$\eta_0 = 9 \times 10^{-6} \, \text{Pa s}, \qquad Q = \frac{CT_K}{B} = 14\,900 \, \text{K}, \qquad T_0 = 436 \, \text{K} \tag{1.18}$$

Using $T = 300 \, K$, $T_f = 816 \, K$ and $G_g = 30$ GPa, this leads to the estimate $\tau_{eq}(T_f) = 2 \times 10^{23}$ years, though considerably smaller than previous 10^{32} years, but still much longer than 10^{10} years, the age of the universe. In particular this implies that the presumed flow of cathedral glasses does not occur on human timescales, their eventual destruction being due to contingent processes such as crizzling (cf. Introduction).

1.6   Fragility index

We have seen that Arrhenius glass formers are said to be strong, whereas glass formers whose viscosity diverges as a VF law are called fragile. We have also understood that the adjective *fragile* is not referred to the material considered in the real space, but to its representation in the configurational space of the degrees of freedom (see discussion in Sec. 1.2). Now, one can wonder about the possibility of quantitatively describing the degree of fragility of a glass former [Angell, 1985], for instance starting from the fit parameters of the temperature laws expressed by Eq. (1.3) for the relaxation time or for the viscosity (kinetic definitions of fragility).

The kinetic fragility describes how fast the relaxation time increases with decreasing temperature as $T \to T_g$. In strong glasses, τ_{eq} has an Arrhenius-like dependence on T, that is the slowest increase detected in glass formers approaching T_g. Since τ_{eq} is not easily accessible experimentally (and it can turn out to be dependent on the technique adopted to measure it), the fragility is usually experimentally inferred from viscosity measurements. Viscosity is related to the relaxation time by the Maxwell relation, Eq. (1.1), involving the infinite frequency shear modulus G_g as proportionality factor. For temperatures around T_g, however, as we mentioned in Sec. 1.2, G_g strongly depends on the material and its value can vary on about three decades, implying that the relaxation time and the viscosity definitions do not coincide.

A further definition can be determined by means of the temperature dependence of the mass diffusion coefficient. Indeed, according to the Stokes-Einstein relation,

$$D = k_B T / (6\pi r \eta) \qquad (1.19)$$

D/T is inversely proportional to the viscosity. We stress that this relation is meant to be valid if the system is at equilibrium and, therefore, might not hold when the system falls out of equilibrium, as in the present case approaching T_g. Moreover, the hydrodynamic radius r can display a temperature dependence varying from material to material. The mobility and the viscosity fragility can, as well, be different. The reader can refer, among others, to the review work of Ruocco *et al.* [2004] for a comprehensive and up-to-date description of the state of the art of the studies on fragility.

One way of defining a *fragility index* K is, e.g., to rewrite Eq. (1.3) as

$$\eta = \eta_0 \exp \frac{1}{K_\eta (T/T_0 - 1)} \qquad (1.20)$$

The larger K_η is, the more fragile the glass former. This is the *global* kinetic definition arising from the functional dependence of the relaxation on temperature. Starting, instead, from the viscosity at a given value of the temperature, without assuming any functional law, a *local* kinetic definition for the fragility can be devised [Angell, 1991]:

$$K_{\eta,\text{loc}} = \left. \frac{d \log \eta(T)/\eta_0}{dT_\text{g}/T} \right|_{T=T_g} \tag{1.21}$$

where $\eta_0 \sim 10^{-3}$ Pa·s for the vast majority of liquids. This fragility takes values going from $K_{\eta,\text{loc}} \simeq 17$ (strong glasses, Arrhenius viscosity behavior) to, as far as we know, $K_{\eta,\text{loc}} = 160$ (for the tri-phenyl-phosphate). For a rather recent and complete list see [Qin & McKenna, 2006].

Starting from the configurational space, considering the degrees of freedom that freeze in as the thermal glass transition is approached, another way of defining a fragility parameter is in terms of the configurational entropy, cf. the AG relation, Eq. (1.14), as

$$K_\text{c} = \frac{T \, T_K \, S_\text{c}(T)}{T - T_K} \tag{1.22}$$

where T_K is the Kauzmann temperature introduced in Sec. 1.4.1. Actually, this is the behavior assumed for the configurational entropy at low temperature in order to match the AG relation and the VF law and identify the fit parameter T_0 with the temperature at which the excess entropy vanishes.[16] The definition is, therefore, global, based once again on a given functional T dependence. The jump in specific heat (as well the jump of thermal expansivity and compressibility), see Fig. 1.4, is proportional to K_c: the larger the variation, the larger the fragility. A local definition can be, otherwise, expressed as in Martinez & Angell [2001]:

$$K_{c,\text{loc}} = T_\text{g} \left. \frac{d \log S_\text{c}(T)}{dT} \right|_{T=T_g} \tag{1.23}$$

the relationship among the two is: $K_{c,\text{loc}} = K_\text{c}/(T_\text{g}S_\text{c}(T_\text{g})) - 1$.

1.7 Kovacs effect

One memory effect that shows up in a one-time observable is the so-called "Kovacs effect" [Kovacs, 1963], which manifests itself under a specific experimental protocol. Even though more than forty years old this effect is still the subject of many investigations, see, e.g., the recent works of [Bellon *et al.*, 2002; Berthier & Bouchaud, 2002; Buhot, 2003; Bertin *et al.*, 2003; Cugliandolo *et al.*, 2004; Mossa & Sciortino, 2004; Arenzon & Sellitto, 2004; Aquino *et al.*, 2006a,b].

---

[16]$S_\text{c} \sim T - T_K$. In this context configurational and the excess entropy are thought to be the same observable. Cf. Sec. 1.4.3 for critical discussion.

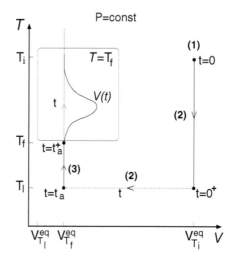

FIGURE 1.7

A pictorial description of the Kovacs protocol. Starting from an equilibrium condition at $T = T_i$ (step 1) at time $t = 0$, the system is quenched to $T = T_l$ and let evolve (step 2). In step 3, the temperature is switched to the final temperature T_f. This is done at the time t_a for which: $V(t_a; T_l) \equiv \bar{V}(T_f)$. In the frame, the typical evolution of the volume $V(t)$ at $T = T_f$, after the temperature switch, is illustrated. Reprinted figure with permission from [Aquino *et al.*, 2006a]. Copyright (2006) by the American Physical Society.

In Chapter 3 we will implement the Kovacs protocol on the harmonic oscillator spherical spin model for fragile glasses that, in spite of its simplicity, captures the phenomenology of the Kovacs effect and allows for a critical discussion on the validity of the definition of effective temperature in the time regimes where memory effects take place in glasses.

The experimental protocol, as originally devised by Kovacs [1963], consists of three main steps, reported in Fig. 1.7.

Step 1 The system is equilibrated at a given high temperature T_i.

Step 2 At time $t = 0$ the system is quenched to a lower temperature T_l, close to or below the glass transition temperature, and it is left to evolve for a time t_a. One then follows the evolution of a given thermodynamic variable. In the original Kovacs experiment this was the volume $V(t)$ of a sample of polyvinyl acetate.

Step 3 After the time t_a, the volume, or other corresponding observable, has reached a value which is, by definition of t_a, equal to the equilibrium value $\bar{V}$ corresponding to an intermediate temperature T_f ($T_l < T_f < T_i$), i.e., such that $V(t_a; T_l)) \equiv \bar{V}(T_f)$. At this time, the bath temperature is switched to the final value T_f. The pressure (or corresponding variable) is kept constant throughout the whole experiment.

Naively one would expect the observable under consideration, after the third step, to remain constant, since it already has reached its equilibrium value at time $t = t_a^+$. The system as a whole, however, has not equilibrated yet and so the observable goes through a non-monotonic evolution before relaxing back to its equilibrium value, showing the characteristic hump plotted in the frame of Fig. 1.7, whose maximum increases with the magnitude of the final jump of temperature $T_f - T_l$ and occurs at a time which decreases with increasing $T_f - T_l$.

2

Two temperature thermodynamics

Puisque tout rétablissement d'équilibre dans le calorique peut être la cause de la production de la puissance motrice, tout rétablissement d'équilibre qui se fera sans production de cette puissance devra être considéré comme une véritable perte: or, pour peu qu'on y réfléchisse, on s'apercevra que tout changement de tempèrature qui n'est pas dû à un changement de volume des corps ne peut être qu'un rétablissement inutile d'équilibre dans le calorique.[1]

Nicolas Léonard Sadi Carnot[2]

Below the temperature T_g, a vitreous liquid becomes solid. Even though its properties seem to be constant in time for any practical use, we have seen that the amorphous structure is actually not in an equilibrium state and we cannot refer to it as a thermodynamic state. Indeed, the glass and the liquid phase cannot be connected by any path in a time independent parameter space: no time independent thermodynamic transformation can *ever* bring a glass former from the liquid phase to the glass phase below T_g. Because of this ever-standing lack of equilibrium, the time will always play a fundamental role in the formation, in the description and in the properties of the glass.

Nonequilibrium thermodynamic theories were worked out in the first half of the last century. They apply to systems close to equilibrium. Typical applications are systems with heat flows, electric currents, and chemical reactions. Key players have been De Donder, who introduced the concept of "rate of reaction" for chemical reactions, his follower Prigogine, who wrote with Defay the book *Chemical Thermodynamics* [Prigogine & Defay, 1954] and received the Nobel prize in chemistry for his contributions to nonequilibrium thermodynamics, and Prigogine's student Mazur, who spent most of his scientific career in Leiden (The Netherlands) and wrote with de Groot (Leiden, Amsterdam) the classical textbook *Nonequilibrium Thermodynamics* [de Groot & Mazur,

---

[1]Since all reestablishment of equilibrium in the caloric can be the cause of production of motive power, any reestablishment of equilibrium occurring without production of such power will be considered as a real loss: or, reflecting a bit on it, one will discern that all changes of temperature that are not due to a change in the volume of the bodies cannot be other than an useless reestablishment of equilibrium in the caloric.

[2]From Carnot [1824].

1962]. The basic assumption in this field is the presence of local thermodynamic equilibrium, and the basic task consists in calculating the entropy
production.[3] For a modern review, we may mention Schmitz [1988].

The nonequilibrium thermodynamics we aim at should apply, instead, to
systems *far* from equilibrium. For example, neither a large nor a small piece of
glass is in local thermodynamic equilibrium; in either cases a very long waiting
time would be needed to reach equilibrium, much longer than the time of the
experiment. Thermodynamics far from equilibrium is yet a mined field, still
waiting for a comprehensive theory that could represent in a thermodynamic
frame the most general aging systems, whose behavior is dominated by nonthermalized processes. Until very recently, some confusion has been caused
by statements implying that thermodynamics had no applicability for glasses
because these are intrinsically out of equilibrium. Actually, this point of view,
that superficially neglects the possibility of nonequilibrium thermodynamics
both near and far from equilibrium, is, in our opinion, arguable.

Thermodynamics started out as a theory on the energy household of steam
engines. The founding paper by Nicolas Léonard Sadi Carnot "Sur la puissance " still stands as a benchmark paper in the field.[4] The theory was born
in the first half of the 19th century as a new way of looking at phenomena
that, in contrast with the Newtonian mechanics approach, was not deterministic nor predictive, and whose goal was to establish the constraints imposed
by nature in the exploitation of its forces, and to control and drive energy
transformations in order to estimate the optimum performance of a thermal
machine. The fact that the theory was later mainly developed at equilibrium
does not mean that the equilibrium hypothesis is the fundamental issue of
thermodynamics. The difficulties met so far in the attempt of using thermodynamics for glasses could be simply related to the unfounded equilibrium
hypothesis.

Many kinds of dynamics can occur in nature. Our aim is to find some
universality in glassy systems, to find at least a subset of variables for which
their dynamics can be encoded in a thermodynamic framework. We should not
want much more than this. The considered systems are, at the basic level, very
intricate and too finely detailed mechanisms cannot expose thermodynamic
behavior. It may very well happen that certain variables display some kind of
thermodynamic behavior, while other variables of the same system do not.[5]

---

[3]In the linear regime, the entropy production is a bilinear functional in generalized thermodynamic forces and the elements of the coupling matrix are called Onsager coefficients.

[4] Indeed, it has even survived the limits of applicability where the target system is small,
but still coupled to a large bath and a large work source [Allahverdyan & Nieuwenhuizen,
2000, 2002a; Scully, 2001]. This field is called "quantum thermodynamics" and has its own
thermodynamic rules.

[5]Likewise, in macroscopic or even mesoscopic systems, certain variables show classical behavior, even though at the basis the system is quantum mechanical. For instance, while
the point of gravity can typically be described classically, to understand the heat capacity
at low temperatures, a quantum description is needed, no matter how large the system is.

This is still a gain, as, in principle, no universality is expected.[6]

Another limitation on our goal is set by thermodynamics itself. The second law, in its Clausius formulation "Heat cannot of itself pass from a colder to a hotter body," is of extreme generality, but it does not quantify the amount of heat that flows in a given system with a given initial state. Likewise, for thermodynamics far from equilibrium we should just try to find some general trends.

Very many decades in time are involved, ranging from the microscopic sub-picosecond regime up to, e.g., for silicate-rich glasses, the age of the solar system, thus covering more than 25 decades. Naively, we expect that processes occurring on a certain time window of a couple of decades could display their own dynamics, basically independent of the dynamics occurring on previous time windows and not influencing the one at later time decades.

Going further, we will address the issue of building a thermodynamics working also for systems out of equilibrium, at least in the long time regime of aging systems, where separation of timescales occurs. In order to do that, we will insert the time dependence of the relaxing observables into effective thermodynamic-like parameters, checking whether or not it is possible to synthesize the system's features into one unique *effective temperature*.

This extra variable is fundamentally a quantity keeping track not simply of the age of the system, but of its whole history, including, e.g., also the cooling rate under which the glass has been formed. In some cases, making use of the effective temperature it is actually possible to connect, in the space of thermodynamic parameters, the liquid and the glass phase like in a standard thermodynamic transformation.

We shall discuss in this and in the following chapters the possibility that, within a yet unknown class of systems, under a fixed dynamical protocol, such as, e.g., cooling at a fixed rate, the glassy state is described by one extra state variable. This relies on having, together with a set of fast processes, that are in instantaneous equilibrium, also a set of slow modes with a much larger characteristic timescale. This timescale can be the age of the system (for isolated aging systems), or else, the inverse of the cooling rate under which the glass has been formed. The slow modes can be so slow that they set out a sort of quasi-equilibrium at some effective temperature T_e, slowly depending on time. In good cases, the same effective temperature describes a variety of different physical phenomena, as if the slow modes carrying the structural relaxation were, indeed, at an equilibrium at that effective temperature. We shall see that in several model systems this effective temperature is a useful extra thermodynamic parameter, at least in the long time regime (cf. Chapters 3, 4, 6). Already in these models, however, we will find limits to the validity of this picture. One can try to go beyond, then, by extending the theory and

---

[6]Universality occurs, for instance, in the problem of turbulence, where all possible dynamical timescales are involved, but there occurs no good separation of timescales. Rather, they should be combined in a renormalization group approach [Falkovich *et al.*, 2001].

involving more extra parameters, e.g., the effective volume or pressure, or, in magnetic systems, an effective magnetic field. The drawback is, however, that, in this case, mechanical (or magnetic) and calorimetric properties get mixed and the physical meaning of the new parameters becomes less clear even from a theoretical point of view (let alone the possible measurement of these parameters!). Moreover, if one ends up needing as many effective parameters as the number of principal observables of the system, then the thermodynamic description loses completely any character of generality, being just a reformulation of the dynamic behavior of each observable (as was the case for the so-called *fictive temperature*, cf. Sec. 2.2).

2.1 Elements of thermodynamics

In order to see how far glass dynamics owns universal properties, such that it can be reshaped into a thermodynamic framework, we first have to recall some basic notions of equilibrium thermodynamics.

2.1.1 First law and second law

The first law of thermodynamics states that the energy change of a system may arise from the heat added to it and the work done on it,

$$dU = đQ + đW \qquad (2.1)$$

We call $đW$ the work done on the system and $đQ$ the heat exchanged from the heat-bath towards the system, using the symbol $đ$ to stress that, generically, they are not expressible as exact differentials. Mathematically, it would generally be impossible that the sum of two non-differentials add up to a differential. Physically, this occurs because of the internal (classical) dynamics of the system, which quickly brings the system to a new equilibrium state.

The second law has many formulations, such as heat goes from high temperatures to low ones, or no work can be extracted from a cyclic change exerted on an equilibrium system. In thermodynamic calculations, it is customary to employ the Clausius inequality

$$đQ \leq TdS \qquad (2.2)$$

For adiabatic changes, the equality sign applies.[7]

---

[7]In quantum thermodynamics the target system is small, so there is no thermodynamic limit. Then each formulation has its own domain of validity [Allahverdyan & Nieuwenhuizen, 2000, 2002a, 2005].

2.1.2   Clausius-Clapeyron relation

Let a standard first order phase transition occur at a transition line $p = p_c(T)$ in the pressure-temperature phase diagram. There, discontinuities appear in the energy (the latent heat) and the volume. These are related to the slope of the transition line via the Clausius-Clapeyron relation. We briefly bring to mind how this comes about.

For adiabatically slow changes the Clausius inequality $đQ \leq TdS$ is an equality and the work shift is $đW = -pdV$. For a normal liquid the first law $dU = đQ + đW$ becomes, then,

$$dU = TdS - pdV \qquad (2.3)$$

One can define the enthalpy $H = U + pV$ and the free enthalpy (or Gibbs free energy)

$$G = H - TS = U + pV - TS \qquad (2.4)$$

whose adiabatic infinitesimal change satisfies the relation

$$dG = -SdT + Vdp \qquad (2.5)$$

This implies that

$$\left.\frac{\partial G}{\partial T}\right|_p = -S, \qquad \left.\frac{\partial G}{\partial p}\right|_T = V \qquad (2.6)$$

The total or co-moving derivative along the transition line is defined as

$$\frac{d}{dT} = \frac{\partial}{\partial T} + \frac{dp_c}{dT}\frac{\partial}{\partial p} \qquad (2.7)$$

One, thus, obtains

$$\frac{dG}{dT} = -S + V\frac{dp_c}{dT} \qquad (2.8)$$

This relation is valid on both sides of the transition line. Let us denote the discontinuity of a macroscopic observable O along the transition line as

$$\Delta O(T, p_c(T)) \equiv O(T^+, p_c(T)) - O(T^-, p_c(T)) \qquad (2.9)$$

and consider Eq. (2.8) across this line. The left-hand side coincides on both sides, in order to keep the free enthalpy continuous, $\Delta G(T, p_c(T)) = 0$, and one finds

$$\Delta S = \Delta V\frac{dp_c}{dT} \qquad (2.10)$$

The entropy can be eliminated using the definition of G, Eq. (2.4), and the fact that $\Delta G = 0$. We end up, this way, with the Clausius-Clapeyron relation

$$\frac{\Delta U + p_c\Delta V}{T} = \Delta V\frac{dp_c}{dT} \qquad (2.11)$$

This relation has been abundantly confirmed in equilibrium systems. In glassy physics, however, it may be violated, since no true equilibrium occurs. We will go further into this point in Sec. 2.4.

2.1.3 Maxwell relation

The strength of thermodynamics lies in certain relations between derivatives, that occur because they derive from a generalized potential. The mathematical consistency relation (Schwarz identity) $\partial^2 G/\partial T \partial p = \partial^2 G/\partial p\,\partial T$ yields the physical relationship known as the Maxwell relation

$$-\frac{\partial S}{\partial p}\bigg|_T = \frac{\partial V}{\partial T}\bigg|_p \qquad (2.12)$$

which relates certain entropy changes to certain volume changes. Though just arising from Eq. 2.4, on a deeper level it goes back to the fact that statistical mechanics is an appropriate tool to describe the physics. The free enthalpy is, indeed, the logarithm of a partition sum, and for, what concerns G, the above Maxwell relation is a mathematical identity.

2.1.4 Keesom-Ehrenfest relations and Prigogine-Defay ratio

In the old days Ehrenfest conjectured that, besides first order phase transitions, there would exist second order phase transitions where second derivatives of the free energy have jumps, third order phase transitions where third derivatives make jumps, etc. Nowadays it is known that the picture is different, namely, besides first order transitions with discontinuities, there exist continuous phase transitions with a diverging correlation length and critical behavior described by scaling power-laws with exponents being identical within a universality class of fairly different physical systems.

The idea of second order phase transitions was based on the analysis of so-called mean-field models, like the Curie-Weiss model for a ferromagnet with long range interactions. For our purposes, this is still a quite useful concept, because, when bringing a substance to the glassy phase, there will occur a smeared discontinuity in the specific heat, the compressibility and the thermal expansivity. They are defined, respectively, as

$$C_p = \frac{\partial(U+pV)}{\partial T}\bigg|_p, \qquad \kappa = -\frac{\partial\log V}{\partial p}\bigg|_T, \qquad \alpha = \frac{\partial\log V}{\partial T}\bigg|_p \qquad (2.13)$$

Even though their jumps are not true discontinuities in reality, the inspection of Ehrenfest was expected to apply to them as well. We shall explain in Sec. 2.4.3 why this is not the case for glasses, where certain nonequilibrium aspects must be taken into account. Let us first reproduce the original result for standard mean-field transitions.

Knowing that, in this case, besides the free enthalpy the volume also is continuous along the transition line, one differentiates the identity $\Delta V(T, p_c(T)) = 0$ with respect to T. This simply yields

$$\Delta\frac{\partial V}{\partial T}\bigg|_p + \frac{\mathrm{d}p_c}{\mathrm{d}T}\,\Delta\frac{\partial V}{\partial p}\bigg|_T = 0 \qquad (2.14)$$

With the above definitions, Eq. (2.13), one gets the so-called first, or *mechanical*, Keesom-Ehrenfest relation [Keesom, 1933; Ehrenfest, 1933; Landau & Lifshitz, 1980]

$$\frac{\Delta \alpha}{\Delta \kappa} = \frac{dp_c}{dT} \tag{2.15}$$

Since the energy also is continuous, one can, otherwise, consider the temperature derivative of $\Delta U(T, p_c(T)) = 0$, to obtain

$$\Delta \frac{\partial U}{\partial T}\Big|_p + \frac{dp_c}{dT} \Delta \frac{\partial U}{\partial p}\Big|_T = 0 \tag{2.16}$$

From the first law (2.1) we have

$$\frac{\partial U}{\partial p}\Big|_T = T\frac{\partial S}{\partial p}\Big|_T - p\frac{\partial V}{\partial p}\Big|_T = -T\frac{\partial V}{\partial T}\Big|_p - p\frac{\partial V}{\partial p}\Big|_T \tag{2.17}$$

In the second equality, the entropy term has been eliminated with the help of the Maxwell relation (2.12).

Now we have collected all elements to express Eq. (2.16) as

$$\Delta \frac{\partial U}{dT}\Big|_p = T\frac{dp_c}{dT}\Delta \frac{\partial V}{\partial T}\Big|_p + p\frac{dp_c}{dT}\Delta \frac{\partial V}{\partial p}\Big|_T \tag{2.18}$$

Inserting the first Ehrenfest relation, Eq. (2.14), this may be written as

$$\Delta \frac{\partial U}{\partial T}\Big|_p + p\Delta \frac{\partial V}{\partial T}\Big|_p = T\frac{dp_c}{dT}\Delta \frac{\partial V}{\partial T}\Big|_p \tag{2.19}$$

and applying the definitions of specific heat and thermal expansivity, Eq. (2.13), this can now be represented as the second, or *calorimetric*, Keesom-Ehrenfest relation

$$\frac{\Delta C_p}{TV \Delta \alpha} = \frac{dp_c}{dT} \tag{2.20}$$

The first Ehrenfest relation is a tautology for second order phase transitions (it is a rewriting of the requirement of continuous volume), while the second one is nontrivial, since it relies on the Maxwell relation expressed by Eq. (2.12). Exactly for this reason, the first Ehrenfest relation will remain to be obeyed in glasses, while the second one may be violated, together with the underlying Maxwell relation. We anticipate that this is against standard belief, according to which the calorimetric relation is considered generally satisfied and the mechanical one is not, see, e.g., the review of Hodge [1994]. We will discuss this key point later on in Sec. 2.4.3.

One can now define the *Prigogine-Defay ratio* involving the variations of specific heat, compressibility and expansivity in a unique expression:

$$\Pi \equiv \frac{\Delta C_p \Delta \kappa}{TV \Delta \alpha^2} \tag{2.21}$$

It is simple to see, from Eqs. (2.15) and (2.20), that in the above-considered equilibrium situation, one simply has

$$\Pi = 1 \qquad (2.22)$$

At equilibrium, the fluctuation-dissipation theorem (FDT) holds (cf. Sec. 2.8), thus allowing the rewriting of Eq. (2.21) as

$$\Pi = \frac{\langle (\Delta S)^2 \rangle \langle (\Delta V)^2 \rangle}{(\langle \Delta S \Delta V \rangle)^2} \qquad (2.23)$$

We may mention that Π was not explicitly introduced in the Prigogine & Defay [1954] treatment of the Keesom-Ehrenfest relations. They did put forward, though, that these relations apply to glasses, with the consequence $\Pi = 1$. To make the experimental findings as clear as possible, since then it became fashionable to consider data for Π, see, e.g., [Davies & Jones, 1953; Gupta & Moynihan, 1976; O'Reilly, 1977; Gupta, 1980; Moynihan & Lesikar, 1981; McKenna, 1989; Richert & Angell, 1998; Hodge, 1994] and the recent books of Donth [2001] and Rao [2002]. Moreover, exploiting Eq. (2.23) or similar expressions, just by applying the Schwarz inequality it was "proved" that $\Pi = 1$ was a lower band edge for the ratio describing the transition between the out-of-equilibrium solid glass and the ergodic liquid. Hereby it was overlooked that, because of lack of equilibrium, on one side the FDT cannot be applied: Eq. (2.23) does not hold for the glass transition.

Typical values of Π were found to lie between 2 and 5 (but rarely also below 1 [O'Reilly, 1977]), leaving a puzzle that remained unsolved till a deeper insight into the definition and applicability of thermodynamics to systems out of equilibrium has been gained. We will dedicate Sec. 2.4 to the topic.

2.2   Fictive temperature

Systems with several temperatures are abundant in nature, we only need to think of our body temperature that exceeds the environment temperature. Nevertheless, the idea to introduce the concept of several temperatures in the field of glasses has met some resistance, and it took a long time before this path was taken up, since its first formulation in literature.

The concept of mapping the nonequilibrium nature of glass relaxation into an extra parameter is an old idea starting, as far as we know, with Tool [1946]. He, then, introduced a *fictive temperature* to account for the nonlinearity of the structural relaxation of given observables across the transition interval (around T_g). In that context, the fictive temperature T_f was a formal object mapping the molecular ordering inside the material in the glass state, that was inhibited with respect to the ordering in the liquid state by the freezing at T_g. The fictive temperature was defined as the temperature at which the glass would have been if the ordering behavior in temperature would have continued below T_g in the same way as it was doing above it. Otherwise said, it was defined as the thermodynamic temperature at which some observed nonequilibrium excess properties, e.g., volume (cf. Sec. 1.7), enthalpy, refractive index, shear viscosity or electrical conductivity, would have been at equilibrium value. It was a way to describe the progress of the structural relaxation. Already in the 1970s, however, (cf., e.g., Moynihan *et al.* [1976]) it was a known fact that measuring the relaxation of different observables for the same glassy state, the related fictive temperatures were not necessarily identical.

In a simple isothermal experiment just below T_g, after a sudden quench at time t', $T_f(t)$ depends on the history of the system for $t > t'$, that is, on nonthermal perturbations. In a more complicated experiment, following a given temperature program $T(t')$ (cooling or heating), the fictive temperature at time $t > t'$, $T_f(t)$, depends both on the temperature history $T(t')$ and on the previous response $T_f(t')$ to that history.

A phenomenological, qualitative formula has been developed to reproduce the structural relaxation, by Narayanaswamy [1971]:

$$T_f(t) = T_f\left[t; T(t'), T_f(t')\right] , \qquad t > t' \qquad (2.24)$$

where $T(t')$ is the reference quantity for the equilibrium fluctuations, while $T_f(t')$ describes the *acceleration* below the glass transition, due to the freezing in of many degrees of freedom of the glass out of equilibrium. Acceleration means the enhancement of the timescales separation [Donth, 2001] and, in particular, of the slowest relaxation time increase extrapolated, below T_g, from the behavior in the region between T_d and T_g (where also a timescale separation between fast and slow processes occurred, see Sec. 1.1.1). We should remark that such a relation should depend on the cooling-heating history of the system, and we believe that, in general, an equation like Eq. (2.24) can only exist within a well-defined class of histories.

Actually, T_f is just a further fit parameter for the relaxation time with respect, e.g., to the two parameter fit of the VF law, Eq. (1.3). Indeed, it allows us to write down the relaxation time dependence on temperature in terms of a parameter that takes into account the out-of-equilibrium relaxation of the glass, the so-called Narayanaswamy-Moynihan (NM) [Narayanaswamy, 1971; Moynihan *et al.*, 1976] equation:

$$\tau_{eq} = \tau_{eq}[T(t), T_f(t)] = K \exp\left[\frac{\Delta h^\star}{R}\left(\frac{x}{T} + \frac{1-x}{T_f}\right)\right] \qquad (2.25)$$

where x and $\Delta h^\star$ are two fit parameters, alternative to, e.g., B and T_0 in the VF law. The constant $R = 8.31447$ J·K^{-1}·mol^{-1} is the gas constant. The fit parameter x depends on the glassy material (it is roughly between 0.1 and 0.5) and $\Delta h^\star$ is the "effective equilibrium activation energy" just above the glass transition at T_g, i.e.,

$$\frac{\Delta h^\star}{N_A} = \left[\frac{d}{d\beta}\left(\lim_{T_f \to T} \log \tau_{eq}\right)\right]_{T=T_g} \qquad (2.26)$$

where $\beta = (k_B T)^{-1}$ and $N_A = 6.022 \cdot 10^{23}$ is the Avogadro number.

To be more quantitative, let us consider a cooling-heating experiment, see Fig. 2.1, looking at a macroscopic observable O of a glassy system. We said that T_f is defined as the temperature at which the equilibrium liquid has a relaxational value of O equal to the one it has in the glassy state at a temperature T_ℓ (lower than T_f if we cool down or higher in case of heating). For a glass former cooled down or heated up across the thermal glass transition we can, thus, write down the equality

$$O(T_\ell) = O_{eq}(T_f(T_\ell;t)) - \int_{T_\ell}^{T_f(T_\ell;t)} dT \left(\frac{\partial O}{\partial T}\right)_g \qquad (2.27)$$

where the subscript "eq" refers to the equilibrium state and "g" to the glass state. Differentiating with respect to the temperature one obtains an expression of the derivative of the fictive parameter in terms of temperature derivatives of the observable under probe:

$$\left.\frac{dT_f}{dT}\right|_{T_\ell} = \left[\frac{dO}{dT} - \left(\frac{\partial O}{\partial T}\right)_g\right]_{T_\ell} \left[\left(\frac{\partial O}{\partial T}\right)_{eq} - \left(\frac{\partial O}{\partial T}\right)_g\right]^{-1}_{T_f} \qquad (2.28)$$

In this approach, the fictive temperature is practically considered as the relaxational part of O in temperature units. Pictorially, see Fig. 2.1, a glass represented by the point A on the cooling curve has a fictive temperature $T_f(A)$ obtained by crossing the straight lines whose slope is the one of the $O(T)$ line deep in the glass phase and the extrapolation of $O(T)$ for the undercooled liquid below T_g.

In terms of the fictive temperature the specific heat of the glassy state phenomenologically reproduces the Tool law [Tool, 1946]

$$C_p(T) = C_p^{glass} + \Delta C_p \frac{dT_f}{dT}, \qquad \Delta C_p = C_p^{liquid} - C_p^{glass} \qquad (2.29)$$

The fictive temperature of the system seen as a time dependent ordering parameter for the description of the thermodynamics of glass has been a long debated issue (see, for instance, Davies & Jones [1953] or, for a more recent report, the work of McKenna [1989] and references therein). To go beyond the purely phenomenological fictive temperature to a more microscopic concept that can, in addition, be partitioned in the time (or frequency) domain (see Sec. 1.3.1 for a discussion about time sectors in the aging regime), one possible way is to identify a connection between structural relaxation and the violation of the FDT far from equilibrium (cf. Sec. 2.8), or with some temperature-like parameter conjugated to the configurational entropy in a thermodynamic

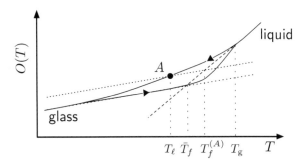

FIGURE 2.1
Pictorial representation of the temperature behavior of an observable O, such as
the internal energy, of a glass former through the glass transition on cooling (upper
curve) and heating (lower curve). The dashed line represents the extrapolation to
low temperature of the relaxation values of O in the liquid phase. $T_f^{(A)}$ is the fictive
temperature relative to the relaxation of O at the point A in the cooling. $\bar{T}_f$ is the
limiting value for the fictive temperature as the glass is cooled down at a given rate.

description (cf. Sec. 2.4). Clearly, in any generalized thermodynamic a
necessary, though not sufficient, requisite will be that all definition of effective
temperature on a given dynamic regime will have to match. This is not
the case for fictive temperature discussed in the present section. We will
tackle in the following sections (and in the following chapters) the problem
of identifying reliable definitions of thermodynamic parameters for the glassy
state and devising measurement procedures.

2.3 Two temperature thermodynamics for a system with separated timescales in contact with two heat baths

Having to deal with a two temperature situation, it is instructive to first consider some aspects of it in an idealized standard setup composed by a subset
at equilibrium in a heat-bath at one temperature and by a remaining part
at equilibrium at a second temperature. Generally, when two independent
thermodynamic systems coupled to two different heat-baths are considered
together, the adiabatic heat exchanges will add up,

$$\dlap dQ = \dlap dQ_1 + \dlap dQ_2 = T_1 dS_1 + T_2 dS_2 \tag{2.30}$$

This relation will appear to carry over to the glassy state, see Eq. (2.45).

Let us consider a system with two subsets of variables: $\{x_1\}$, at equilibrium in a thermal bath at temperature T_1, and $\{x_2\}$, at equilibrium in a

bath at temperature T_2, whose generic Hamiltonian is $\mathcal{H}_0(\{x_2\}, \{x_1\})$. Following Allahverdyan & Nieuwenhuizen [2002b] we now consider the statistical mechanics of such a system. Our starting point will be the case where two external fields $H_{1,2}$ linearly couple to its observables $M_{1,2}$. The Hamiltonian reads:

$$\mathcal{H}(\{x_2\}, \{x_1\}; H_2, H_1) = \mathcal{H}_0(\{x_2\}, \{x_1\}) - H_2 M_2(\{x_2\}) - H_1 M_1(\{x_1\}) \quad (2.31)$$

Before computing thermodynamic functions, though, we introduce the time, in the form of characteristic timescales over which the two apart sets of variables evolve. We will call $\tau_{1,2}$ the typical timescale of the variables $\{x_{1,2}\}$ and we will assume that $\tau_2 \gg \tau_1$. We are, therefore, imposing that the variables 1 are "fast," i.e., they equilibrate in a short time with respect to the relaxation time of the variables 2, the "slow" variables. The behavior of the fast variables $\{x_1\}$ at any time depends on the value of $\{x_2\}$. The values of the slow variables stay constant on the characteristic time of $\{x_1\}$ dynamics: with respect to the motion of the subset of variables 1, the subset 2 is *quenched*. On the contrary, with respect to the motion of $\{x_2\}$, the $\{x_1\}$ variables just create a noise that can be integrated over in the partition function. The partition sum over the fast variables reads, indeed,

$$Z_1(\{x_2\}) = e^{\beta_1 H_2 M_2(\{x_2\})} \int \mathcal{D}x_1 \; e^{-\beta_1 \mathcal{H}_0(\{x_2, x_1\}) + \beta_1 H_1 M_1(\{x_1\})} \quad (2.32)$$

while the total one, over all variables is

$$Z_2 = \int dx_2 e^{-\beta_2 F_1(\{x_2\})} = \int dx_2 \left(\int dx_1 e^{-\beta_1 \mathcal{H}(\{x_2, x_1\}; H_2, H_1)} \right)^{\beta_2/\beta_1} \quad (2.33)$$

$$= \int dx_2 e^{\beta_2 H_2 M_2(\{x_2\})} \left(\int dx_1 e^{-\beta_1 \mathcal{H}_0(\{x_2, x_1\}) + \beta_1 H_1 M_1(\{x_1\})} \right)^{\beta_2/\beta_1}$$

where $\beta_{1,2} = 1/T_{1,2}$ (the Boltzmann constant is set equal to one). We shall define the related free energies by

$$Z_1(\{x_2\}) = e^{-\beta_1 F_1(\{x_2\})}, \qquad Z_2 = e^{-\beta_2 F} \quad (2.34)$$

The magnetizations take their expected forms,

$$\langle M_1 \rangle = -\frac{\partial F}{\partial H_1} = \frac{1}{Z_2} \int \mathcal{D}x_2 \; e^{-\beta_2 F_1} \frac{1}{Z_1(\{x_2\})} \int \mathcal{D}x_1 \; M_1 \; e^{-\beta_1 \mathcal{H}} = \langle\langle M_1 \rangle_1 \rangle_2$$

$$(2.35)$$

$$\langle M_2 \rangle = -\frac{\partial F}{\partial H_2} = \frac{1}{Z_2} \int \mathcal{D}x_2 \; M_2 \; e^{-\beta_2 F_1} = \langle M_2 \rangle_2 \quad (2.36)$$

where $\langle \cdots \rangle_1$ stands for the $\{x_2\}$ dependent average over the fast processes at their temperature T_1, and $\langle \cdots \rangle_2$ for the average over the slow processes at temperature T_2.[8]

The susceptibilities read simplest in terms of the fluctuations

$$\delta M_2 \equiv M_2 - \langle M_2 \rangle_2, \quad \delta M_1 \equiv M_1 - \langle M_1 \rangle_1, \quad \delta \langle M_1 \rangle_1 \equiv \langle M_1 \rangle_1 - \langle \langle M_1 \rangle_1 \rangle_2 \tag{2.37}$$

Some straightforward steps lead to

$$\frac{\partial \langle M_2 \rangle}{\partial H_2} = \beta_2 \langle (\delta M_2)^2 \rangle_2 \tag{2.38}$$

$$\frac{\partial \langle M_2 \rangle}{\partial H_1} = \frac{\partial \langle M_1 \rangle}{\partial H_2} = \beta_2 \langle \delta M_2 \delta \langle M_1 \rangle_1 \rangle_2 \tag{2.39}$$

$$\frac{\partial \langle M_1 \rangle}{\partial H_1} = \beta_2 \langle (\delta \langle M_1 \rangle_1)^2 \rangle_2 + \beta_1 \langle \langle (\delta M_1)^2 \rangle_1 \rangle_2 \tag{2.40}$$

If $H_2 = 0$, $M_1 = M$ and $H_1 = H$, or, equivalently, choosing $M = M_2 + M_1$ and $H_1 = H_2 = H$, the above relations add up to

$$\frac{\partial \langle M \rangle}{\partial H} = \frac{\langle (\delta \langle M \rangle_1)^2 \rangle_2}{T_2} + \frac{\langle \langle (\delta M)^2 \rangle_1 \rangle_2}{T_1} \tag{2.41}$$

We stress that $\langle (\delta \langle \ldots \rangle_1)^2 \rangle_2$ is the contribution to fluctuations of the slow variables ($\rightarrow \langle (\delta \ldots)^2 \rangle_{\text{slow}}$), whereas $\langle \langle (\delta \ldots)^2 \rangle_1 \rangle_2$ is the contribution to the fluctuations imputable to the fast variables ($\rightarrow \langle (\delta \ldots)^2 \rangle_{\text{fast}}$).

The fluctuation formula Eq. (2.41) will appear to be applicable to glasses that can be described by one extra parameter, the effective temperature (cf. Sec. 2.7).

For systems with more than two well-separated timescales, the presentation can be generalized directly. For infinitely many well separated timescales, the partition sum will be infinitely nested and it will, then, coincide with the replicated partition sum, as it is known in the theory of spin glasses [Mézard *et al.*, 1987; van Mourik & Coolen, 2001].

2.3.1 Two temperature thermodynamics for glassy systems

Let us now consider a glass. The temperature at which the fast variables are thermalized, T_1, will be the temperature T of the environment (or of the experimental set-up) at which all β (fast) processes are in equilibrium on timescales much smaller than the observation time ($\tau_1 \ll t_{\text{obs}}$). According to the picture just exposed, we introduce a second, *effective*, thermal bath at

---

[8]Let us make a remark about the Bayes formula. It derives the conditional probability for the process x_1, given the value of the process x_2, from their joint probability, viz. $P(x_1|x_2) = P(x_1, x_2)/P(x_2)$. When the timescale τ_2 of the processes x_2 equals τ_1, the one of the x_1, this relation makes sense, and it continues to make sense when $\tau_2 \gg \tau_1$. But, in the latter situation the definition $P(x_2|x_1) = P(x_1, x_2)/P(x_1)$ is devoid of any physical meaning, as the time τ_1 is too short to make a sensible statement about the long time process x_2.

temperature $T_2 = T_e$ in which the slow, α, modes behave *as if they were* at equilibrium. They are actually far from equilibrium at the room temperature T, but the fact that their relaxation time is $\tau_2 \gg t_{\text{obs}}$ can induce us to consider them as stuck in a quasi-equilibrium situation at a different temperature.

This is a very strong assignment and we will see that, actually, life is more complicated than this. However, a relevant contribution to the analysis of the glass behavior comes even from this simplification. Eq. (2.41), for instance, becomes now the contributions of both fast and slow fluctuations to the susceptibility (cf. Sec. 2.7)

$$\chi = \frac{\partial \langle M \rangle}{\partial H} = \frac{\langle (\delta \langle M \rangle)^2 \rangle_{\text{slow}}}{T_e} + \frac{\langle (\delta M)^2 \rangle_{\text{fast}}}{T} \qquad (2.42)$$

In more complicated situations, one might even need two effective parameters, an effective temperature and an effective field. For instance, one can define $H_1 = H$ and the effective field $H_2 = H_e(H, T)$ [or $H_2 = 0$ and introduce the effective fields in $H_1 = H + H_e(H, T)$]. In general, the susceptibility expressions will depend on the choice made and will be more complicated than Eq. (2.41), involving terms proportional to $\partial H_e / \partial H$. We will see an alternative, self-consistent, introduction of an effective field in the specific case of the harmonic oscillator spherical spin model presented in Chapter 3.

Still another case is where $M_2 = 0$, $M_1 = M$, but $H_1 = H + H_e \neq H$. Then one gets

$$\frac{\partial \langle M \rangle}{\partial H} = \left(1 + \frac{\partial H_e}{\partial H} \Big|_T \right) \left[\frac{\langle \delta \langle M_1 \rangle^2 \rangle_{\text{slow}}}{T_e} + \frac{\langle \langle \delta M_1^2 \rangle_{\text{fast}}}{T} \right] \qquad (2.43)$$

Finally, one may consider $M_1 = 0$, $M_2 = M$, but $H_e \neq H$. Then one gets

$$\frac{\partial \langle M \rangle}{\partial H} = \frac{\partial H_e}{\partial H} \Big|_T \frac{\langle \delta M^2 \rangle_2}{T_e} \qquad (2.44)$$

The latter two connections may be useful in situations where an effective field has to be taken into account.

2.4 Laws of thermodynamics for off-equilibrium systems

Finally, we will start exploring the possibility that, within a certain class of systems, the glass state can be described by an effective temperature, that maps the history of the out-of-equilibrium system into a thermodynamic frame.

In this section, we follow a scheme in which the nonequilibrium state at time t can be represented, e.g., by the parameter T, p, and $T_e(t)$, provided t

is long enough. A necessary check for this approach will be that the results of cooling, heating and aging experiments are consistently described by this set of parameters.

The assumption of separation of timescales and the concept of configurational entropy (cf. Sec. 1.4) allows for the formulation of first and second law of thermodynamics in terms of two temperatures. In order for the conservation of energy to be satisfied in a glass-forming liquid that has fallen out of equilibrium at low temperature, the heat variation has to take the same form as in the standard two temperature picture described in Sec. 2.3:

$$đQ = T \, dS_{ep} + T_e \, dS_c \tag{2.45}$$

where now T is the bath temperature and S_{ep} is the entropy of equilibrium processes coupled to this heat-bath, while T_e is the effective temperature and S_c the configurational entropy. The equality holds in the absence of currents.

The first law, then, reads

$$dU = đQ + đW = T \, dS_{ep} + T_e \, dS_c - p \, dV \tag{2.46}$$

and the second law, expressed as the Clausius inequality, as

$$đQ \leq T \, dS = T \, d(S_{ep} + S_c) \tag{2.47}$$

becomes

$$(T_e - T) \, dS_c \leq 0 \tag{2.48}$$

In this framework, Maxwell relations can be generalized to out-of-equilibrium expressions, as well as the Clausius-Clapeyron relation for first order phase transitions. Furthermore, for continuous, or second order, phase transitions, the Keesom-Ehrenfest relations between variations of thermal expansivity, $\Delta\alpha$, compressibility, $\Delta\kappa$, and specific heat, ΔC_p, can be rewritten showing that the first relation, Eq. (2.15), always holds because it is a trivial thermodynamic equality, while the second one, Eq. (2.20), contains corrections in terms of T_e and its derivatives. As a consequence, dogmas about the Prigogine-Defay ratio being larger than one (cf. Sec. 2.1.4) are unveiled, and proved inconsistent. The Prigogine-Defay ratio between the caloric and the mechanical relations - equal to one for true, equilibrium, second order phase transitions - can take any value at the kinetic, far-from-equilibrium, smeared liquid-glass transition. We will now analyze this topic in detail. See also [McKenna, 1989; Hodge, 1994; Angell, 1995; Nieuwenhuizen, 1997a, 1998c, 2000] in the Bibliography for further widening of the subject.

As a first observation, we notice that in the nonequilibrium thermodynamic formalism just introduced, the specific heat at constant pressure can be written as

$$C_p \equiv \left.\frac{\partial(U+pV)}{\partial T}\right|_p = T\left(\left.\frac{\partial S_{\mathrm{ep}}}{\partial T}\right|_{p,T_e} + \left.\frac{\partial T_e}{\partial T}\right|_p \left.\frac{\partial S_{\mathrm{ep}}}{\partial T_e}\right|_{p,T}\right)$$

$$+T_e\left(\left.\frac{\partial S_{\mathrm{c}}}{\partial T}\right|_{p,T_e} + \left.\frac{\partial S_{\mathrm{c}}}{\partial T_e}\right|_{T,p} \left.\frac{\partial T_e}{\partial T}\right|_p\right)$$

$$= C_0 + C_1 \left.\frac{\partial T_e}{\partial T}\right|_p \qquad (2.49)$$

where, in the last equality, we evaluated all prefactors at $T_e = T$. This is what Tool conjectured for its fictive temperature and it is a relationship usually verified in experiments [cf. Eq. (2.29)].

2.4.1   Maxwell relation for aging systems

We define the generalized Gibbs free enthalpy with an extra temperature

$$G = U - TS_{\mathrm{ep}} - T_e S_{\mathrm{c}} + pV \qquad (2.50)$$

whose differential form is

$$dG = -S_{\mathrm{ep}}dT - S_{\mathrm{c}}dT_e + V dp \qquad (2.51)$$

Within this setup we have now three Maxwell relations instead of one [cf. Eq. (2.12)]:

$$-\left.\frac{\partial S_{\mathrm{ep}}}{\partial p}\right|_{T_e,T} = \left.\frac{\partial V}{dT}\right|_{T_e,p} \qquad (2.52)$$

$$-\left.\frac{\partial S_{\mathrm{ep}}}{\partial T_e}\right|_{T,p} = -\left.\frac{\partial S_{\mathrm{c}}}{\partial T}\right|_{T_e,p} \qquad (2.53)$$

$$-\left.\frac{\partial S_{\mathrm{c}}}{\partial p}\right|_{T,T_e} = \left.\frac{\partial V}{\partial T_e}\right|_{T,p} \qquad (2.54)$$

The last two set constraints on the newly introduced configurational contribution to the entropy: S_{c}. These relations cannot be used immediately, since it is unknown how to control T_e at fixed T and p. The basic step is to specify a protocol for a smooth set of experiments to be discussed [Nieuwenhuizen, 1997a]. For a smooth sequence of cooling procedures at a set of nearby pressures of a glass-forming liquid, Eq. (2.51) implies a modified Maxwell relation between macroscopic observables, such as, e.g., energy $[U(t; T, p) \to U(T, p) = U(T, T_e(T, p), p)]$ and volume. This solely occurs since $T_e = T_e(T, p)$ is a nontrivial function of temperature and pressure, within the set of experiments under consideration. The Schwarz identity $\partial^2 G/\partial T \partial p = \partial^2 G/\partial p\, \partial T$ now yields

$$-\left.\frac{\partial S_{\mathrm{ep}}}{\partial p}\right|_T - \left.\frac{\partial S_{\mathrm{c}}}{\partial p}\right|_T \left.\frac{\partial T_e}{\partial T}\right|_p = \left.\frac{\partial V}{\partial T}\right|_p - \left.\frac{\partial S_{\mathrm{c}}}{\partial T}\right|_p \left.\frac{\partial T_e}{\partial p}\right|_T \qquad (2.55)$$

The differential relations, Eqs. (2.46, 2.51), do not invoke the functional dependence $T_e(T, p)$, since they hold for any functional dependence, and even in the absence of it. However, their interdependence does become relevant when taking the time, i.e., the T_e, derivatives as it was done to obtain Eq. (2.55).

As before, we eliminate S_{ep} from these relations. The first law, Eq. (2.46), implies

$$T\frac{\partial S_{ep}}{\partial p}\bigg|_T = \frac{\partial U}{\partial p}\bigg|_T - T_e\frac{\partial S_c}{\partial p}\bigg|_T + p\frac{\partial V}{\partial p}\bigg|_T \qquad (2.56)$$

Combining equations (2.55) and (2.56) one obtains

$$\frac{\partial U}{\partial p}\bigg|_T + p\frac{\partial V}{\partial p}\bigg|_T + T\frac{\partial V}{\partial T}\bigg|_p = T\frac{\partial S_c}{\partial T}\bigg|_p \frac{\partial T_e}{\partial p}\bigg|_T - T\frac{\partial S_c}{\partial p}\bigg|_T \frac{\partial T_e}{\partial T}\bigg|_p + T_e\frac{\partial S_c}{\partial p}\bigg|_T \qquad (2.57)$$

This is the modified Maxwell relation between observables U and V. The effects of aging are represented by the terms in the right-hand side and all involve the configurational entropy. In equilibrium one has $T_e(T, p) = T$, so that the right-hand side vanishes, and the standard form is recovered.

2.4.2 Generalized Clausius-Clapeyron relation

Let us consider a first order transition between two glassy phases A and B. An example could be the transition from low-density amorphous ice to high-density amorphous ice [Mishima *et al.*, 1985] or the coordination transformation occurring in strong glasses such as, for instance, germanium dioxide (from fourfold to sixfold coordination raising the pressure) and silica (from tetrahedral to octahedral coordination) [Tsiok *et al.*, 1998]. Another kind of pressure-induced amorphous-amorphous transition takes place in densified porous silicon, where the high-density amorphous packing transforms into a low-density amorphous packing upon decompression [Deb *et al.*, 2001]. A similar transition also takes place in undercooled water [Poole *et al.*, 1992].[9]

---

[9]Theoretical models have been recently introduced to describe an amorphous-amorphous transition. As, for instance, a model of hard-core repulsive colloidal particles subject to a short-range attractive potential that induces the particles to stick to each other [Dawson *et al.*, 2000; Zaccarelli *et al.*, 2001; Sciortino, 2002]. In the framework of the mode coupling theory (MCT) it has been shown that the interplay of the attractive and repulsive mechanisms results in the existence of a high(er) temperature "repulsive" glass, where the hard-core repulsion is responsible for the freezing in of many degrees of freedom and the kinetic arrest, and a low(er) temperature "attractive" glass that is energetically more favored than the other one but only occurs when the thermal excitation of the particles is rather small. Such theoretical and numerical predictions seem to have been successfully tested in recent experiments [Chen *et al.*, 2003; Eckert & Bartsch, 2002; Pham, 2002]. Other models where amorphous-amorphous transitions are found are the spherical p-spin model on lattice gas of Caiazzo *et al.* [Caiazzo *et al.*, 2004] where an off-equilibrium Langevin dynamics is considered, and the $s + p$-spin models [Crisanti & Leuzzi, 2004, 2006], both going beyond the MCT assumption of equilibrium.

For the standard Clausius-Clapeyron relation one observes that the free enthalpy G is continuous along the first order phase transition line $p_g(T)$, and it is commonly assumed to hold for the glass transition as well.[10] The phases A and B under probe, will have their own T_e, S_{ep} and S_c. In principle, they are all discontinuous along the first order transition line. Let us denote the discontinuities of an observable O between the two phases as

$$\Delta O(T, p_g(T)) \equiv O_A - O_B \qquad (2.58)$$

Taking $O = G$ and differentiating the identity $\Delta G = 0$ one finds

$$\left[\Delta V - \Delta \left(\frac{\partial T_e}{\partial p} \Big|_T S_c \right) \right] \frac{dp_g}{dT} = \Delta S_{ep} + \Delta \left(\frac{\partial T_e}{\partial T} \Big|_p S_c \right) \qquad (2.59)$$

The entropy of equilibrated processes, S_{ep}, can be eliminated by means of Eq. (2.50). Using again $\Delta G = 0$, Eq. (2.59) becomes

$$\Delta V \frac{dp_g}{dT} = \frac{\Delta U + p_g \Delta V}{T} + \Delta \left(\frac{dT_e}{dT} S_c - \frac{T_e}{T} S_c \right) \qquad (2.60)$$

where $d/dT = \partial/\partial T + (dp_g/dT)\partial/\partial p$ is the "total" derivative, i.e., the derivative along the transition line. This is the modified Clausius-Clapeyron equation. One possible application of the above formulation could be tested on physical systems undergoing amorphous-amorphous transitions presenting first order discontinuities. No experimental tests have been attempted so far.[11]

If phase A is an equilibrium undercooled liquid, and phase B is a glass, $T_e = T$ in the A phase and its S_c-terms will cancel from Eq. (2.60) that reduces to

$$\Delta V \frac{dp_g}{dT} = \frac{\Delta U + p \Delta V}{T} + \left(\frac{T_e}{T} - \frac{dT_e}{dT} \right) S_c \qquad (2.61)$$

where T_e and S_c are properties of the glassy phase B [Nieuwenhuizen, 1997a].[12]

2.4.3 Keesom-Ehrenfest relations and Prigogine-Defay ratio out of equilibrium

For standard glass-forming liquids, there are, actually, no discontinuities in U and V. It holds that, along the glass transition line, $T_e(T, p_g(T)) = T$,

---

[10]Since $T_e \neq T$, it is actually not obvious that G should still be continuous there. But if there exists a statistical mechanics description by some type of partition sum, it follows again automatically.

[11]For ice, Mishima & Stanley [1998] have presented a thermodynamic construction of the free enthalpy G. This is, however, based on equilibrium ideas and does not involve the effective temperature in the amorphous phases. In particular, they assumes the validity of the original Clausius-Clapeyron relation.

[12]Notice that Eq. (7) by Nieuwenhuizen [1997a] contains a misprint in the prefactor of S_c.

implying $dT_e/dT = 1$, which, indeed, removes the S_c terms from Eqs. (2.60), (2.61) and reduces them to trivial statements. To describe what happens at the thermal glass transition we can, then, use a generalization of the Keesom-Ehrenfest relations reported in Sec. 2.1.4.

When deriving the mechanical Keesom-Ehrenfest relation, we may just copy the derivation of Sec. 2.1.4 and again arrive at Eq. (2.15). It is a rather trivial equality,

$$\frac{\Delta\alpha}{\Delta\kappa} = \frac{dp_g}{dT} \tag{2.62}$$

that is not related to the thermodynamics of the system. Actually, in many experiments, this equation appears as violated. In boron trioxide, for instance, the left-hand side amounts to 8.010^{-7} K/Pa [Richert & Angell, 1998] whereas the right-hand side is 1.9710^{-7} K/Pa [O'Reilly, 1962], four times smaller. Analogous discrepancies are detected in selenium, PVC and polyvinyl acetate (PVAC).

Why is a mathematical identity, just stating that the volume decreases continuously at the glass transition, with no discontinuities, so often violated by some of the most common glassy materials? The answer is in the experimental procedures adopted to measure the variations of the compressibility and, to a lesser extent, of the thermal expansivity. Since one extreme of the variations of κ, α and C_p is out of equilibrium, any techniques relying on the equilibrium hypothesis will provide strongly biased results. Indeed, this is what happens when, as was often done, the compressibility is measured by the sound waves in the amorphous media. Sound waves are fast modes, that cannot pick up slow contributions to κ and, thus, yield a too small value. This is by itself not a problem, but it forbids replacement of $\Delta\kappa$ in Eq. (2.62) by the one obtained from speed of sound measurements.[13] For this reason, one should carefully review all experimental protocols adopted to determine where the source of deviation from identity is in the cited data (and many more). Fundamentally, the point is that out of equilibrium the only way to derive the right compressibility and thermal expansivity is by their definitions, Eq. (2.13). In order to do this, however, the whole phase diagram p,V,T must be reconstructed in the glass phase as well, and the $p_g(T)$ line must be obtained by direct measurements. As an instructive instance we show in Fig. 2.2 the re-elaboration of the data obtained by the experiments of Rehage & Oels [1977] on atactic polystyrene. In the figure, the specific volume, the pressure and the temperature are reported both in the liquid phase and in the glass phase. In both phases they lie on a surface. The two surfaces meet at the glass transition, across which the volume continuously decreases

---

[13] In spin-glass theory with one step of replica symmetry breaking (cf. Sec. 7.2), the long time susceptibility would be the field cooled susceptibility, while the analogue of speed of sound measurements would yield the zero-field-cooled susceptibility, which is smaller, as the slow processes do not contribute to it.

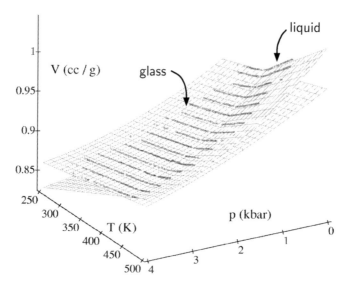

FIGURE 2.2

Data of the glass transition for cooling atactic polystyrene at a rate of 18 K/h, scanned from the paper of Rehage & Oels [1977]: specific volume V (cm^3/g) versus temperature T (K) at various pressures p (kbar). As confirmed by a polynomial fit, the data in the liquid essentially lie on a smooth surface, and so do the data in the glass. The first Keesom-Ehrenfest relation [see Eq. (2.15)] describes no more than the intersection of these surfaces, and is, therefore, automatically satisfied. The values for the compressibility derived in this manner by its definition $\partial \log V/\partial p|_T$, will generally differ from results obtained via other experimental procedures. Reprinted figure with permission from [Nieuwenhuizen, 2000]. Copyright (2000) by the American Physical Society.

as temperature decreases. The equation of this intersection is nothing else than the mechanical Keesom-Ehrenfest relation. Compressibility and thermal expansivity, both in the glass and liquid phase, can be obtained from the p and T derivatives of the volume curves, respectively.

For what concerns the calorimetric Keesom-Ehrenfest relation, it was already observed in Sec. 2.1 that the thermodynamics of the system does, instead, play a role, since the Maxwell relation is employed in its derivation. Outside equilibrium, the Maxwell relation is modified, see Eq. (2.57). We thus obtain the generalized relation

$$
\begin{aligned}
\frac{\Delta C_p}{T_{\mathrm{g}} V \Delta \alpha} &= \frac{\mathrm{d} p_g}{\mathrm{d} T} + \frac{1}{V \Delta \alpha}\left(1 - \left.\frac{\partial T_e}{\partial T}\right|_p\right)\left(\left.\frac{\partial S_{\mathrm{c}}}{\partial T}\right|_p + \frac{\mathrm{d} p_g}{\mathrm{d} T}\left.\frac{\partial S_{\mathrm{c}}}{\partial p}\right|_T\right) \\
&= \frac{\mathrm{d} p_g}{\mathrm{d} T} + \frac{1}{V \Delta \alpha}\left(1 - \left.\frac{\partial T_e}{\partial T}\right|_p\right)\frac{\mathrm{d} S_{\mathrm{c}}}{\mathrm{d} T}
\end{aligned}
\tag{2.63}
$$

where dS_c/dT is the total derivative of the configurational entropy along the glass transition line. The last term is new and vanishes only at equilibrium, where $T_e = T$.

Combining the two Keesom-Ehrenfest relations one may eliminate the slope of the transition line and compute the Prigogine-Defay ratio. For equilibrium transitions it should be equal to unity. Assuming that at the glass transition a number of unspecified parameters undergo a phase transition, Davies & Jones [1953] showed that $\Pi \geq 1$, while Di Marzio [1974] showed that in that case the correct value is $\Pi = 1$. To explain the glass property $\Pi > 1$ in terms of the framework for relaxation presented in Sec. 2.2, Gupta [1980] introduced a "fictive pressure" next to the fictive temperature. It was, hence, generally expected that $\Pi \geq 1$ is a strict inequality. As we have shown in Sec. 2.1, however, these results completely rely of the validity of the FDT or just on the equilibrium hypothesis. In glasses, typical experimental values are reported in the range $2 < \Pi < 5$. Although, in order to give credit to all these numbers, the experimental protocol should be known as well.

Being conscious of the out-of-equilibrium nature of the glassy state, we have pointed out that, as the first Ehrenfest relation is satisfied, the second one is not, implying

$$\Pi = 1 + \frac{1}{V\Delta\alpha}\left(1 - \frac{\partial T_e}{\partial T}\bigg|_p\right)\frac{dS_c}{dT}\left(\frac{dp_g}{dT}\right)^{-1} = 1 + \frac{1}{V\Delta\alpha}\left(1 - \frac{\partial T_e}{\partial T}\bigg|_p\right)\frac{dS_c}{dp} \tag{2.64}$$

Depending on the set of experiments to be chosen, dp_g/dT can be small or large, and Π can also be below unity, and even be negative. [Rehage & Oels, 1977] found $\Pi = 1.09 \approx 1$ at $p = 1\,kbar$, using a short-time value for κ. Indeed, though these authors understand very well that speed of sound measurements do not yield a proper long time compressibility, they still try to measure it independently, by inserting extra temperature steps in the cooling protocol. However, this method also is bound to fail in reproducing the value of κ from Eq. (2.62) that is needed for the Keesom-Ehrenfest relations, derived without accounting for such steps. Reanalyzing their data one computes, from Eq. (2.64), a Prigogine-Defay ratio $\Pi = 0.77$ [Nieuwenhuizen, 1997a, 2000]. Surprisingly, this lies below unity. As this would be impossible in equilibrium, it stresses, once more, that a nonequilibrium approach is needed for a proper description of the glass transition.

In summary, even though the definition (2.23) of Π looks like a combination of equilibrium quantities this is not the case when the glass state is involved in the transition. The Prigogine-Defay ratio is, indeed, a misleading expression, since κ_{glass} sensitively depends on how the experiment is carried out and the commonly accepted inequality $\Pi \geq 1$, based on equilibrium or stationary assumptions, is incorrect. In particular, this rules out the Gibbs & Di Marzio [1958] model as a principally correct model for the glassy state, simply because it is an equilibrium model and, hence, leads to predictions in contradiction to experiments (such as $\Pi = 1$).

2.5 Laws of thermodynamics for glassy magnets far from equilibrium

In this section we give, without derivation, the analogous relations for glassy magnets, i.e., collections of magnetic moments (spins). The work, at constant volume, is

$$\mathrm{d}W = -M\mathrm{d}H \qquad (2.65)$$

The generalized Helmholtz free energy is

$$F = U - TS_{\mathrm{ep}} - T_e S_c \qquad (2.66)$$

and the differential comes out to be

$$\mathrm{d}F = -S_{\mathrm{ep}}\mathrm{d}T - S_c\mathrm{d}T_e - M\mathrm{d}H \qquad (2.67)$$

The modified Maxwell relation for a glassy magnet reads

$$\left.\frac{\partial U}{\partial H}\right|_T + M - T\left.\frac{\partial M}{\partial T}\right|_H = T_e\left.\frac{\partial S_c}{\partial H}\right|_T + T\left(\left.\frac{\partial T_e}{\partial H}\right|_T\left.\frac{\partial S_c}{\partial T}\right|_H - \left.\frac{\partial T_e}{\partial T}\right|_H\left.\frac{\partial S_c}{\partial H}\right|_T\right) \qquad (2.68)$$

Instead of compressibility and thermal expansivity, for magnets we have the susceptibility χ and we, further, define a thermal "magnetizability" α_M,

$$\chi = \frac{1}{N}\frac{\partial M}{\partial H}, \qquad \alpha_M = \frac{1}{N}\frac{\partial M}{\partial T} \qquad (2.69)$$

where N is the number of spins. The specific heat is now taken at constant volume: C_V.

The first, *magnetic* Keesom-Ehrenfest relation is written, in this case, as

$$\frac{\Delta\alpha}{\Delta\chi} = \frac{\mathrm{d}H_g}{\mathrm{d}T} \qquad (2.70)$$

whereas the modified calorimetric Keesom-Ehrenfest relation becomes

$$\frac{\Delta C_V}{NT\Delta\alpha} = \frac{\mathrm{d}H_g}{\mathrm{d}T} + \frac{1}{N\Delta\alpha}\left(1 - \left.\frac{\partial T_e}{\partial T}\right|_H\right)\left(\left.\frac{\partial S_c}{\partial T}\right|_H + \frac{\mathrm{d}H_g}{\mathrm{d}T}\left.\frac{\partial S_c}{\partial H}\right|_T\right) \qquad (2.71)$$

Along the glass transition line, the equality $T_e(T, H_g(T)) = T$ implies

$$\frac{\mathrm{d}T_e}{\mathrm{d}T} = \left.\frac{\partial T_e}{\partial T}\right|_H + \left.\frac{\partial T_e}{\partial H}\right|_T\frac{\mathrm{d}H_g}{\mathrm{d}T} = 1 \qquad (2.72)$$

Using Eq. (2.72) and dividing Eq. (2.71) by Eq. (2.70), the Prigogine-Defay ratio takes the form

$$\Pi \equiv \frac{\Delta C\Delta\chi}{NT(\Delta\alpha)^2} = 1 + \frac{1}{N\Delta\alpha}\left(1 - \left.\frac{\partial T_e}{\partial T}\right|_H\right)\frac{\mathrm{d}S_c}{\mathrm{d}H} \qquad (2.73)$$

where $\mathrm{d}/\mathrm{d}H$ is the total derivative along the transition line $T_g(H)$.

2.6 Effective temperature in thermal cycles: a universal nonlinear cooling-heating procedure

In the field of glass physics, there are not many universal or exact results. Experiments are usually performed at linear cooling or heating, e.g., in [Moynihan *et al.*, 1976; Torell, 1982; Angell & Torell, 1983] for boron trioxide, borosilicate crown glass (strong) and $Ca^{++}K^{+}NO_3^{-}$ (fragile) (cf. also Fig. 1.5). We denote the thermal history imposed on the glass sample by $T(t)$. A general feature is that, both in cooling and heating, the glass transition temperature, that is, the "transformation range" in which the specific heat sensibly changes, shifts towards a lower temperature as the rate of change, $\dot{T}$, decreases. In the heating processes, furthermore, a hump in the specific heat is observed, its height depending on the cooling rate with which the glass was formed.

Let us now discuss the behavior of the specific heat near the glass transition for a different kind of protocol, which appears to bear universal features. Inspired by the study of analytical models for glassy behavior, like the harmonic oscillator model of Bonilla *et al.* [1998] (to be discussed in the next chapter) and a spherical spin model [Nieuwenhuizen, 1998c], Nieuwenhuizen [2000] considered a *nonlinear* cooling-heating process with universal aspects.

Let us begin considering, for simplicity, a strong glass, the relaxation of which is described by an Arrhenius law,

$$\tau_{eq}(T) = \tau_0 \exp \frac{A}{T}, \tag{2.74}$$

whose inverse function is

$$\tilde{T}_{eq}(\tau) = \frac{A}{\log(\tau/\tau_0)} \tag{2.75}$$

The specific nonlinear cooling or heating setup $T(t)$ exploits this inverse Arrhenius law and is defined as

$$T(t) = (1 - \tilde{Q})T_{g} + \tilde{Q}\tilde{T}_{eq}(t) = (1 - \tilde{Q})T_{g} + \tilde{Q}\frac{A}{\ln(t/\tau_0)} \tag{2.76}$$

It involves two parameters: the thermal glass transition temperature T_{g} and the dimensionless parameter $\tilde{Q}$. Cooling is described by $\tilde{Q} > 0$ and heating by $\tilde{Q} < 0$. A nonlinear cooling experiment of this form could be performed in any system with a quickly diverging relaxation time.

We first show that a glass transition occurs around the timescale and temperature

$$t_{g} = \tau_0 e^{A/T_{g}}, \qquad T_{g} = \frac{A}{\ln(t_{g}/\tau_0)} \tag{2.77}$$

where one has $T(t) = T_{\mathrm{g}}$. The timescale during which the system basically remains at temperature T during the temperature program $T(t)$ is

$$\tau_{\mathrm{cool/heat}} = \frac{T(t)}{|\dot{T}(t)|} \sim t \tag{2.78}$$

where we neglected logarithmic corrections. Writing $t = s\, t_g$, we are around the glass transition for $s = \mathcal{O}(1)$ and Eq. (2.76) yields

$$T = T_{\mathrm{g}} + \tilde{Q}A \left(\frac{1}{\ln t_g + \ln s} - \frac{1}{\ln t_g} \right) \approx T_{\mathrm{g}} - \frac{\tilde{Q}A \ln s}{\ln^2 t_g} \tag{2.79}$$

where we used $\ln t_g \gg \ln s$. In the same approximation we may write this as

$$\beta = \beta_g + \frac{\tilde{Q}}{A} \ln s \tag{2.80}$$

so it yields

$$\tau_{\mathrm{eq}}(T(t)) = \tau_0 e^{\beta_g A} s^{\tilde{Q}} = t_g \left(\frac{t}{t_g} \right)^{\tilde{Q}} \tag{2.81}$$

The ratio of the cooling-heating timescale to the momentary equilibrium timescale is

$$\frac{\tau_{\mathrm{cool/heat}}(T(t))}{\tau_{eq}(T(t))} \sim \left(\frac{t_g}{t} \right)^{\tilde{Q}-1} \tag{2.82}$$

We can discriminate three cases depending on the value of $\tilde{Q}$.

- $\tilde{Q} > 1$. Cooling, $\dot{T} < 0$: for $t \ll t_g$ there is equilibrium at the instantaneous temperature $T(t)$, whereas for $t \gg t_g$ the instantaneous equilibration time τ_{eq} is larger than the cooling timescale $\tau_{\mathrm{cool/heat}}$, and the system becomes glassy.

- $0 < \tilde{Q} < 1$. Slow (adiabatic) cooling: Eq. (2.76) describes cooling *inside* a glassy state ($\dot{T} < 0$) but occurring so slowly, that equilibrium is reached around time t_g. Indeed, for $t \ll t_g$, one has $\tau_{\mathrm{cool/heat}} \ll \tau_{eq}$ and for $t \gg t_g$, $\tau_{\mathrm{cool/heat}} \gg \tau_{eq}$.

- $\tilde{Q} < 0$. Heating from the glassy state: as time goes by, and temperature increases, equilibrium is reached around time t_g.

The parameter $\tilde{Q}$ can, then, be considered as a dimensionless "nonlinear" cooling parameter (with negative values corresponding to heating).

Let us now introduce the effective temperature T_e and write an expression for the specific heat depending on its derivatives, in the spirit of Tool's pioneering observations [cf. Eqs. (2.29) and (2.49)]. Generally, in an approach where $T_e(t)$ is close to $T(t)$, we may define the effective temperature by

$$U(t) = \bar{U}(T(t)) + C_2[T_e(t) - T(t)] \tag{2.83}$$

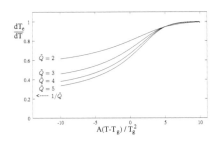

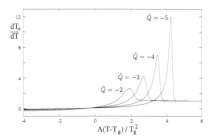

FIGURE 2.3

Specific heat factor $\partial T_e/\partial T$ as a function of reduced temperature in nonlinear, universal cooling obtained by solving Eqs. (2.87)-(2.90), with different values of the speed parameter $\tilde{Q}$. The asymptotic values are 1 to the right and $1/\tilde{Q}$ to the left. Reprinted figure with permission from [Nieuwenhuizen, 2000]. Copyright (2000) by the American Physical Society.

FIGURE 2.4

Specific heat factor $\partial T_e/\partial T$ as a function of reduced temperature in nonlinear heating, Eq. (2.90), with different values of the parameter $\tilde{Q}$ controlling the increase of the temperature ($\tilde{Q} < 0$ for heating). Reprinted figure with permission from [Nieuwenhuizen, 2000].

This implies for the specific heat

$$C_V = \frac{\dot{U}}{\dot{T}} = \bar{C}_V + C_2\left[\frac{\dot{T}_e}{\dot{T}} - 1\right] \equiv C_1 + C_2\frac{\mathrm{d}T_e}{\mathrm{d}T} \qquad (2.84)$$

where the barred functions are the equilibrium ones. Now, from the general relaxation for energy fluctuations near equilibrium at temperature T, we have

$$\dot{U} = -\frac{U - \bar{U}(T)}{\tau_{\mathrm{eq}}(T)} \qquad (2.85)$$

It seems reasonable to assume that, in the glassy regime, the slow part of the left-hand side of Eq. (2.83) is ruled by the slow part of the right-hand side, i.e.,

$$\dot{T}_e = -\frac{T_e - T}{\tau_{\mathrm{eq}}(T_e)} \qquad (2.86)$$

Here we have inserted for the characteristic time $\tau_{\mathrm{eq}}(T_e)$, which, in the aging regime, is basically $\tau_{\mathrm{eq}}(T_e) \approx t$, while close to equilibrium it will just be $\tau_{\mathrm{eq}}(T)$. In the next chapter, we shall find support for these assumptions in the harmonic oscillator model, where they are exact properties of the dynamics.

Inverting and deriving Eq. (2.76) with respect to time one obtains

$$\dot{T} = \frac{\tilde{Q}}{\tau'_{\mathrm{eq}}(T_{\mathrm{g}} - (T_{\mathrm{g}} - T)/\tilde{Q})} \qquad (2.87)$$

where the prime denotes the derivative with respect to the whole argument. We may combine Eqs. (2.86) and (2.87) in a thermodynamic shape, where time does not appear explicitly,

$$\frac{dT_e}{dT} \equiv \frac{\dot{T}_e}{\dot{T}} = \frac{T - T_e}{\tilde{Q}} \frac{\tau'_{eq}(T_g - (T_g - T)/\tilde{Q})}{\tau_{eq}(T_e)} \tag{2.88}$$

and we may, then, go to dimensionless variables by putting

$$T = T_g + \frac{T_g^2}{A} x, \qquad T_e = T_g + \frac{T_g^2}{A} y \tag{2.89}$$

and obtain (recalling that $T_g \ll A$),

$$\frac{dy}{dx} = \frac{y - x}{\tilde{Q}} e^{y - x/\tilde{Q}} \tag{2.90}$$

This equation, valid in the linear regime $|T - T_g| \ll A$, has some universality and can be extended to fragile glasses, as well. Indeed, when τ_{eq} satisfies a generalized Vogel-Fulcher law $\tau_{eq} \sim \exp(B^\gamma (T - T_0)^{-\gamma})$, and a glass transition occurs in a narrow range around some T_g, with $T_g - T_0 \ll B$, one may define the deviations x and y as

$$T = T_g + \frac{(T_g - T_0)^{\gamma+1}}{\gamma B^\gamma} x, \qquad T_e = T_g + \frac{(T_g - T_0)^{\gamma+1}}{\gamma B^\gamma} y \tag{2.91}$$

and verify that, to the leading order in $T - T_g$, one arrives at the same Eq. (2.90). The derivation still holds for $T_0 = 0$, where the timescale satisfies an enhanced Arrhenius law. This finding invites us to carry out the nonlinear cooling protocol Eq. (2.76) for any good glass former. In all cases, the specific heat will be

$$C = C_1 + C_2 \frac{dy}{dx} \tag{2.92}$$

where $C_{1,2}$ are material parameters.

Eq. (2.90) can be solved perturbatively for large negative and large positive values of $x = (T - T_g)A/T_g^2$, the scaled temperature distance from the glass transition. The derivation of the solution is displayed in Appendix 2.A.

In the intermediate regime, the differential Eq. (2.90) can be solved numerically. In Figs. 2.3-2.7 we present the universal line-shapes for the specific heat factor $(C - C_1)/C_2 = dT_e/dT$ for several values of cooling-heating speed parameter $\tilde{Q}$. The labels along the x-axis refer to the Arrhenius situation, but the same plots would apply to the Vogel-Fulcher situation. They exhibit the features known from experiments: upon cooling (in the figures: $\tilde{Q} > 0$ and going from right to left) there occurs a smeared downward jump, while upon heating (in the figures: $\tilde{Q} < 0$ and going from left to right) an overshoot shows up. The latter is related to a rather late, steep increase of the internal energy from the glassy value to the equilibrium value.

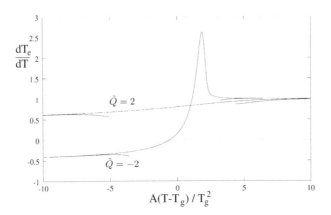

FIGURE 2.5

Specific heat factor $\partial T_e/\partial T$ as a function of reduced temperature in a nonlinear cooling with $\tilde{Q} = 2$ and in a nonlinear heating experiment with $\tilde{Q} = -2$. Dashed lines are asymptotes (2.A.4) and (2.A.9), obtained from the analytic solution of Eq. (2.90) for large positive and large negative x. Reprinted figure with permission from [Nieuwenhuizen, 2000].

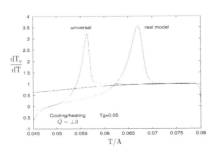

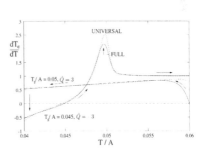

FIGURE 2.6

Specific heat factor $\partial T_e/\partial T$ in nonlinear universal cooling ($\tilde{Q} = 3$) and heating ($\tilde{Q} = -3$). The "universal" curve is obtained solving Eq. (2.90). The full curve is computed by solving the differential-integral Eq. (3.31) for the dynamics of the harmonic oscillator model (next chapter). Reprinted with permission from [Nieuwenhuizen, 2000].

FIGURE 2.7

Factor $\partial T_e/\partial T$ as a function of reduced temperature in a nonlinear universal cooling followed by heating. The jump from cooling to heating is adjusted such that peaks overlap. "Full" denotes the solution of Eq. (3.31). The parameter $\tilde{Q}$ is switched from 3 to -3 after reaching $T/A = 0.045$. Reprinted with permission from [Nieuwenhuizen, 2000].

The analysis of this section thus shows that cooling in systems with an Arrhenius law or a Vogel-Fulcher law, leads to glassy behavior quite similar to that expected for realistic glasses. More specifically, following Nieuwenhuizen [1998c, 2000] we have considered a cooling protocol that predicts universal behavior near typical glass transitions. To implement it in practice, one should first determine the equilibrium timescale $\tau_{eq}(T)$, and then impose a cooling or heating protocol as defined by the first equality in Eq. (2.76).

2.7   Fluctuation formula and effective temperatures

The basic result of nonequilibrium statistical physics is the capability of relating fluctuations in macroscopic variables to response of the averages of these variables caused by changes in external fields or temperature.

One can try to generalize such relations to the glassy phase. Susceptibilities appear to have a nontrivial decomposition, that look to be general in form, even though, every susceptibility (each one bound to a different activity) can, in principle, bring quantitatively different coefficients of the decomposition (see also Sec. 2.8.2).[14]

Let us consider, as exemplification, a model with an external field H and the conjugated coordinate, M. In slow-cooling experiments at a set of fixed fields, according to the framework presented in Sec 2.4, it holds that $M = M(T(t), T_e(t, H), H)$. For thermodynamics one eliminates time to express $T_e(t, H)$ as $T_e(T, H)$, implying $M = M(T, T_e(T, H), H)$. One may then expect that the susceptibility will decompose as

$$\chi \equiv \frac{1}{N} \frac{\partial M}{\partial H}\bigg|_T = \frac{1}{N} \frac{\partial M}{\partial H}\bigg|_{T,T_e} + \frac{1}{N} \frac{\partial M}{\partial T_e}\bigg|_{T,H} \frac{\partial T_e}{\partial H}\bigg|_T \qquad (2.93)$$

In a cooling (heating) experiment, the time t on which the susceptibility depends is the time of the cooling (heating) program $T(t)$. In the above decomposition such cooling is considered slow enough to allow for the definition of an effective temperature $T_e(t)$, depending on time both through the cooling program applied to the system and through its out-of-equilibrium aging relaxation.

In order to better identify the different components, we have to consider a pure aging relaxation without cooling, thus fixing T. The system will continue to age, as expressed by $T_e = T_e(t; T, H)$.

---

[14]If this always would be the case, it would be something dangerously similar to the observable dependence of the fictive temperature measured experimentally [Moynihan *et al.*, 1976; Donth, 2001], that brought about the downfall of the idea of fictive temperature as ... a temperature (cf. Sec. 2.2).

We may, then, use the equality

$$\frac{\partial M}{\partial H}\bigg|_{T,t} = \frac{\partial M}{\partial H}\bigg|_{T,T_e} + \frac{\partial M}{\partial T_e}\bigg|_{T,H} \frac{\partial T_e}{\partial H}\bigg|_{T,t} \tag{2.94}$$

Exploiting the assumption of separation of timescales, it can be conjectured, cf. Sec. 2.3, that the left-hand side of Eq. (2.94) is the sum of fluctuation terms for fast and slow processes,

$$\chi_{\text{fluct}}(t) = \frac{1}{N} \frac{\partial M}{\partial H}\bigg|_{T,t} = \frac{\langle \delta M(t)\delta M(t) \rangle_{\text{fast}}}{NT} + \frac{\langle \delta M(t)\delta M(t) \rangle_{\text{slow}}}{NT_e(t)} \tag{2.95}$$

where the first "fast" term in the right-hand side corresponds to the fluctuations in the stationary regime preluding to the aging regime: it is just the standard equilibrium expression for the β equilibrium processes. The second term is the contribution given by the fluctuations of those degrees of freedom of the partially frozen, out-of-equilibrium, glassy system. Notice that slow processes enter with their own temperature, the effective temperature, cf. Eq. (2.41).

The fluctuation terms are instantaneous, and they are, thus, the same for aging and cooling. Looking back to Eq. (2.94), one can, then, identify the last term on the right-hand side with a corrective loss term, related to an aging experiment:

$$\chi_{\text{loss}} = -\frac{1}{N} \frac{\partial M}{\partial T_e}\bigg|_{T,H} \frac{\partial T_e}{\partial H}\bigg|_{T,t} \tag{2.96}$$

It measures the decrease of fluctuations below the glass transition, due to the partial freezing in of the degrees of freedom of the glass. In actual situations that we will discuss in next chapter, it will be seen that this loss term decays very fast in the glassy regime,[15] so that it can basically be neglected.

The first term in Eq. (2.93) can be, then, decomposed in a fluctuation contribution and a loss term:

$$\chi_{\text{fluct}} + \chi_{\text{loss}} = \frac{1}{N} \frac{\partial M}{\partial H}\bigg|_{T,T_e} \tag{2.97}$$

The second term in Eq. (2.93) is a new, configurational, term:

$$\chi_{\text{conf}} = \frac{1}{N} \frac{\partial M}{\partial T_e}\bigg|_{T,H} \frac{\partial T_e}{\partial H}\bigg|_T \tag{2.98}$$

different from zero as far as $T_e \neq T$. This allows us to write Eq. (2.93) as

$$\chi = \chi_{\text{fluct}} + \chi_{\text{conf}} + \chi_{\text{loss}} \approx \chi_{\text{fluct}} + \chi_{\text{conf}} \tag{2.99}$$

---

[15]Notice that the quantity χ_{loss} is conceptually different from the loss part of the susceptibility discussed in Chapter 1, that is the imaginary part of the susceptibility in the frequency domain.

Thermodynamics of the glassy state

TABLE 2.1

Conjugated field-variable couples are shown together with the name of the associated "susceptibility" for some of the most typical response experiments on glasses.

Fluctuating variable	Perturbing field	Susceptibility
Entropy	Temperature	Entropy compliance
Volume	Pressure	Compressibility
Volume	Temperature	Thermal expansivity
Shear angle	Shear stress	Shear compliance
Dielectric polarization	Electric field	Dielectric susceptibility
Magnetization	Magnetic field	Magnetic susceptibility

The configurational contribution originates from the difference in the system's structure for cooling experiments at nearby fields. For glass-forming liquids such a term occurs in the compressibility. Its existence was anticipated in some earlier works. Goldstein [1973] pointed out that the volume of the glass phase depends more strongly on the pressure of formation than on the pressure eventually exerted on the system, after that a partial release of pressure has taken place. Jäckle [1989] also assumed an extra parameter and argued the existence of a configurational term, but he limited himself to the case of infinitely slow cooling and took as an extra parameter only the formation pressure, predicting eventually a Prigogine-Defay ratio Π larger than one. Not restricting ourselves to adiabatically slow cooling, ignoring any constraint on Π, that would be meaningless as seen in Sec. 2.1.4, the approach we have just presented allows, in principle, us to find the configurational term (2.98) for typical cooling procedures from the construction of the function $M(T, T_e, H)$ in full (T, T_e, H)-space.

2.8   Fluctuation and dissipation out of equilibrium

The fluctuation-dissipation theorem states that the connection between the thermodynamic fluctuation of a generic *variable* v and the response of such a variable to a small perturbation in its conjugated *field* f are simply connected by a linear relation, whose proportionality factor is the temperature of the heat-bath in which the system under probe is embedded. Examples of conjugated variables are shown in Table 2.1.

The first proof of the theorem, based on the second law of thermodynamics, was given by Nyquist [1928]. Further on, more complete proofs for generalizations of the theorem were provided by Callen & Welton [1951], Takahasi [1952] and Kubo [1957]. It is a very considerable finding, since it states an equivalence between quite different concepts such as reversible fluctuation and irreversible response. We define the *correlation function* $C_{eq}(t-t')$ as the product of fluctuations of v at time t and at time $t' < t$ averaged over the ensemble of all configurations at equilibrium:

$$C_{eq}(t - t') \equiv \langle \delta v(t) \, \delta v(t') \rangle \tag{2.100}$$

and the *response function*

$$G_{eq}(t - t') \equiv \left. \frac{\delta \langle v(t) \rangle}{\delta f(t')} \right|_{f=0} \tag{2.101}$$

The brackets $\langle \ldots \rangle$ stand for the ensemble average. At equilibrium, time translational invariance (TTI) holds and, therefore, the statistical ensemble at time t is the same as the ensemble at t'. In other words, the system forgets its previous history. For what concerns quantities depending on two times, such a property implies that they are actually only functions of the difference of times. The susceptibility is an integrated response:

$$\chi_{eq}(t - t') = \int_{t'}^{t} dt'' G(t - t'') \tag{2.102}$$

The FDT can, then, be written both in terms of time- and frequency-dependent observables. In the time domain it reads:

$$\chi_{eq}(t - t') = \frac{C_{eq}(0) - C_{eq}(t - t')}{T} \tag{2.103}$$

whereas in the frequency domain it becomes:

$$\chi''(\omega) = \frac{\pi \omega}{T} \tilde{C}(\omega) \tag{2.104}$$

where χ'' is the imaginary part (loss part), of the Fourier transform of the integrated response function and $\tilde{C}$ is the spectral density, usually denoted by $S(\omega)$.[16]

If instead of χ we use G, Eq. (2.103) becomes

$$G_{eq}(t - t') = -\frac{1}{T} \frac{\partial C_{eq}(t - t')}{\partial t} = \frac{1}{T} \frac{\partial C_{eq}(t - t')}{\partial t'} \tag{2.105}$$

---

[16] An example widely used in experiments is the spectral density of voltage noise (see experiments mentioned later in this section), conjugated to the impedance of the experimental setup.

What we said up to now is valid as long as TTI holds,[17] i.e., as far as systems are at equilibrium or in a quasi-equilibrium state.[18] For up-to-date reviews of the FDT in connection with glassy systems, the reader can refer to [Crisanti & Ritort, 2003; Cugliandolo, 2003]. For what concerns glasses, that are intrinsically never at true thermodynamic equilibrium, the situation is sensitively different. The correlation and the response functions depend both on the waiting time and on the observation time. Moreover, the statistical ensemble on which the average $\langle \ldots \rangle$ is computed changes with time. Since TTI does not hold anymore, the off-equilibrium average has to be performed over a set of different time trajectories, each starting from different, random, initial conditions. For the typical correlation and response function behavior in a glass, look at Figs. 1.2, 1.3.

2.8.1   Fluctuation-dissipation ratio

Out of equilibrium, the contribution of the frozen states to the fluctuations and to the response will, in general, be different from the one in the warm liquid state (in particular, the contribution to the response will be damped with respect to the one to the fluctuations) and no theorem can be enunciated to connect the ratio of fluctuation and response to the heat-bath temperature. In jargon, one speaks of "FDT violation," even though no theorem actually exists when the hypothesis of equilibrium (or stationarity) is not satisfied. Contrary to what is written in Eqs. (2.103), (2.104), the relation between fluctuation and response will not display the time- and frequency-independent heat-bath temperature T as a proportionality factor [Cugliandolo & Kurchan, 1993, 1994; Cugliandolo *et al.*, 1997; Parisi, 1997b].[19]

Consider structural relaxation after a quench from high temperature to $T < T_g$ and, after a time t_w, probe this state by linear response in the time (frequency) domain. Let us call the probing time t (or the probing frequency $\omega \sim 1/t$). Assuming that we are able to measure the correlation functions $C(t, t_w)$ [or spectral densities $\tilde{C}(\omega, t_w)$] in the frozen phase and the corresponding susceptibilities $\chi(t, t_w)$ [loss parts $\chi''(\omega, t_w)$] depending on t_w, we can investigate their relation as it is brought about in the FDT.

---

[17]Even in chaotic systems FDT holds, provided a suitable generalization is implemented [Falcioni & Vulpiani, 1995].

[18]The fact that many glasses are quite stable at low temperature (e.g., room temperature for window glass), displaying extremely long relaxation times to equilibrium, very low energy dissipation rates (else called small entropy production), stable mechanical, electrical, optical and, possibly, magnetic properties, may induce us to think that they are in a quasi-equilibrium state. This is not the case, however, as it is observed, cf. Chapter 1, that measurements yield constant results only if performed on timescales much smaller than the waiting time, otherwise the phenomenon of aging becomes evident.

[19]In the time domain, the "violation" is usually described by the multiplicative fluctuation-dissipation *violation factor* $X(t, t_w)$ entering in Eq. (2.109) as $\chi(t, t_w) = [C(t_w, t_w) - C(t, t_w)]X(t, t_w)/T$. The further assumption is, then, made that $X(t, t_w) = X[C(t, t_w)]$.

Indeed, one can still consider the fluctuation-dissipation ratio (FDR) $(C_0 - C)/\chi$, (or $\tilde{C}/\chi''$, else $\partial_{t'}C/G$) as an indication of how far the system is out of equilibrium, at least for what concerns the property whose fluctuations and response are measured or computed.

A step forward is, then, to regard such a FDR as an effective temperature, i.e., as the temperature of the slow processes occurring in the aging system. This has been put forward in several ways, starting from the mean-field approximation for spin models with quenched disordered exchange interactions,[20] e.g., by Sompolinsky [1981]; Sompolinsky & Zippelius [1982]; Sommers & Dupont [1984], by Horner [1992b,a]; Crisanti *et al.* [1993] or by [Cugliandolo & Kurchan, 1993, 1994; Cugliandolo *et al.*, 1997] (see also [Bouchaud *et al.*, 1998; Cugliandolo & Kurchan, 1999]) or analyzing spatially extended chaotic systems [Hohenberg & Shramian, 1989].

The study of FDR in slowly relaxing systems, either aging isolated or driven out of equilibrium by external forces, is a field on its own and we cannot give a complete overview of the state of the art. We will just concentrate on the possibility of interpreting the ratio as an effective temperature, limiting ourselves to glassy systems. For an up-to-date review we indicate the work of Crisanti & Ritort [2003].

In the time domain, a "FD" effective temperature is, then, introduced either through the ratio

$$T_e^{\text{FD}}(t, t_{\text{w}}) \equiv \frac{C(t_{\text{w}}, t_{\text{w}}) - C(t, t_{\text{w}})}{\chi(t, t_{\text{w}})} \qquad (2.106)$$

or by means of

$$T_e^{\text{FD}}(t, t_{\text{w}}) \equiv \frac{1}{G(t, t_{\text{w}})} \frac{\partial C(t, t_{\text{w}})}{\partial t_{\text{w}}} \qquad (2.107)$$

where we have put $t' = t_{\text{w}}$, i.e., the waiting time elapsed between the preparation[21] of the sample and the beginning of the measurements (the switching on of the external perturbative field). The relation between susceptibility and response function out of equilibrium is

$$\chi(t, t_{\text{w}}) = \int_{t_{\text{w}}}^{t} dt' G(t, t') \qquad (2.108)$$

---

[20]Some of these models, indeed, very well describe the phenomenology of structural glasses, e.g., the spherical p-spin glass model or other models whose statics is computable by means of a one step replica symmetry breaking [Kirkpatrick & Thirumalai, 1987a,b; Kirkpatrick *et al.*, 1989; Crisanti & Sommers, 1992; Crisanti *et al.*, 1993; Crisanti & Sommers, 1995; Franz & Parisi, 1997; Nieuwenhuizen, 1998a; Mézard & Parisi, 1999b,a; Coluzzi *et al.*, 2000b], cf. Secs. 7.2-7.3. The verified conjecture that p-spin like spin-glass models and structural glasses are in the same universality class has been the basis to extend analytical results on spin-glass [Cugliandolo & Kurchan, 1993, 1994; Franz & Parisi, 1997; Cugliandolo *et al.*, 1997; Nieuwenhuizen, 1997b; Exartier & Peliti, 2000] to the study of structural glasses in the context of complex systems.

[21]By preparation we mean not only the growth of the sample but also the setting of the thermodynamic parameters at which the experiment has to take place.

We notice that Eqs. (2.106) and (2.107) are equivalent only at the long times, in the aging regime.

In the frequency domain, the effective parameter $T_e^{FD}(\omega, t_w)$ can, as well, be defined as

$$T_e^{FD}(\omega, t_w) \equiv \frac{\pi \omega \tilde{C}(\omega, t_w)}{\chi''(\omega, t_w)} . \qquad (2.109)$$

Numerous numerical tests of this FDT violation and of the model dependence of the FDR have been made and some evidence has been found in favor of its interpretation as a thermodynamic-like temperature. Among those, besides the ones already cited, some remarkable analyses have been carried out on binary mixtures of soft particles [Parisi, 1997b; Grigera *et al.*, 2004] and on Lennard-Jones interacting particles [Barrat & Kob, 1999; Berthier & Barrat, 2002] with size ratios preventing crystallization, as well as on a mono-atomic Lennard-Jones glass provided with a many-body interaction term inhibiting crystallization [di Leonardo *et al.*, 2000]. More will be considered in Chapters 3, 4 and 6.

The usual picture is that, in as far as the observation time is less than the waiting time of the system, the ratio between response and correlation behaves as if it is in a stationary state, i.e., FDT seems to be satisfied. As the time goes by and t becomes of the order of t_w, this apparent stationarity is, however, unmasked and aging takes over: $T_e^{FD} \neq T$. Making glass by cooling down a liquid, the effective temperature is higher than the heat-bath temperature. One can understand this by thinking that slow processes were stuck out of equilibrium at a higher temperature with respect to that of the environment in which the glass lives and they, somehow, "remember" it.

In mean-field models for glasses, the effective temperature turns out to depend exclusively on time through the value of the correlation function. A rather standard way to visualize all this is, mostly by theoreticians, to plot the $\chi(C)$ parametric curve, see Fig. 2.8. This is the shape usually occurring in models for structural glasses (mean-field, certainly, but also short-range, see the above-mentioned examples and Figs. 2.9-2.10). Other out-of-equilibrium aging systems, such as models of coarsening and domain growth [Cugliandolo & Kurchan, 1995][22] and spin-glasses [Marinari *et al.*, 1998; Ricci-Tersenghi, 2003] display other characteristic shapes (see insets in Fig. 2.8).

In order for T_e^{FD} to be a well-defined temperature, a necessary condition is that it is the same for any measurable observable reacting to a perturbation in a small conjugate field. A remarkable analysis on the insensitivity of the FDR to the observable chosen is represented in Fig. 2.9 for a Lennard-Jones binary mixture (LJBM), see also Appendix 6.A.2, a computer model with a van der Waals-like interaction [Berthier & Barrat, 2002].

Experimental evidence has been reached, as well, for the existence of such a kind of effective parameter. A violation of FDT has been measured in the

---

[22]Read further for criticism on this behavior: Sec. 2.8.2.

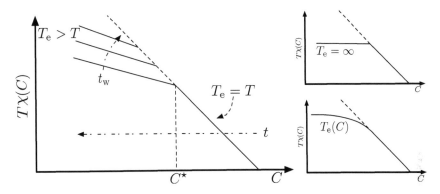

FIGURE 2.8

Susceptibility times heat-bath temperature versus correlation function. As the correlation decreases, that is, as time increases from t_w, the slope passes from minus one (i.e., FDR= T) to something less than one (i.e., FDR= $T_e > T$). The value $C^\star$ at which the departure from equilibrium occurs is the plateau value of the correlation function as plotted in Fig. 1.2 and it depends on the waiting time. In the insets on the right-hand side the typical behaviors expected for coarsening systems (top) and spin-glasses (bottom) are shown.

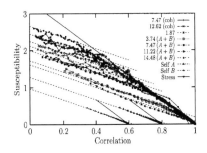

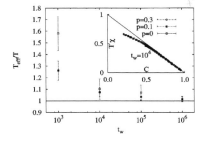

FIGURE 2.9

Plot of $\chi(C)$ for the LJBM. Several measures done in numerical simulations are reported at the heat- temperature of $T = 0.3$ (the slope of the stationary part is $1/0.3$) and all curves display the same slope as they fall out of equilibrium $(1/T_e^{\mathrm{FD}} = 1/0.65)$: the FD effective temperature is observable independently. Reprinted with permission from [Berthier & Barrat, 2002]. Copyright (2002) by American Institute of Physics.

FIGURE 2.10

Effective temperature versus waiting time in the soft sphere binary mixture obtained by numerical simulations with different Monte Carlo algorithms at $T = 0.89T_d$. The $\chi(C)$ plot is shown in the inset for $t_w = 10^4$ Monte Carlo steps at a lower temperature, $T = 0.53T_d$. Reprinted with permission from [Grigera *et al.*, 2004]. Copyright (2004) by the American Physical Society.

fragile glass former glycerol by Grigera & Israeloff [1999], via electric noise applying a small AC field of fixed angular frequency ω. For frequencies slightly larger than 1 Hz one can probe the long waiting time regime, where aging and violation of FDT is expected to occur, within the measurement time. Instead of formula (2.104), the equivalent relation initially given by Nyquist [1928] between electrical resistance and voltage noise has been used to get the noise power spectral density

$$S_V(\omega) = \frac{2}{\pi} k_B T_e^{\mathrm{FD}}(\omega, t_w) \mathrm{Re} Z \qquad (2.110)$$

where Z is the (aging) impedance of the circuit employed in the measurements. The glassy behavior is signaled whenever the proportionality factor $T_e(\omega, t_w)$ is different from the heat-bath temperature T. The deviation from T is evident, though weak, even for $\omega t_w \lesssim 10^5$ at $T = 180K$, cf. Fig. 2.11 where, for this material, the glass temperature is $T_g \sim 196K$.[23] Grigera & Israeloff [1999] also measured the fictive temperature from the excess enthalpy [Simon & McKenna, 1997]. Its behavior versus heat bath temperature turns out, however, to be different from the one of the effective temperature as can be seen in Fig. 2.12. As a matter of fact, there would have been no reason to account for their coincidence. T_e is relative to voltage fluctuations (and dielectric response), while T_f rationalizes the excess enthalpy relaxation and we know that, for what concerns the latter, every observable relaxation is described by a different fictive temperature (Sec. 2.2).

Another experiment has been carried out on a colloidal material, Laponite RD, [24] a synthetic clay made of disc shaped charged particles [Bellon *et al.*, 2001; Bellon & Ciliberto, 2002]. At low temperature this material forms a solid-like solution that, for low concentration, turns out to be glassy. General characteristics of Laponite RD are a fast dispersion in water, solidification even for a very low mass fraction, aging behavior below the transition and, at low concentration, a structure function very similar to the one of glass formers [Kroon *et al.*, 1996]. The fluctuation-dissipation relation between the voltage noise spectral density $S_V(\omega)$ and the electrical impedance Z of the experimental setup has been analyzed by Bellon *et al.* [2001]; Bellon & Ciliberto [2002], again using the generalization of the Nyquist formula given in Eq. (2.110). Different from the previous experiment, measurements are not done at one fixed frequency, but a two dimensional map of T_e^{FD} is built both in time and frequency (see Figs. 2.13 and 2.14). The effective temperature measured in this way turns out to be a decreasing function of time and frequency (as expected) and, in the time domain, T_e decays faster to the

---

[23]This persistence of the FDT violation is actually unexpected since the aging regime sets in as $t - t_w \sim 1/\omega \sim t_w$, whereas on shorter timescales, $t - t_w \sim 1/\omega \ll t_w$, the local equilibrium of the stationary regime is expected to occur, and FDT holds (cf. the right part of the line plotted in Fig. 2.8).

[24]Laponite RD is a registered trademark of Laporte Absorbents - P.O Box 2, Cheshire, U.K.

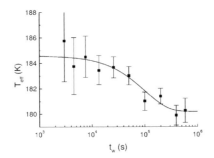

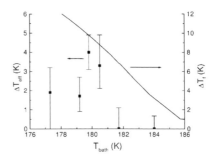

FIGURE 2.11

Effective temperature dependence on the waiting time at a fixed observation time in glycerol. The heat-bath temperature is $T = 180K$. Reprinted figure with permission from [Grigera & Israeloff, 1999]. Copyright (1999) by the American Physical Society.

FIGURE 2.12

Distance of T_e and T_f from T. The barred points are for dielectric ΔT_e^{FD}, whereas the full curve represents the enthalpy ΔT_f. Reprinted figure with permission from [Grigera & Israeloff, 1999]. Copyright (1999) by the American Physical Society.

heat-bath temperature T at high frequency than a low frequency, as shown in Fig. 2.13.[25] Moreover, as it was observed also from the measurements of the density fluctuations in polymers in the glassy phase [Wendorff & Fischer, 1973], the violation of FDT at a given angular frequency ω lasts for $t_w \omega \gg 1$, in agreement with the experiment on glycerol. What is peculiar on Laponite, however, is that, for relatively long waiting times, the T_e^{FD} at low frequencies is several orders of magnitude larger than the heat-bath temperature.

This very large FDR, as well as the persistence of aging for long times (at low frequencies), has been confirmed by Buisson *et al.* [2003b,a] by comparing noise and dielectric response functions in a polymer glass (Makrofol DE 1-1C, a bisphenol A polycarbonate). Here, the glass transition occurs lowering the temperature, unlike Laponite where the vitrification occurs varying the density of the discoidal particles. In this material $T_g = 419$ K and, for low frequencies and t_w of the order of hours, the FDR at $T = 333$ K is still a couple of orders of magnitude larger than T_g. One has to reach an aging waiting time of a day to measure an FDR similar to T_g, that is, the temperature where the α processes (low frequency) fall out of equilibrium with the heat-bath. Buisson *et al.* [2003b] bring the large FDR back to a highly intermittent dynamics characterized by large fluctuations (similar, in some ways, to the trap model, cf. Sec. 2.8.2), but the physical origin of these large fluctuations is yet to be understood. All in all, however, it seems that, at least in the time-frequency

---

[25]In Chapter 4 we discuss an exactly solvable model, the disordered backgammon model, where modes at higher energy, in absolute value, thermalize much faster than modes at low energy.

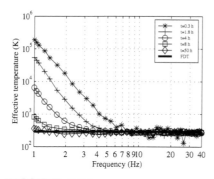

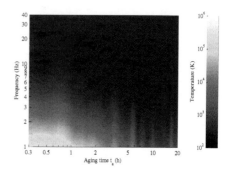

FIGURE 2.13

Effective temperature dependence on frequency in Laponite at different waiting times, expressed in hours: $t \in [0.3h : 50h]$. As the t_w is increased the effective temperature tends to the heat-bath temperature for lower and lower frequencies. Reprinted with permission from [Bellon *et al.*, 2001]. Copyright (2001) by the European Physical Society.

FIGURE 2.14

Effective temperature as defined in Eq. (2.110) is measured in Laponite varying both time and frequency. The height of T_e is indicated by the tone of grey in the plot. T_e is large for short times/small frequencies and decays as they increase. Reprinted figure with permission from [Bellon *et al.*, 2001]. Copyright (2001) by the European Physical Society.

range explored by these experiments, the FDR cannot be taken as a plausible definition of effective temperature.

Very recently, Jabbari-Farouji *et al.* [2007] observed no deviation from the FDT in Laponite, on the basis of microrheology experiments, rather than voltage noise, performed using optical tweezers to study the motion of silica probe particles. They compare the directly (actively) measured loss part $\chi''(\omega, t_w)$ with the response obtained from the spectral density by means of Eq. (2.104) (passive measurement), i.e., assuming the validity of the FDT. The displacement power spectral density $\tilde{C}(\omega)$ is obtained as the Fourier transform of the time correlation function of the silica bead position x:

$$2\pi\tilde{C}(\omega) = \langle |x(\omega)|^2 \rangle = \int_0^\infty dt \; \langle x(t)x(0) \rangle \, e^{i\omega t}$$

In Fig. 2.15 the two loss parts are compared at two different waiting times. The real part of the response function is computed, as well, using the principal-value Kramers-Kroning integral

$$\chi'(\omega) = \frac{2}{\pi} P \int_0^\infty d\xi \, \frac{\xi \, \chi''(\xi)}{\xi^2 - \omega^2}$$

Even though both correlation and response display aging, the actively and passively measured data sets collapse on each other at different waiting times, implying the validity of the FDT also out of equilibrium, for the timescales and the range of frequencies considered.

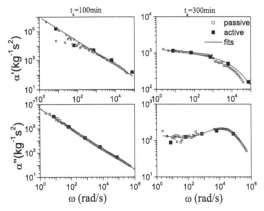

FIGURE 2.15

Real and imaginary parts of the response $\chi(\omega) \equiv \alpha(\omega)$ at $t_w \equiv t_a = 100$ and 300 min, obtained from active (solid symbols) and passive (open symbols) microrheology performed on the same 1.16 μm silica bead in the Laponite sample. In "passive" measurements the response is computed from the spectral density via the FDT, cf. Eq. (2.104). α' is calculated from α'' using a Kramers-Kronig relation (see text). Reprinted with permission from [Jabbari-Farouji *et al.*, 2007]. Copyright (2007) by the American Physical Society.

Going back to Fig. 2.8, it is interesting to note that even if very spread among theoreticians, the parametrization $\chi(C)$ is very rarely probed by experimentalists and mostly in experiments on spin-glasses, see, e.g., [Herisson & Ocio, 2002] and also the indirect determination of $\chi(C)$ curves from existing experimental data performed by Cugliandolo *et al.* [1999]. As far as we know, if truth be told, no analysis of this kind has ever appeared for real glass materials. The only example we know is the just-mentioned measurements of Buisson *et al.* [2003a] and a preliminary study on Laponite [Maggi, 2006], where, however, the occurrence of an aging regime describable by means of a unique T_e, at least for a long time window, is far from being evident.

2.8.2 Limits to the role of FDR as a temperature

The framework above reported is, actually, not valid for any aging system. In particular, drawbacks on the definition of an effective temperature by means of Eq. (2.106) have been found in model systems for which the mean-field (or else the "mean-bath," cf. Sec. 2.9) approximation does not hold. Most of these inconsistencies have been noticed in coarsening systems, for which the mean-field expectation is an infinite effective temperature as the system falls out of equilibrium (see top inset in Fig. 2.8). In one and two dimensional ferromagnetic Ising spin chains with Glauber dynamics [Godrèche & Luck, 2000; Sollich *et al.*, 2002; Mayer *et al.*, 2003; Pleiming, 2004; Mayer *et al.*, 2004; Mayer & Sollich, 2005] different couples of conjugated field-variable provide different FDRs in the same aging time sector; in plaquette spin models the FDR has been found to be nonuniversal, depending on the observation wave-vector, and even takes negative values in probes at small wave numbers [Buhot & Garrahan, 2002; Jack et al., 2006]; in the Ising spin dynamically facilitated model introduced by Fredrickson & Andersen [1984], a negative FDR is found both in one and three dimensions [Crisanti et al., 2000; Mayer et al., 2006]; in two dimensional ferromagnets with Kawasaki dynamics [Krzakala, 2005] for temperatures not much lower than the critical one and not too long waiting times, the response function is not monotonic as a function of the correlation. The FDR in coarsening systems has been systematically studied in phase-ordering systems by Lippiello & Zannetti [2000], Godrèche & Luck [2000] and Corberi et al. [2001a,b, 2002b,a, 2003b,a, 2004] that have analyzed systems with conserved and nonconserved order parameters (both scalar and vectorial) in different space dimensions, reaching the conclusion that above the lower critical dimension of a model the FDR is flat (as in the top inset of Fig. 2.8), whereas at the lower critical dimension a finite FDR occurs.

For what concerns out-of-equilibrium aging systems that resemble glasses, a generalized version of the trap model of Bouchaud [1992] (see also [Bouchaud & Dean, 1995; Monthus & Bouchaud, 1996] for a description of its glass phenomenology) has been studied, where, besides the energy, arbitrary observables (in the form of energy biased probability distributions) can be considered. Under the assumption that dynamics is Markovian, i.e., that the transition rate for the evolution from one initial state to a final one only depends on the initial state, Fielding & Sollich [2002] have shown that FDR is observable dependent.[26]

Also in dense granular materials,[27] where the heat-bath temperature actually plays no role, an effective temperature can be defined as FDR and also in this field a negative FDR has been found in numerical simulations of lattice gas models for vibrated dry granular media [Nicodemi, 1999]. However, using an Einstein relation for sheared granular matter between the diffusion of tracer particles and the response function to the shear (the mobility), recent experiments by D'Anna et al. [2003] and Song et al. [2005] have measured an effective temperature that, besides being positive, is also independent of the tracer and of the packing.

All in all, the impression that the indirect FDR measure of effective temperature is ill-defined or nonuniversal for the particular class of off-equilibrium aging systems called glasses (*stricto sensu*, cf. Chapter 1) does not gather much evidence. Indeed, going to more realistic computer models the FDR seems to hold the right properties to be a valid effective temperature candidate (observable independent, constant on a long time window), cf. the

---

[26]The reader can refer to Chapter 4 for a discussion on the validity of Markovian approximation for the relaxation of glassy systems.

[27]For the definition of temperature in diluted granular systems (granular gases) we refer, instead, to Puglisi et al. [2002]; Barrat et al. [2004]; Baldassarri et al. [2005].

Lennard-Jones and the soft spheres binary mixtures mentioned above (cf. Figs. 2.9, 2.10). Certainly, to understand the problem, a much wider inspection is necessary both in more and more realistic models and in experiments on real materials. We will see, however, in the next section and in the next chapters, that, possibly, other questions have to be addressed to express a quantity with the properties of a temperature that can encode the history of a glass on a wide timescale and describe the glass state by means of a two temperature thermodynamics.

2.9 Direct measurement of the effective temperature

We have looked, so far, at different proposals for a generalization of thermodynamics out of equilibrium involving a time dependent effective temperature next to the standard thermodynamic parameters such as heat-bath temperature, pressure, volume, magnetic field (or any other kind of external field).

In particular, we have concentrated on a two temperature picture that, assuming separation of timescales, implements a description of the out-of-equilibrium thermodynamics by means of the conjugated variables effective temperature and configurational entropy. Furthermore, we have seen that, both theoretically and experimentally, some evidence exists that the effective temperature defined as the FDR is a well-defined thermodynamic parameter for long time windows (longer than or comparable with the observation timescale). At least, this seems to hold in a certain subclass of systems, including not only mean-field models but also microscopically realistic computer glass models and real glass substances such as glycerol and Laponite, while, in general, even limiting ourselves to the class of out-of-equilibrium aging systems, some counterexamples have been found, above all in domain growth models (Sec. 2.8.2).

We will see in the following chapters how most of the approaches presented in this chapter to build a thermodynamics of the glassy state can be carried out in exactly solvable models (Chapters 3 and 4). In Chapter 6, we will also introduce the concept of *inherent structure* and, there, we will see how an effective temperature can be defined, associated with the symbolic dynamics occurring through inherent structures (rather than through the actual configurations of the system).

In Table 2.2 we report all the effective temperatures that we inspect in the book, together with the chapters where they are considered and discussed.

In order to call T_e a temperature, however, it has to satisfy some minimal thermologic requirements.

1. It should be possible to measure it *directly* with some kind of thermometer. We stress that, for instance, the FDR is an indirect measure of the

TABLE 2.2
Different possible definitions of effective temperature considered in the book.
The Boltzmann constant k_B is set equal to one.

Effective temperature	Out-of-equilibrium generalized formulas	Definition equation	Section
Fictive	$T_f(t) = \mathcal{F}_{t>t'}\left[T(t'), T_f(t')\right]$	2.24	2.2
Thermodynamic configurational	$đQ = T_e dS_c$ $T_e = \partial\Phi/\partial S_c\vert_T$	2.45, 3.143	2.4 3.3.2
Fluctuation	$\chi^{\text{fluct}} = \frac{\langle(đM)^2\rangle_{\text{slow}}}{NT_e^{\text{fl}}}$	2.95	2.7, 3.3.6
FDR	$T_e^{\text{FD}}(t, t_w) = \frac{\partial_{t_w} C(t,t_w)}{G(t,t_w)}$	2.106	2.8 3.3.4
FDR (integrated)	$T_e^{\text{FD}}(t, t_w) = \frac{C_{\text{eq}}(0) - C(t,t_w)}{\chi(t,t_w)}$	2.107	2.8, 6.3.2 6.5.1
Quasi-static	$\omega(T, T_e, H_e)$ $= e^{-\mathcal{H}_{\text{eff}}(\{x_i\}, T, H_e)/T_e}$	3.127	3.3.1
Transition rate	$\frac{W(\Delta E)}{W(-\Delta E)} = e^{\Delta E/T_e}$	3.148	3.3.3
Adiabatic	$u(t)\, z(T^\star(t))$ $= -e^{1/T^\star(t) - z(T^\star(t))}$	4.36	4.1.3
PEL equilibrium matching	$\bar{\phi}(T_e^{\text{eq}}(t)) = \bar{\phi}(T, t)$	6.37	6.3.1, 6.5.1 6.6.1
PEL configurational	$T_e^{\text{int}}(\phi, T)$ $= \left(1 + \frac{\partial f_{\text{vib}}}{\partial \phi}\right)\left(\frac{\partial s_c}{\partial \phi}\right)^{-1}$	6.39	6.3.1
PEL quasi-static	$\omega(0, T_{\text{is}}, H_{\text{is}})$ $= e^{-\Phi(\{x_i\}, H_{\text{is}})/T_{\text{is}}}$	6.78	6.6.1

T_e by means of fluctuations and responses.

2. There should be a heat flux from modes whose temperature is T_e to
   modes whose temperature is T (if the glass is formed by cooling $T_e > T$).
   Moreover, if different off-equilibrium processes are evolving and they
   are at different effective temperatures, there has to be a heat exchange
   between them.

3. Processes evolving on similar timescales should have the same effective
   temperature (constant for the whole time window considered). In other
   words, the zeroth law of thermodynamics is expected to hold on a given
   timescale.

Actually, it is not clear, so far, which properties of a thermodynamic temperature are shared by the effective temperature(s) defined in the previous
sections (e.g., Gibbs-like, conjugated to the configurational entropy, FDR).
Let us consider a glass former, a silica-based one, for instance, and suppose
that it vitrified at about 2000 Kelvin degrees and has been cooled down to

room temperature and let relax for a long time t_w (long enough to overcome the stationary regime and reach the aging regime). Following the formal definitions of two temperature thermodynamics, or the experimental measurements of FDR, its effective temperature will be larger (much larger in the present example) than the heat-bath temperature. Yet, if we touch a window at room temperature we do not burn ourselves. If we measure the temperature inside the glass with a standard thermometer the temperature detected is the room one. This is more than obvious, according to our experience. The problem, then, arises of the compatibility of the definition of an effective temperature as the "temperature of the slow processes" and what we experience in everyday life.

We know that, in a glass, fast processes occur that have thermalized on a short timescale. One can, hence, hypothesize that a standard thermometer couples right to these modes and our hand exclusively perceives these modes at equilibrium. What about the off-equilibrium α modes? They are there, by definition of a glass, and are responsible for the structural relaxation. If T_e can be identified as their temperature there must be some reason why it is not detected at all by ordinary instruments (hands included). A first reason might be that a thermometer should have response time comparable with the characteristic timescales of structural relaxation, therefore very long. A second one might be that the thermal conductivity of slow modes decays very fast so that the heat exchange with the environment is negligible. This, we notice, would have further implications on the zeroth law of thermodynamics. These reasonable assumptions would explain the compatibility of the existence of the effective temperature with the impossibility of a direct measure of it. They would, if we could verify them experimentally, or, at least, prove them theoretically.

Cugliandolo, Kurchan and Peliti [1997] tried a comprehensive treatment of the measurability problem and designed a consistent framework based on the FDR definition of effective temperature and on the observation that a standard thermometer just measures the room temperature. In that theoretical framework the effective temperature controls the direction of the heat flux between different processes (provided they evolve on comparable timescales) and effective thermalization between different processes occurs in the aging regime. Moreover, they predicted that a thermometer with long response time would measure an effective temperature larger than the room one in a glass cooled down and left at room temperature for a long time. We have seen that this last property has been indirectly verified by Grigera & Israeloff [1999]; Bellon *et al.* [2001] who measured correlations and response functions in the aging regime and calculated the corresponding evolution of their ratio (Sec. 2.8).

Following Kurchan [2005], one can try to give a pictorial idea of a direct measurement, taking the simplest thermometer, a harmonic oscillator, Θ, and coupling it to an observable O_A of a system A, yielding a perturbation as small as possible. To deal with a very small coupling, imagine first duplicating many

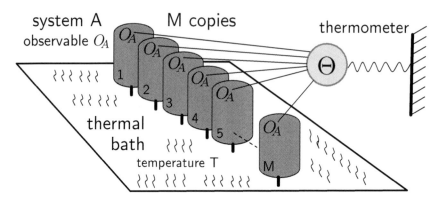

FIGURE 2.16

Thermometer Θ (a harmonic oscillator) coupled to the observable O in M copies of a system A embedded in a heat-bath at temperature T.

times our slow relaxing system at some point of its evolution and couple the same thermometer to each one of the copies (see Fig. 2.16). The more the copies, the smaller the perturbation. The observable O fluctuates around its mean value and reacts to the presence of Θ, being slightly modified. This is described, respectively, by the correlation function Eq. (2.100) and by the response function (2.101). The change in O_A causes a feedback in Θ, dissipating energy. The thermometer, thus, interacts with the noise due to the fluctuations of the observable it is measuring and is sensitive to the reaction of O_A to its own presence. At equilibrium, the two effects maintain energy equipartition, for any observable and any thermometer and in order for this to occur, the FDT, Eq. (2.103), or Eq. (2.104), must be satisfied.

Always sticking to this *Gedanken* experiment we can couple Θ to an observable O_B of yet another system, B. By the same mechanisms by which the thermometer system thermally interacts with the system A, the system B will feel the noise of O_A in the system A and the feedback of O_A to the presence of Θ, through the coupling to Θ itself (and vice versa).

What happens if we are not at equilibrium? If we are in a specific subclass on nonequilibrium, namely, in the system A, processes take place on two different, well-separated, timescales, and if each type of process behaves as if at equilibrium at a different temperature (i.e., we are considering a glass) the propagation from system A to system B will work the same way as in the equilibrium case: fluctuation and dissipation will occur in the same way also in B, even if, this time, they will yield both a fast and a slow contribution and there will be two different proportionality factors, corresponding to two different temperatures of the system.

Instead of many copies we can consider a decomposition of a single large system in many thermally coupled components. One component will, then, play the role of Θ. This scheme is self-consistent if all components are supposed to behave in the same way in the average (from the point of view of correlation functions), that is, if we can neglect the fluctuations of the fluctuations. This reminds us of the mean-field approximation, but now we deal with mean fluctuations. Kurchan [2005] calls this a "mean-bath" approximation. In this approximation, an out-of-equilibrium system belonging to the class of aging systems, whose processes evolve on well-separated timescales, consistently owns a two temperature state.

So far, however, no direct measurement procedure has been devised, nor has a comprehensive theory involving differently defined T_e been proposed. We will see in Chapter 3 how a step forward in the latter issue can be made using exactly solvable model glasses, but in general the question remains wide open.

2.A Asymptotic solution in nonlinear cooling

Let us recall Eq. (2.90),

$$\frac{dy}{dx} = \frac{y-x}{\tilde{Q}}e^{y-x/\tilde{Q}} \tag{2.A.1}$$

We shall analyze this here for large positive and large negative x, that is, for temperatures far above and far below the glass transition. We introduce $w = (1/\tilde{Q} - 1)x$. For large negative w we set $y = x - \ln z(v)$ and $v = e^w = e^{-(\tilde{Q}-1)x/\tilde{Q}}$, to obtain the differential equation

$$(\tilde{Q} - 1)v^2 z'(v) + \ln z(v) + \tilde{Q}vz(v) = 0 \tag{2.A.2}$$

This allows a series expansion in the small variable v,

$$z = 1 \ -\tilde{Q}v + \frac{\tilde{Q}}{2}(5\tilde{Q} - 2)v^2 - \frac{\tilde{Q}}{3}(29\tilde{Q}^2 - 27\tilde{Q} + 6)v^3$$

$$+ \frac{\tilde{Q}}{24}(1181\tilde{Q}^3 - 1812\tilde{Q}^2 + 900\tilde{Q} - 144)v^4 \tag{2.A.3}$$

$$- \frac{\tilde{Q}}{5}(1529\tilde{Q}^4 - 3345\tilde{Q}^3 + 2690\tilde{Q}^2 - 940\tilde{Q} + 120)v^5 + \cdots$$

and it implies for the specific heat factor an exponential approach to equilibrium

$$\left.\frac{\partial T_e}{\partial T}\right|_H = \frac{dy}{dx} = -\frac{\ln z(v)}{\tilde{Q}vz(v)}$$

$$= 1 + (\tilde{Q} - 1)[-v + 2(2\tilde{Q} - 1)v^2 - \frac{3}{2}(5\tilde{Q} - 2)(3\tilde{Q} - 2)v^3$$

$$+ \frac{4}{3}(4\tilde{Q} - 3)(29\tilde{Q}^2 - 27\tilde{Q} + 6)v^4 + \cdots] \tag{2.A.4}$$

When $\tilde{Q} > 1$ or $\tilde{Q} < 0$, this result applies for a large positive reduced temperature x, while for $0 < \tilde{Q} < 1$, it applies for large negative x. In all these cases, w is large negative and $v = e^w$ is exponentially small.

For large positive w we set

$$s = \frac{1}{w} = -\frac{\tilde{Q}}{(\tilde{Q}-1)x} \tag{2.A.5}$$

and

$$y = -\frac{1}{(\tilde{Q}-1)s} + \ln s - \ln u(s) \tag{2.A.6}$$

which yields the differential equation

$$u(s) = 1 + s\ln s - s\ln u(s) - (\tilde{Q}-1)su(s) + (\tilde{Q}-1)s^2 u'(s) \tag{2.A.7}$$

By iteration there arises an expansion in powers of s and $\Lambda = \ln s$

$$
\begin{aligned}
u(s) = \ & 1 + (\Lambda - \tilde{Q} + 1)s + (2\tilde{Q} - 2 - \Lambda)s^2 + \left(-8\tilde{Q} + \frac{11}{2} + 3\Lambda + \frac{1}{2}\Lambda^2\right. \\
& \left. -2\Lambda\tilde{Q} + \frac{5}{2}\tilde{Q}^2\right)s^3 + \left(-\frac{57}{2}\tilde{Q}^2 + \frac{16}{3}\tilde{Q}^3 + 45\tilde{Q} + 18\Lambda\tilde{Q} + 2\Lambda^2\tilde{Q}\right. \\
& \left. -5\Lambda\tilde{Q}^2 - \frac{1}{3}\Lambda^3 - \frac{7}{2}\Lambda^2 - 14\Lambda - \frac{131}{6}\right)s^4 + \cdots
\end{aligned}
\tag{2.A.8}
$$

This implies for the specific heat factor an algebraic approach to the low temperature value $1/\tilde{Q}$, with logarithms in the sub-leading terms,

$$\left.\frac{\partial T_e}{\partial T}\right|_H = \frac{dy}{dx} = \frac{1 + s\Lambda - s\ln u(s)}{\tilde{Q}u(s)} \tag{2.A.9}$$

$$
\begin{aligned}
= \ & \frac{1}{\tilde{Q}} + \frac{\tilde{Q}-1}{\tilde{Q}}\left\{ s + (\tilde{Q} - \Lambda - 2)\,s^2 + \left[\tilde{Q}^2 - (2\Lambda + 7)\tilde{Q} + \Lambda^2 + 5\Lambda + 7\right]s^3 \right. \\
& + \left[\tilde{Q}^3 - \left(\frac{35}{2} + 3\Lambda\right)\tilde{Q}^2 + (23\Lambda + 46 + 3\Lambda^2)\,\tilde{Q}\right. \\
& \left.\left. -26\Lambda - \frac{17}{2}\Lambda^2 - \frac{61}{2} - \Lambda^3\right]s^4\right\}
\end{aligned}
$$

When $\tilde{Q} > 1$ or $\tilde{Q} < 0$ this applies for large negative x; if $0 < \tilde{Q} < 1$ it applies for large positive x. In all these cases, the variable s is small.

It is evident that both Eqs. (2.A.4) and (2.A.9) go the correct value $\partial T_e/\partial T = 1$ in the limit $\tilde{Q} \to 1$. Indeed, in that limit the system cools down, but remains in equilibrium at all times, because the procedure, Eq. (2.76), reduces to $T(t) = \tau_{\text{eq}}^{-1}(t)$, just imposing a sliding state, always in good enough equilibrium.

3

Exactly solvable models for the glassy state: a dynamic approach to a dynamic phenomenon

The study of exactly solvable models has been always an active area of research in the field of statistical physics. They help us to grasp general principles governing the physical behavior of realistic systems that, due to the complicated interactions among their different constituents, cannot be predicted using standard perturbative techniques. Glasses in general are systems falling into this category. The slow relaxation of glasses observed in the laboratory is a consequence of the simultaneous interplay of its components that yields a very complex and rich phenomenology.

Firstly, we want to get more insight into the glassy dynamics, in its various aspects, exploiting the analytical solubility of the models we will discuss. Indeed, thanks to their simplicity, the features of the glassy materials can be connected in a direct correspondence with given elements of the models. We can even switch on and off certain properties or certain dynamic behaviors, tuning the model parameters or implementing a given - *facilitated* or *constrained* - dynamics in alternative ways. Furthermore, in some cases, the thermodynamic state functions, including the configurational entropy, can be computed as functions of the dynamic variables of the model. Our second goal is to check the principal applicability and generality of the concept of *effective temperature*, very often discussed in literature in many different approaches (see Chapter 2) and to verify whether the possibility exists of inserting such a parameter into the construction of a consistent out-of-equilibrium thermodynamic theory. The question whether there exists a single effective temperature encoding the aging dynamics and whether this coincides with the fluctuation-dissipation ratio is still controversial (see Sec. 2.8.2). We believe that this and other related questions can be better addressed by analyzing simple models. Recent reviews of kinetically constrained models for glass are, e.g., Workshop [2002], Ritort & Sollich [2003], Leonard *et al.* [2007].

In this chapter, we propose a set of dynamically facilitated models undergoing a Monte Carlo parallel dynamics. For these models the statics is trivial and, nevertheless, the exactly solvable dynamics exhibits interesting glassy aspects. Basically all of the relevant features of much more complicated real glasses are displayed, such as aging, diverging relaxation time to equilibrium (either Arrhenius or Vogel-Fulcher), configurational entropy, the Adam-Gibbs relation between relaxation time and configurational entropy,

Kauzmann transition, off-equilibrium fluctuation-dissipation relation and Kovacs effect.

In the next chapter, we present *urn models* whose dynamics proceeds through *entropic barriers* and that can be exactly computed in a certain adiabatic approximation. The dynamic behavior describes the slow relaxation and many related properties typical of glasses. Also there we will face the case where the dynamical relaxation of different energy modes can be made explicitly clear. In Chapter 5 we present another solvable model based on directed polymers, that is simple enough to be analytically worked out and yet displays glassy behavior.

The possibility of computing an exact solution for the dynamics allows for a precise formulation of the two temperature picture presented in Secs. 2.3, 2.4, 2.5. Even though the physics of the class of models that we are going to discuss in this chapter is simple, we shall formulate general aspects of the results by analyzing them in thermodynamic language. This thermodynamic formulation also incorporates the interpretation of the fluctuation-dissipation ratio (FDR) as an effective temperature, as exposed in Sec. 2.8. Indeed, the relation between thermal correlation functions and responses to external drivings has become a central point in investigating out-of-equilibrium systems. The approach of Sec. 3.3, will show that the effective temperature that occurs in thermodynamics and the one that occurs in the FDR can be equal in the aging regime only if the relaxation is *slow enough*. We will discuss and quantify the latter expression in terms of the parameters of the models that will be presented in the following.

The working hypothesis at the basis of the simplified models we will deal with is shaped on one particular physical property of glasses: the exponential divergence of timescales around the glass transition.[1] This induces an *asymptotic decoupling of the time-decades* (cf. Sec. 1.1). The reasonable assumption is then made that, in a glass system that has aged a long time t, all processes with equilibration time much less than t are in equilibrium (the β processes) while those evolving on timescales much larger than t (if existing) are still quenched, leaving the processes with a timescale of the order t (i.e., the α processes) as the only interesting ones. The asymptotic decoupling of timescales is the input for the family of models we are going to analyze in Sec. 3.2 and could be the basis for a generalization of equilibrium thermodynamics to systems out of equilibrium. We will see, specifically in Sec. 3.3, how this approach involves systems in which one extra variable (the effective temperature) describes the nonequilibrium physics and if and up to which extent the addition of this single variable is sufficient to yield a consistent thermodynamic theory, able to describe even typical off-equilibrium phenomena such as memory effects (Sec. 3.5). Always employing harmonic oscillators, in Sec. 3.6 we will discuss the problem of the direct measurement of an effective temperature and, in Sec. 3.7, its dependence on different frequency modes.

---

[1] As opposed to the algebraic divergence in standard continuous phase transitions.

3.1 Harmonic oscillator model

In order to introduce the dynamic approach, we will first consider the study of a simplified model, the harmonic oscillator (HO) model of Bonilla *et al.* [1996a], to which an external field contribution is added [Nieuwenhuizen, 1998c]. The aim is to be very clear about the procedure involved, presenting the principal steps in the derivation of the equations of motion, before generalizing the same dynamics to a class of models representing both strong and fragile glasses.

The Hamiltonian of the harmonic oscillator model is

$$\mathcal{H}_{\rm ho}[\{x_i\}] = \sum_{i=1}^{N} \left(\frac{K}{2} x_i^2 - H x_i \right) \tag{3.1}$$

where K is the Hooke constant, H an external field and $\{x_i\} \in] - \infty, \infty[$ the harmonic oscillator positions.

We also define

$$m_1 \equiv \frac{M_1}{N} \equiv \frac{1}{N} \sum_{i=1}^{N} x_i; \qquad m_2 \equiv \frac{M_2}{N} \equiv \frac{1}{N} \sum_{i=1}^{N} x_i^2 \tag{3.2}$$

The statics of the model is trivial. The free energy is simply given by

$$\frac{F}{N} = -\frac{1}{2} \frac{H^2}{K} - \frac{T}{2} \log \frac{2\pi T}{K} \tag{3.3}$$

and the equilibrium value of the harmonic oscillator position is

$$x_i = \frac{H}{K} \qquad \forall i \tag{3.4}$$

independent of the temperature.

The entropic contribution to Eq. (3.3), $S = \frac{1}{2} \log T/K + \text{const}$, is clearly ill-defined at zero temperature, but this is a known artifact of the fact that the HO variables are continuous even at zero temperature. No quantization is considered and the Nernst principle (vanishing entropy at zero temperature) is, thus, not expected to be satisfied.

Many generalizations of this model have been studied for gaining insight into the properties of slowly relaxing materials. Besides the two timescales generalization [Nieuwenhuizen, 1999; Leuzzi & Nieuwenhuizen, 2001a,b, 2002] that we will extensively consider in Sec. 3.2, further examples are

1. the homogeneous potential model [Ritort, 2004]:

$$\mathcal{H}[\{x_i\}] = \frac{K}{p} \sum_{i=1}^{N} x_i^p \qquad p > 2 \tag{3.5}$$

2. the wedge potential model [Ritort, 2004]:

$$\mathcal{H}[\{x_i\}] = K \sum_{i=1}^{N} |x_i| \tag{3.6}$$

3. the disordered harmonic oscillator model [Garriga & Ritort, 2005]:

$$\mathcal{H}[\{x_i\}; \{\epsilon_i\}] = \frac{K}{2} \sum_{i=1}^{N} \epsilon_i x_i^2 \tag{3.7}$$

where $\{\epsilon_i\}$ takes random values with a certain distribution (Sec. 3.7 is dedicated to it)

4. spherical spins in a random and an external field [Nieuwenhuizen, 1998c, 2000]

$$\mathcal{H}[\{S_i\}] = -J \sum_{i=1}^{N} x_i S_i - L \sum_{i=1}^{N} S_i \tag{3.8}$$

where the x_i are quenched random Gaussian variables with zero average and unit variance and the spherical spins have arbitrary lengths under the constraint $\sum_{i=1}^{N} S_i^2 = N$.

3.1.1   Analytically solvable Monte Carlo dynamics

The dynamics that we apply to the system is a parallel Monte Carlo dynamics, as first introduced by Bonilla *et al.* [1996a]. The thus-obtained dynamical model composed by the simple local Hamiltonian (3.1) and the dynamic rules that we are going to present in the following is analytically solvable.

In a Monte Carlo step, a random updating of the harmonic oscillator variables is performed,

$$x_i \to x_i' = x_i + \frac{r_i}{\sqrt{N}} \tag{3.9}$$

where the variables $\{r_i\}$ are independent random variables with a Gaussian distribution of zero mean and a variance Δ^2. We indicate by δE the energy between the proposed new configuration and the initial one, viz., $\delta E \equiv \mathcal{H}(\{x_i'\}) - \mathcal{H}(\{x_i\})$. The dynamics is set by the Metropolis probability W of performing a randomly proposed update of all oscillator positions:

$$W(\delta E) = \begin{cases} 1 & \text{if } \delta E \leq 0 \\ \exp(-\beta \delta E) & \text{if } \delta E > 0 \end{cases} \tag{3.10}$$

If the energy of the new configuration is higher than the energy of the initial configuration ($\delta E > 0$) the move is accepted with a probability $\exp(-\beta \delta E)$; if δE is negative, the proposed move is always accepted.

We look at how the quantities M_1 and M_2 are updated in a Monte Carlo (MC) step. Let us denote the changes induced by Eq. (3.9) as

$$y_1 \equiv M_1' - M_1 = \sum_{i=1}^{N} \frac{r_i}{\sqrt{N}} \qquad (3.11)$$

$$y_2 \equiv M_2' - M_2 = \sum_{i=1}^{N} N \left(\frac{2}{\sqrt{N}} r_i x_i + \frac{r_i^2}{N} \right) \qquad (3.12)$$

As a consequence, the distribution function of y_1 and y_2, for prescribed values of m_1 and m_2, is:

$$p(y_1, y_2 | m_1, m_2) \equiv \int \prod_{i=1}^{N} \frac{dr_i}{\sqrt{2\pi \Delta^2}} \exp\left(-\frac{r_i^2}{2\Delta^2} \right) \qquad (3.13)$$

$$\times \int \prod_{a=1}^{2} \delta \left(\sum_{i=1}^{N} (x_i')^a - \sum_{i=1}^{N} (x_i)^a - y_a \right)$$

$$= \frac{1}{4\pi\Delta^2 \sqrt{m_2 - m_1^2}} \exp\left(-\frac{y_1^2}{2\Delta^2} - \frac{(y_2 - \Delta^2 - 2y_1 m_1)^2}{8\Delta^2 (m_2 - m_1^2)} \right)$$

As a function of the M's updates, the energy difference reads

$$x \equiv \delta E = \frac{K}{2} y_2 - H\, y_1 \qquad (3.14)$$

In terms of x and $y \equiv y_1$ the distribution function can be formally written as the product of two Gaussian distributions:

$$p(y_1, y_2 | m_1, m_2)\, dy_1\, dy_2 \qquad (3.15)$$

$$= \frac{1}{\sqrt{2\pi \Delta_x}} \exp\left(-\frac{(x - \bar{x})^2}{2\Delta_x} \right) \frac{1}{\sqrt{2\pi \Delta_y}} \exp\left(-\frac{[y - \bar{y}(x)]^2}{2\Delta_y} \right) dx\, dy$$

$$= \qquad p(x|m_1, m_2) \qquad p(y|x, m_1, m_2) \qquad dx\, dy$$

where

$$\bar{x} = \frac{\Delta^2 K}{2}, \qquad\qquad \Delta_x = \Delta^2 K^2 \left(\mu_2 + \mu_1^2 \right), \qquad (3.16)$$

$$\bar{y}(x) = \frac{\mu_1}{\mu_2 + \mu_1^2} \frac{\bar{x} - x}{K}, \qquad \Delta_y = \frac{\Delta^2 \mu_2}{\mu_2 + \mu_1^2} \qquad (3.17)$$

In the above expressions we have introduced the abbreviations

$$\mu_1 \equiv \frac{H}{K} - m_1 \qquad (3.18)$$

$$\mu_2 \equiv m_2 - m_1^2 \qquad (3.19)$$

They are just the deviations from the equilibrium values of $m_{1,2}$ at zero temperature [cf. Eqs. (3.2, 3.4)]. As equilibrium is approached, $\mu_1 \to 0$ and $\mu_2 \to T/K$.

We now derive the equation of motion of the dynamic observables $m_{1,2}$. After a MC update from time t to time $t + dt$ the probability distribution of the new values M'_1, M'_2 after the update evolves like

$$
P(M'_1, M'_2, t + dt) = \int \prod_{a=1}^{2} dM_a \, P(M_1, M_2, t) \int \prod_{a=1}^{2} dy_a \, p(y_1, y_2|m_1, m_2)
$$

$$
\times \left[W(\beta x) \prod_{a=1}^{2} \delta(M'_a - M_a - y_a) + (1 - W(\beta x)) \prod_{a=1}^{2} \delta(M'_a - M_a) \right]
$$

$$
= P(M'_1, M'_2, t) + \int \prod_{a=1}^{2} dM_a \, dy_a \, P(M_1, M_2, t) \, p(y_1, y_2|m_1, m_2)
$$

$$
\times W(\beta x) \left[\prod_{a=1}^{2} \delta(M'_a - M_a - y_a) - \prod_{a=1}^{2} \delta(M'_a - M_a) \right] \tag{3.20}
$$

where we have used the closure property

$$
\int \prod_{a=1}^{2} dy_a \, p(y_1, y_2|m_1, m_2) = 1 \tag{3.21}
$$

This induces, in the evolution of the mean values of the Ms,

$$
\langle M_b(t + dt) \rangle = \int \prod_{a=1}^{2} dM'_a \, M'_b \, P(M'_1, M'_2, t + dt) \tag{3.22}
$$

$$
= \langle M_b(t) \rangle + \int \prod_{a=1}^{2} dM_a \, P(M_1, M_2, t) \int \prod_{a=1}^{2} dy_a \, p(y_1, y_2|m_1, m_2) \, y_b \, W(\beta x)
$$

The time interval dt of a MC step is then set equal to $dt = 1/N$, that is, it becomes infinitesimal in the thermodynamic limit. As $N \to \infty$ the probability distribution of the Ms is strictly peaked around their mean values, i.e.,

$$
\lim_{N \to \infty} P(M_1, M_2, t) = \prod_{a=1}^{2} \delta(M_a - \langle M_a(t) \rangle) \tag{3.23}
$$

and one has

$$
\frac{d \langle m_a \rangle}{dt} = \lim_{N \to \infty} \frac{1}{N} \frac{\langle M_a(t + dt) \rangle - \langle M_a(t) \rangle}{dt} \tag{3.24}
$$

$$
= \int \prod_{k=1}^{2} dy_k \, y_a \, p(y_1, y_2| \langle m_1(t) \rangle, \langle m_2(t) \rangle) W(\beta x)
$$

Let us denote $\langle m_i(t) \rangle = m_i(t)$ from now on, to ease the notation.

Going to integration variables x, y, the integration over y in Eq. (3.24) can be carried out, and we obtain the MC equations of motion:[2]

$$\dot{m}_1(t) = \int dx \ W(\beta x) \ \bar{y}(x) \ p(x|m_1(t), m_2(t)) \tag{3.25}$$

$$\dot{m}_2(t) = \frac{2}{K} \int dx \ W(\beta x) \left[x + H \ \bar{y}(x)\right] p(x|m_1(t), m_2(t)) \tag{3.26}$$

where $\bar{y}$ is defined in Eq. (3.17). Let us define the parameter

$$\alpha \equiv \frac{\bar{x}}{\sqrt{2\Delta_x}} = \frac{\Delta}{\sqrt{8(\mu_2 + \mu_1^2)}} \tag{3.27}$$

Dynamically speaking this quantity is large when equilibrium is approached and temperature is small. We also introduce two basic MC integrals. The first one is the acceptance rate of the proposed MC move,

$$A_0(t) = \int dx \ W(\beta x) \ p(x|m_1(t), m_2(t)) \tag{3.28}$$

For large times, A_0 is proportional to the inverse of the relaxation time to equilibrium. The second integral is the average of the change in energy in the MC move,

$$A_1(t) = \int dx \ W(\beta x) \ x \ p(x|m_1(t), m_2(t)) \tag{3.29}$$

In terms of A_0 and A_1 one can rewrite Eqs. (3.25)-(3.25)

$$\dot{m}_1(t) = 4\alpha^2(t)\mu_1(t) \left[A_0(t) - \frac{2}{\Delta^2 K} A_1(t)\right], \tag{3.30}$$

$$\dot{m}_2(t) = 8\alpha^2(t)\mu_1(t)\frac{H}{K}A_0(t) + \frac{2}{K}\left[1 - \frac{8\alpha^2(t)\mu_1(t)}{\Delta^2}\frac{H}{K}\right] A_1(t) \tag{3.31}$$

It is useful, to identify the long time contributions in the equations, to compute $A_0(t)$ and $A_1(t)$ as functions of the parameter $\alpha(t)$. The acceptance rate is

$$A_0(t) = \int_{-\infty}^{0} \frac{dx}{\sqrt{2\pi\Delta_x}} \exp\left[-\frac{(x - \bar{x})^2}{2\Delta_x}\right] + \int_{0}^{\infty} \frac{dx}{\sqrt{2\pi\Delta_x}} e^{-\beta x} \exp\left[-\frac{(x - \bar{x})^2}{2\Delta_x}\right] \tag{3.32}$$

---

[2]These equations hold provided that spring constant, external field and temperature are kept constant. If they change in time, extra contributions appear, that can be identified in a straightforward manner, starting with the proper expression $\delta E = \frac{1}{2}Ky_2 - Hy_1 + dt\left(\dot{K}m_2 + \dot{H}m_1\right)|_{m_1,m_2}$, generalizing Eq. (3.14) accordingly. We shall not go into this any further here.

Making the transformations

$$z = \frac{(x - \bar{x})}{\sqrt{2\Delta_x}}, \qquad \zeta = \frac{x}{\sqrt{2\Delta_x}} + \sqrt{\frac{\Delta_x}{2}}\left(\beta - \frac{\bar{x}}{\Delta_x}\right) \qquad (3.33)$$

in the first and the second integral, respectively, $A_0(t)$ is rewritten as

$$A_0(t) = \frac{1}{2}\text{erfc}[\alpha(t)] + \frac{b(\alpha(t))}{2} \qquad (3.34)$$

where erfc is the complementary error function [defined in Eq. (3.A.2)] and we have introduced the function

$$b(\alpha(t)) \equiv \exp\left[\frac{\beta^2 \bar{x}^2}{4\alpha^2(t)} - \beta\bar{x}\right]\text{erfc}\left[\beta\frac{\bar{x}}{2\alpha(t)} - \alpha(t)\right] \qquad (3.35)$$

In a similar way one also can express $A_1(t)$ as:

$$A_1(t) = \frac{\bar{x}}{2}\left[\text{erfc}(\alpha(t)) + b(\alpha(t))\left(1 - \frac{\beta\bar{x}}{2\alpha^2(t)}\right)\right] \qquad (3.36)$$

The case of our interest is the slow relaxation regime, thus large α. Therefore, one can expand the above expressions for large α, see Eq. (3.A.3).

Since everything in the long time regime is more clearly expressed in terms of μ_1, μ_2, it is more useful to work with the equations of motion for these observables, connected to Eqs. (3.30)-(3.31) by

$$\dot{\mu}_1(t) = -\dot{m}_1(t), \qquad (3.37)$$
$$\dot{\mu}_2(t) = \dot{m}_2(t) - 2m_1(t)\,\dot{m}_1(t) \qquad (3.38)$$

We will come back to these equations in Sec. 3.2.2, in the framework of an interesting generalization of the HO model.

3.1.2   Parallel Monte Carlo versus Langevin dynamics

We have so far presented the formalism for an analytical treatment of the parallel MC dynamics that leads, as we will see in the next sections, to slow relaxation and aging. Before going into the details of the resolution of the MC equation, we would like to explain why the Langevin dynamics has not been applied in this case. In the field of disordered systems, indeed, the dynamics is usually computed following Langevin's approach, that is, embedding the (disordered) Hamiltonian dynamics into a heat-bath, represented by a white noise of variance T. Why are Langevin dynamics not applied in our simplified models? What are the differences between the two, both widely used, dynamic approaches?

The Monte Carlo-Metropolis algorithm is the kernel of very many numerical simulations of glassy materials of any kind, among which are coarsening systems, spin-glasses, structural glasses, colloids and polymers. We saw in the previous section - and we will verify in detail later on - that a main role in the MC equations of motion is played by the acceptance ratio A_0: the frequency by which the proposed updates of the harmonic oscillator positions are accepted. In the infinite time limit the acceptance rate is the inverse of the relaxation time to equilibrium τ_{eq}.

The nature of the off-equilibrium MC dynamics strongly depends on the behavior of A_0. Indeed, in models like HO (and its generalizations in the next sections), the very slow evolution dynamic, including the Arrhenius temperature behavior of the relaxation time are connected to the fact that acceptance rates are very small. In Langevin dynamics, no analogue of the acceptance rate is present, though, with the consequence that no slow relaxation, no glass-like behavior, no aging, no Arrhenius law ever arise. Thus, our choice for parallel updating is lastly a compromise between having a simple model for a system with glassy dynamics and demanding analytic solvability.

Exactly at $T = 0$, furthermore, the MC dynamics is intrinsically non-ergodic because only moves decreasing the energy value can be accepted. A dynamic regime is, thus, reached that cannot be traced back to any regime displayed in Langevin dynamics.

Langevin dynamics for the HO model

The Langevin equations for the HO model in an external field H are

$$\frac{\partial x_i(t)}{\partial t} = -\frac{\partial \mathcal{H}}{\partial x_i} + \eta_i(t) \qquad \forall i \tag{3.39}$$

where the white noise $\eta_i(t)$ is such that

$$\langle \eta_i(t)\eta_j(t')\rangle = 2T\delta_{ij}\delta(t - t') \tag{3.40}$$

Manipulating the above equations and averaging over the noise one obtains the equations of motion

$$\dot{\mu}_1(t) = -K\mu_1(t), \tag{3.41}$$
$$\dot{\mu}_2(t) = 2\left[T - K\mu_2(t) - H\mu_1(t)\right] \tag{3.42}$$

whose solution is

$$\mu_1(t) = \mu_1(0)\, e^{-Kt}, \tag{3.43}$$
$$\mu_2(t) = \frac{T}{K} + \left[\mu_2(0) - \frac{T}{K}\right] e^{-2K\,t}$$
$$- \left\{\frac{\mu_2(0)}{K} + 2\left[\mu_2(0) - \frac{T}{K}\right]\right\} e^{-Kt}\left(1 - e^{-Kt}\right) \tag{3.44}$$

i.e., an exponential decay with a temperature-independent relaxation time that is a straightforward generalization of the Brownian harmonic oscillator [van Kampen, 1981].

Parallel MC and Langevin coincide as $\Delta \to 0$.

Bonilla *et al.* [1996a,b, 1998] showed, for the spin-glass spherical version of the Sherrington-Kirkpatrick model and for the HO model that, when the variance of the parallel MC update move is vanishingly small, the MC equations of motion, e.g., Eqs. (3.30, 3.31), can be reduced to the Langevin equations, provided the scaling

$$t_{MC} \to t_L = t_{MC} \beta \Delta^2 / 2 \qquad (3.45)$$

is imposed.

In the model of harmonic oscillators $\{x_i\}$ of spring constant K, embedded in an external field H linearly coupling to x_i, Eqs. (3.34, 3.36, 3.30) and (3.31) tend, for small Δ, to

$$A_0 = 1 \qquad (3.46)$$

$$A_1 = \frac{\Delta^2 K}{2} \left[1 - \beta K \left(\mu_2 + \mu_1^2\right)\right] \qquad (3.47)$$

$$\dot{m}_1 = \frac{\beta \Delta^2}{2} K \mu_1 \qquad (3.48)$$

$$\dot{m}_2 = \Delta^2 \left[1 - \beta K \left(\mu_2 + \mu_1^2\right)\right] \qquad (3.49)$$

Imposing the rescaling Eq. (3.45) one recovers Eqs. (3.41, 3.42), together with their exponential relaxation to equilibrium. When, instead, the variance of distribution of the update change $\{r_i\}$ is finite, the acceptance ratio decreases with time. The larger is Δ, the faster A_0 decreases.

When using the MC dynamics in the HO model and generalizations thereof, even though there is no energy barrier to be overcome, the system spends a long time looking for a configuration of lower energy because the available configurational space decreases as the system evolves toward equilibrium. This regime can be referred to as an *entropic barrier* regime and takes over when the value of the acceptance rate A_0 becomes small. It has no counterpart in Langevin relaxation dynamics.

The MC updating that we implement is parallel and it is this particular feature that yields the collective behavior leading to exponentially divergent timescales, notably in a model with no interactions between particles. A sequential MC updating, like the Langevin dynamics, does not produce any glassy effect. In this sense, there is an analogy with facilitated Ising models [Fredrickson & Andersen, 1984], and with the kinetic lattice-glass model with contrived dynamics of Kob & Andersen [1993], where the transition probabilities depend dynamically on the neighboring configuration; this dynamics may induce glassy behavior in situations where ordinary Glauber [1963] or Langevin dynamics would not. Models of this type may give valuable insight into the long time dynamics, at least within a class that exhibits some longtime universality.

3.2 Kinetic models with separation of timescales: harmonic oscillator spherical spin models

In the HO model in an external field, as well as in any other model of the class recalled in Sec. 3.1, only variables evolving on one, long, timescale were considered, uncoupled to any fast relaxing process. Having determined the essential formalism for these models we move now to study a related class of models possessing both "fast" and "slow" variables, the latter still undergoing the same MC dynamics as before. The existence of fast modes, and the nonlinearity that they induce, is the source, e.g., of the memory effects reported in Sec. 3.5, that would not be reproducible otherwise.

The model we are going to analyze is described by the Hamiltonian

$$\mathcal{H}[\{x_i\}, \{S_i\}] = \sum_{i=1}^{N} \left(\frac{1}{2}Kx_i^2 - H\,x_i - J\,x_iS_i - L\,S_i \right) \tag{3.50}$$

where N is the size of the system and $\{x_i\}$ and $\{S_i\}$ are continuous variables, the last satisfying a spherical constraint: $\sum_i S_i^2 = N$. We call them respectively *harmonic oscillators* and *spherical spins* (accordingly, we will use the abbreviation "HOSS model"). The parameter J is the coupling constant between $\{x_i\}$ and $\{S_i\}$ on the same site i and L is the external field acting on the spherical spins. Relying on the hypothesis of asymptotic decoupling of timescales, we assume that the $\{S_i\}$ relaxes to equilibrium on a timescale much shorter than the one of the harmonic oscillators. In other words, the spins have fast dynamics, which keeps them in instantaneous equilibrium, while the oscillators are slow and may exhibit glassy dynamics.

This model is a drastic simplification of a system of interacting particles with an inner degree of freedom, varying on a timescale shorter than the particles' motion. The potential is harmonic and each particle independently interacts with the medium, encoded in the spring constant K. The inner degree of freedom, the spherical spin S_i, is coupled exclusively to the position of the particle (x_i), and the only global interaction is provided by the spherical constraint.

The model owns a very simple statics and evolves with the parallel dynamics exposed in Sec. 3.1, that retains the fundamental collective nature of the glassy dynamics. We will see how this dynamics undergoes a huge slowing down as the system is cooled down and which kind of aging dynamics the system sets out (Secs. 3.2.2-3.2.4). We can also implement the occurrence of an underlying Kauzmann transition (cf. Sec. 1.4.2). This allows us to perform a dynamic probe also below the Kauzmann temperature, thus getting information in a regime where experimental results on relaxation dynamics are not achievable (Secs. 3.2.4, 3.4 and Appendix 3.A). Memory effects taking place when the system is cooled down and then reheated are eventually considered and satisfactorily reproduced (as reported in Sec. 3.5).

As we said, the spins represent the fast modes and the harmonic oscillators the slow ones. From the point of view of the motion of the $\{x_i\}$, the spins are just a noise. To describe the long time regime of the $\{x_i\}$ we can average over this noise by performing the computation of the $\{S_i\}$ partition function, obtaining an effective Hamiltonian depending only on the $\{x_i\}$, that determines the dynamics of these variables.[3]

Using the saddle point approximation for large N, we find the following partition function for the subsystem of spins at a given $\{x_i\}$ configuration:

$$Z_S(\{x_i\}) = \int \left(\prod_{i=1}^{N} dS_i \right) \exp\{-\beta \mathcal{H}\left[\{x_i\}, \{S_i\}\right]\} \, \delta \left(\sum_{i=1}^{N} S_i^2 - N \right)$$

$$\simeq \exp \left[-\beta N \left(\frac{K}{2} m_2 - H m_1 - w + \frac{T}{2} \log \frac{w + \frac{T}{2}}{2\pi T} \right) \right] \quad (3.51)$$

where $\mathcal{D}S \equiv \prod_i dS_i$. The explicit dependence on $\{x_i\}$ is expressed in the argument of Z_S and coded in the $m_{1,2}$, defined in Eq. (3.2). We have, further, introduced the function of the harmonic oscillator positions

$$w(\{x_i\}) \equiv \sqrt{J^2 m_2 + 2JL m_1 + L^2 + \frac{T^2}{4}} \quad (3.52)$$

We stress that $m_{1,2}$ are just abbreviations for the x_i-dependent expressions. We define, then, the effective Hamiltonian $\mathcal{H}_{\text{eff}}(\{x_i\}) \equiv -T \log Z_S(\{x_i\})$, which is the free energy for a given configuration of $\{x_i\}$:

$$\mathcal{H}_{\text{eff}}(\{x_i\}) = \frac{K}{2} m_2 N - H m_1 N - w N + \frac{TN}{2} \log \frac{w + \frac{T}{2}}{T} \quad (3.53)$$

This can also be written in terms of the internal energy $U(\{x_i\})$ and of the entropy $S_{\text{ep}}(\{x_i\})$ of the equilibrium processes:

$$\mathcal{H}_{\text{eff}}(\{x_i\}) = U(\{x_i\}) - T S_{\text{ep}}(\{x_i\}) \quad (3.54)$$

$$\frac{U(\{x_i\})}{N} = \frac{K}{2} m_2 - H \, m_1 - w + \frac{T}{2} \quad (3.55)$$

$$\frac{S_{\text{ep}}(\{x_i\})}{N} = \frac{1}{2} - \frac{1}{2} \log \frac{w + T/2}{T} \quad (3.56)$$

The function U is the internal energy of the fast processes given a blocked, quenched, $\{x_i\}$ configuration, i.e., it is the average over the spins of the Hamiltonian (3.50). The quantity S_{ep} is the entropy of the spins. We stress that, depending on the timescale of interest, $\mathcal{H}_{\text{eff}}(\{x_i\})$ is both an effective Hamiltonian, for what concerns the x_i variables (mimicking α processes), and a free energy, for the spins (β-like processes).

---

[3]This approach is fully in line with Ginzburg-Landau theory, where fast modes are integrated out to yield the Ginzburg-Landau free energy for the slow modes. Also there, the role of the fast modes is coded in the temperature dependence of prefactors.

To shorten the notation and to allow for a straightforward connection with the simpler case of Sec. 3.1, we define here the effective "spring constant" $\tilde{K}$ and the effective "external field" $\tilde{H}$, viz.

$$\tilde{K}(\{x_i\}) = K - \frac{J^2}{w(\{x_i\}) + T/2}, \qquad \tilde{H}(\{x_i\}) = H + \frac{JL}{w(\{x_i\}) + T/2} \quad (3.57)$$

We also define the (truly) constant combination

$$D \equiv HJ + KL = \tilde{H}J + \tilde{K}L \quad (3.58)$$

If the interaction between x_i and S_i is absent (complete decoupling, $J = 0$), it holds that $\tilde{K} = K$ and $\tilde{H} = H$, without any nonlinear contributions, and Eq. (3.53) reduces to Eq. (3.1) (apart from an irrelevant constant related to the free energy of the uncoupled spins).

Before introducing the dynamics, we first derive the statics of the model.

3.2.1 Statics and phase space constraint

The partition function of the whole system at equilibrium is:

$$Z(T) = \int \mathcal{D}x \mathcal{D}S \exp\left[-\beta \mathcal{H}(\{x_i\}, \{S_i\})\right] \delta\left(\sum_i S_i^2 - N\right) \quad (3.59)$$

$$= \int dm_1 dm_2 \exp\left\{-\beta U(m_1, m_2) + S_{\text{ep}}(m_1, m_2) + S_{\text{c}}(m_1, m_2)\right\}$$

where $\mathcal{D}x \equiv \prod_i dx_i$. Besides U [cf. Eq. (3.55)] and S_{ep} [cf. (3.56)], the additional object S_{c} that appears in the exponent is the contribution to the total entropy of the $\{x_i\}$ configurations, i.e., the configurational entropy:

$$S_{\text{c}}(m_1, m_2) \equiv \frac{N}{2}\left[1 + \log(m_2 - m_1^2)\right] \quad (3.60)$$

The configurational entropy of the HOSS model will be widely discussed in Sec. 3.3 in the framework of the two temperature thermodynamic picture, after having explicitly computed the dynamics. Here it simply comes out from the Jacobian e^{S_c} of the transformation of variables $\mathcal{D}x \to e^{S_c} dm_1 dm_2$. We can compute the large N limit of the partition function using, once again, the saddle point approximation. The saddle point equations are found by minimizing with respect to the variables m_1 and m_2 the function

$$\frac{\beta}{N} F(T, m_1, m_2) \equiv \beta\left(\frac{K}{2} m_2 - H m_1 - w\right) \quad (3.61)$$

$$+ \frac{1}{2}\left[\log\frac{w + T/2}{T} - 1 - \log(m_2 - m_1^2)\right]$$

Denoting the saddle point values of m_1 and m_2 as $\bar{m}_1$ and $\bar{m}_2$, their self-consistent equations are:

$$\bar{m}_1 = \frac{\tilde{H}(\bar{m}_1, \bar{m}_2)}{\tilde{K}(\bar{m}_1, \bar{m}_2)} \qquad (3.62)$$

$$\bar{m}_2 = \bar{m}_1^2 + \frac{T}{\tilde{K}(\bar{m}_1, \bar{m}_2)} \qquad (3.63)$$

The form of the solutions $\bar{m}_1(T)$, $\bar{m}_2(T)$ is quite complicated because each of these equations is actually a fourth order equation, but they can be explicitly computed analytically and easily solved numerically. In terms of the equilibrium values $\bar{m}_k$, we find the following expression for the equilibrium free energy:

$$F(T, \bar{m}_1, \bar{m}_2) = U(T, \bar{m}_1, \bar{m}_2) - T\ [S_{\mathrm{ep}}(T, \bar{m}_1, \bar{m}_2) + S_{\mathrm{c}}(T, \bar{m}_1, \bar{m}_2)]\ \ (3.64)$$

When the spin-oscillator coupling J is zero, one has $\tilde{K} \to K$, $\tilde{H} \to H$, $\bar{m}_2 - \bar{m}_1^2 \to T/K$, while S_{c} simply corresponds to the entropic contribution in Eq. (3.3) (the fast process entropy S_{ep} is zero if one sets also $L = 0$, but for $J = 0$ it is irrelevant anyhow).

Constraint on configurations: modeling a fragile glass

Another possible ingredient for the model is a constraint on the slow processes configuration space introduced to prevent the existence of a single global minimum, thus implementing a large degeneracy of the allowable lowest states. The constraint is, hence, taken on the $\{x_i\}$, concerning the long time regime. It reads

$$m_2 - m_1^2 \geq m_0 \qquad (3.65)$$

where m_0 is a fixed, but arbitrary, positive constant (a null m_0 would provide no constraint at all). The model glass obtained this way, with m_0 strictly larger than zero, has no "crystalline" state. In other words the constraint prevent the system from evolving towards the lowest energy configurations (the equilibrium state) in phase space, irrespective of the initial conditions.

This constraint applied to the harmonic oscillator dynamics is a way to reproduce the behavior of fragile glass formers, (e.g., $K^+Ca^{2+}NO_3^-$, $K^+Bi^{3+}Cl^-$, OTP, toluene and chlorobenzene, cf. Sec. 1.2).

If the constraint, Eq. (3.65), is absent, the dynamics at very low T will still be glassy even though for $t \to \infty$ the system will eventually relax to its global minimum, without having to overcome energy barriers.

As we will explain in detail in the next section, either we can implement a dynamics that satisfies this constraint, thus introducing an extra global coupling among $\{x_i\}$ in a dynamic way, or we can adopt a dynamics that is not influenced by the constraint ($m_0 = 0$), but parallel and, thus, approximately reproducing the collective motion of cooperative regions in the glass. The dynamical model that one obtains with the very simple local Hamiltonian

(3.50) and the parallel MC dynamics appear to yield typical examples of glass relaxation. With no constraint, the relaxation time will follow an Arrhenius law as a function of the temperature, Eq. (1.2). Introducing the constraint in the dynamic rule, instead, the relaxation time will take a VF form, Eq. (1.3).

If temperature is large enough, the constraint practically plays no role. More precisely, the constraint is basically inactive above a temperature T_0, consistently defined as the temperature at which the system reaches the phase space constraint in the asymptotic limit $t \to \infty$:

$$T_0 \equiv m_0 \, \tilde{K}(\bar{m}_1(T_0), \bar{m}_2(T_0)) \tag{3.66}$$

Indeed, for the fragile glass case at $T < T_0$, when the constraint is reached at finite t, the saddle point equation (3.63) becomes $\bar{m}_2 - \bar{m}_1^2 = m_0$, no matter what the temperature of the thermal bath is. The equilibrium values of m_1 and m_2 are, thus, expressed by the solution of equation (3.62) and of a modified Eq. (3.63), taking into account the existence of the constraint:

$$\bar{m}_2 - \bar{m}_1^2 = \begin{cases} T/\tilde{K}(\bar{m}_1, \bar{m}_2) & \text{if } T > T_0 \\ m_0 & \text{if } T < T_0 \end{cases} \tag{3.67}$$

If $m_0 = 0$ the constraint does not exist and $T_0 = 0$.

When Eq. (3.65) is first satisfied, at T_0, in the infinite time limit, the configurational entropy S_c [cf. Eq. (3.60)] goes to its minimal value

$$S_c^{(0)} \equiv S_c(T_0) = \frac{N}{2}(1 + \log m_0) \tag{3.68}$$

Zero configurational entropy would mean that only one configuration is allowed for the system, but here we are not using discrete, "quantum," variables and entropies are, therefore, ill-defined at low T, cf. Sec. 3.1. Indeed, the Nernst principle is violated because the configurational entropy counts all the multiple ways in which the *continuous* harmonic oscillators can arrange themselves in order to satisfy the constraint (3.65). As a consequence, the value $S_c(T_0)$ can take any value different from zero in our model. Since we are dealing with classical variables, we can bypass this inconvenience by just subtracting from S_c the constant $S_c^{(0)}$ in order to have $S_c(T \in [0, T_0]) = 0$.[4] Coming from high temperature there would thus be a transition from a many (metastable) states phase to a phase in which the system is stuck forever in one single minimum (or a subextensive number of minima). T_0 is, hence, the Kauzmann temperature (cf. Sec. 1.4.2).

We will look at the Kauzmann transition later on, in Sec. 3.4, in terms of the effective temperature of the model, that we will introduce in Sec. 3.3.

---

[4]From a dynamic point of view the entropy value $S_c^{(0)}$ is related to the dynamics on timescales where all the degenerate minima are sampled. These are much longer than the scales of our interest, and, for our purposes, the value of S_c at T_0 (and below) is irrelevant.

3.2.2 Parallel Monte Carlo dynamics of the HOSS model: equations of motion

We make use of the MC dynamic formalism presented in Sec. 3.1.1 to build the equations of motion. The energy difference between the new and the old states is, in the HOSS model case,

$$\delta E \equiv \mathcal{H}_{\text{eff}}(\{x_i'\}) - \mathcal{H}_{\text{eff}}(\{x_i\}) \simeq \frac{\tilde{K}}{2} y_2 - \tilde{H} y_1 \qquad (3.69)$$

where the last expression is obtained upon neglecting the variations of m_1 and m_2 that are $\mathcal{O}(y_{1,2}^2/N) \sim \Delta^2/N$. If $J = 0$, that is, if we consider only slow modes (no coupling with fast spherical spins), it is exactly $\delta E = K/2\, y_2 - H\, y_1$, as we already discussed [cf. Eq. (3.14)].

In terms of the energy difference $x \equiv \delta E$ and of $y \equiv y_1$, the distribution function can, once again, be written as the product of two Gaussian distributions, as in Eq. (3.15), with averages and variances formally very similar to those presented in Eqs. (3.16)-(3.17), provided the definitions (3.57) are adopted:

$$\overline{x} = \Delta^2 \tilde{K}/2, \qquad\qquad \Delta_x = \Delta^2 \tilde{K}^2 \left(\mu_2 + m_0 + \mu_1^2 \right) \quad (3.70)$$

$$\overline{y}(x) = \frac{\mu_1}{\mu_2 + m_0 + \mu_1^2} \frac{\overline{x} - x}{\tilde{K}}, \qquad \Delta_y = \frac{\Delta^2(\mu_2 + m_0)}{\mu_2 + m_0 + \mu_1^2} \qquad (3.71)$$

The abbreviations $\mu_{1,2}$, inspired by the static self-consistency Eqs. (3.62)-(3.63), are now

$$\mu_1 \equiv \frac{\tilde{H}}{\tilde{K}} - m_1 \qquad\qquad (3.72)$$

$$\mu_2 \equiv m_2 - m_1^2 - m_0 \qquad\qquad (3.73)$$

Notice that Eq. (3.73) is different from the definition of the same quantity in the HO model [Eq. (3.19)], since now we allow for a strictly positive m_0. We will study the dynamics for these two specific combinations of the variables m_1 and m_2. The first variable is defined, starting from the saddle point equation (3.62), as the deviation from the instantaneous equilibrium state. When equilibrium is reached, μ_1 is zero, whereas outside equilibrium $\mu_1 \neq 0$ and its magnitude can be considered as a measure of how far the system is from relaxation.

For μ_2 there are two basic cases:

- $T_0 = 0$. The variable μ_2 is the distance from the ground state of the model (when the constraint constant m_0 is zero there is one unique global minimum). In equilibrium, at $T = 0$, from Eq. (3.67) we know that $\mu_2 = 0$, while for $T > 0$, $\bar{\mu}_2 = T/\tilde{K}(T)$ with the denominator taking the following expression for low T:

$$\tilde{K}_\infty(T) = \lim_{t\to\infty} \tilde{K}(m_1(t), m_2(t); T) = \frac{KD}{D+J^2} + \frac{T}{2}\frac{J^2 K^2}{(D+J^2)^2} \quad (3.74)$$
$$+ \frac{T^2}{8}\frac{J^6 K^3 (J^2 - 3D)}{D(D+J^2)^5} + \mathcal{O}(T^3)$$

For simplicity and for physical interest, we will limit ourselves to the sets of interaction parameters yielding a positive $\tilde{K}$ (and $D = HJ+KL > 0$). Such a choice guarantees that $\bar{\mu}_2$ is always nonnegative. Moreover, it implies that at $T = 0$ the system reaches its minimum

$$x_i = \frac{H+J}{K} \qquad \forall i \qquad (3.75)$$

- $T_0 > 0$. When the constraint constant is different from zero, the asymptotic value of μ_2 is $T/\tilde{K}(T) - m_0$ if $T \geq T_0$, or else it is zero.

In terms of μ_1, μ_2 the square root w, introduced in Eq. (3.52), becomes

$$w(\mu_1, \mu_2) = \sqrt{J^2(m_0 + \mu_2) + \left(\frac{D}{\tilde{K}(\mu_1, \mu_2)} - J\tilde{K}(\mu_1, \mu_2)\mu_1\right)^2 + \frac{T^2}{4}} \quad (3.76)$$

In the following sections we will present separately the two kinds of dynamics corresponding to the relaxation of a strong glass (no constraint) and to that of a fragile glass (configurational constraint plus dynamics depending on constraint).

The equations of motion for the one time variables $m_{1,2}$ are given by Eqs. (3.30)-(3.31), provided $H \to \tilde{H}$, $K \to \tilde{K}$. The long time regime is more clearly expressed in terms of μ_1, μ_2, whose equations of motion are now:

$$\dot{\mu}_1 = \frac{\dot{\tilde{H}}}{\tilde{K}} - \frac{\dot{\tilde{K}}\tilde{H}}{\tilde{K}^2} - \dot{m}_1 \qquad (3.77)$$
$$\dot{\mu}_2 = \dot{m}_2 - 2m_1 \dot{m}_1 \qquad (3.78)$$

Introducing the abbreviations

$$Q(m_1, m_2) \equiv \frac{J^2 D}{\tilde{K}^3 w \left(w + T/2\right)^2}, \qquad \bar{Q} \equiv Q(\bar{m}_1, \bar{m}_2) \qquad (3.79)$$

$$P(m_1, m_2) \equiv \frac{J^4(m_2 - m_1^2)}{2\tilde{K} w \left(w + T/2\right)^2}, \qquad \bar{P} \equiv P(\bar{m}_1, \bar{m}_2) \qquad (3.80)$$

where $w(\mu_1, \mu_2)$ is expressed by Eq. (3.76), after a lengthy but straightforward manipulation one obtains

$$\dot{\mu}_1 = -4\alpha^2 \mu_1 (1 + Q\,D) \left\{ A_0(t) - \frac{2}{\Delta^2 \tilde{K}} A_1(t) \right\} - QJA_1(t) \qquad (3.81)$$

$$\dot{\mu}_2 = 8\alpha^2 \mu_1^2 \left\{ A_0(t) - \frac{2}{\Delta^2 \tilde{K}} A_1(t) \right\} + \frac{2}{\tilde{K}} A_1(t) \qquad (3.82)$$

The acceptance rate $A_0(t)$ and the average energy shift $A_1(t)$ are defined in Eqs. (3.28)-(3.29) and the parameter α is the one introduced in Eq. (3.27), provided $\mu_2 \to \mu_2+m_0$ to deal with the possible presence of the configurational constraint; $\alpha^2 = \bar{x}^2/(2\Delta_x) = 8/\Delta^2(\mu_2 + m_0 + \mu_1^2)$.

The dependence of A_0 and A_1 on μ_1 and μ_2 is hidden in the variables α, $\bar{x}$ and Δ_x. Their functional dependence on μ_1, μ_2 passes through $\tilde{K}$ and can be obtained, e.g., by solving the equation (see definition 3.57)

$$(\tilde{K} - K)(w + T/2) = -J^2 \tag{3.83}$$

It results in a complicated but analytically solvable fourth order equation for $\tilde{K}$,

$$(\tilde{K}-K)^2(D-J\tilde{K}\mu_1)^2+(\tilde{K}-K)^2\tilde{K}^2J^2(m_0+\mu_2)-J^4\tilde{K}^2+TJ^2\tilde{K}^2(\tilde{K}-K) = 0 \tag{3.84}$$

The explicit solution appears not to be very appealing but is easily computable.

3.2.3   Dynamics of the *strong* glass model

To model a strong glass we will consider the dynamics without imposing any constraint on the configuration space and making use of the above-introduced MC dynamics, with a variance Δ^2 for the randomly chosen updating $\{r_i\}$ of the slow variables $\{x_i\}$. We will see that this dynamics displays an Arrhenius relaxation near zero temperature. This happens for similar models, e.g., the oscillator model of Sec. 3.1 or the spherical spin model of Nieuwenhuizen [1998c, 2000], where exactly the same dynamics is applied. The difference is that now the formalism includes two types of dynamical processes: fast and slow.

Zero temperature dynamics

We first present the dynamics exactly at zero temperature. For long times, $\alpha(t)$ is large and diverging and we can expand $A_0(t)$ and $A_1(t)$ [cf., e.g., Eq. (3.27)], obtaining

$$A_0(t) \simeq \frac{e^{-\alpha^2(t)}}{2\alpha(t)\sqrt{\pi}} \tag{3.85}$$

$$A_1(t) \simeq -\frac{e^{-\alpha^2(t)}}{2\alpha(t)\sqrt{\pi}} \frac{\Delta^2 \tilde{K}(t)}{4\alpha^2(t)} \tag{3.86}$$

First of all, we solve the equation of motion for μ_2, Eq. (3.82) neglecting terms of order μ_1^2 with respect to those of order μ_2, i.e., assuming $\alpha^2 \simeq \Delta^2/(8\mu_2)$. Using Eq. (3.86), Eq. (3.82) becomes

$$\dot{\mu}_2 \simeq -2\,(2\mu_2)^{3/2}\,\frac{\exp\left(-\frac{\Delta^2}{8\mu_2}\right)}{\sqrt{\pi}\Delta^2} \tag{3.87}$$

or, equivalently, $\dot{\alpha} \simeq e^{-\alpha^2}/\sqrt{\pi}$, yielding the implicit solution of the dynamics:

$$\frac{\mathrm{erf}[i\,\alpha(t)]}{i} = \frac{2}{\pi}t + \mathrm{const} \tag{3.88}$$

The error function erf is defined in Eq. (3.A.1). At $T = 0$, the solution in the aging regime turns out to be:

$$\mu_2(t) \simeq \frac{\Delta^2}{8} \frac{1}{\log \frac{2t}{\sqrt{\pi}} + \frac{1}{2} \log \log \frac{2t}{\sqrt{\pi}}} \tag{3.89}$$

Combining Eq. (3.85) with Eq. (3.86), Eq. (3.81) takes the form

$$\dot{\mu}_1 = \frac{e^{-\alpha^2}}{2\alpha\sqrt{\pi}} \left\{ \frac{JQ\Delta^2\tilde{K}}{4\alpha^2} - 2\mu_1(1 + DQ)(2\alpha^2 + 1) \right\} \tag{3.90}$$

Dividing Eq. (3.90) by Eq. (3.87) we can write down a differential equation for μ_1 as a function of μ_2:

$$\frac{d\mu_1}{d\mu_2} \simeq 8(1 + DQ)\frac{\alpha^4}{\Delta^2}\mu_1 - \frac{JQ\tilde{K}}{2} \tag{3.91}$$

where we have neglected terms of order $1/\alpha^2 \sim \mu_2$ with respect to those of $\mathcal{O}(1)$. To obtain the leading order behavior we can neglect the left-hand side of the equation with respect to the right-hand side, i.e., we can perform an *adiabatic approximation* stating that the relative variation of μ_1 with respect to a change μ_2 is very slow in the aging regime (it has to be verified afterward). This amounts to assuming that the evolution of μ_1 mainly occurs on a hypersurface of configurations satisfying the constraint $\mu_2 = \mathrm{const}$.

In such an adiabatic approximation the solution to Eq. (3.91) turns out to be

$$\mu_1(t) \simeq \frac{4J^3K}{\Delta^2(D + J^2)^2}\mu_2^2(t) \tag{3.92}$$

where we have inserted the infinite time limits $\tilde{K}_\infty = KD/(D + J^2)$ and $\bar{Q} = J^2/D^2$. This result is both consistent with the initial assumption of neglecting $\mu_1^2 \sim \mu_2^4$ with respect to μ_2 and with the adiabatic approximation of neglecting $d\mu_1/d\mu_2 \sim \mu_2$ with respect to the right-hand side of Eq. (3.91), which is of $\mathcal{O}(1)$. At zero temperature and for long times, one, thus, obtains $\mu_1 \sim \mu_2^2 \ll \mu_2$.

Dynamics at $T > 0$

If T is above zero, but still small, the leading order of the expansion of A_0 and A_1 for large times ($\alpha \gg 1$) reads:

$$A_0 \simeq \frac{e^{-\alpha^2}}{2\alpha\sqrt{\pi}} \frac{1}{1 - \frac{4T\alpha^2}{\Delta^2 \tilde{K}}} \tag{3.93}$$

$$A_1 \simeq -\frac{e^{-\alpha^2}}{2\alpha\sqrt{\pi}} \frac{\Delta^2 \tilde{K}}{4\alpha^2} \frac{1 - \frac{8T\alpha^2}{\Delta^2 \tilde{K}}}{\left(1 - \frac{4T\alpha^2}{\Delta^2 \tilde{K}}\right)} \tag{3.94}$$

where the terms $T\alpha^2$ are of $\mathcal{O}(1)$ (in order for α to be large T must be small). Indeed, introducing the difference $\delta\mu_2(t) = \mu_2(t) - \bar{\mu}_2$, the $\mathcal{O}(T\alpha^2)$ term can be written as

$$\frac{8T\alpha^2}{\Delta^2 \tilde{K}} = \frac{8T}{\Delta^2 \tilde{K}} \frac{\Delta^2}{8\mu_2} = \frac{\bar{\mu}_2}{\mu_2} = \frac{1}{1 + \frac{\delta\mu_2}{\bar{\mu}_2}} \tag{3.95}$$

[see the definition of α, Eq. (3.27)] so that $\lim_{t\to\infty} 8T\alpha^2/(\Delta^2 \tilde{K}) = 1$.

In the latter notation the MC equations (3.81, 3.82) are rewritten as

$$\dot{\mu}_1 \simeq \frac{e^{-\alpha^2}}{\alpha\sqrt{\pi}} \frac{1 + \frac{\delta\mu_2}{\bar{\mu}_2}}{1 + 2\frac{\delta\mu_2}{\bar{\mu}_2}} \left[\frac{JQ\Delta^2\tilde{K}}{2\alpha^2} \frac{\delta\mu_2}{\bar{\mu}_2} \frac{1}{1 + 2\frac{\delta\mu_2}{\bar{\mu}_2}} - 4\alpha^2(1 + DQ)\mu_1 \right] \tag{3.96}$$

$$\dot{\mu}_2 = \dot{\delta\mu}_2 \simeq \frac{e^{-\alpha^2}}{\alpha\sqrt{\pi}} \frac{\Delta^2}{\alpha^2} \frac{\delta\mu_2}{\bar{\mu}_2} \frac{1 + \frac{\delta\mu_2}{\bar{\mu}_2}}{\left(1 + 2\frac{\delta\mu_2}{\bar{\mu}_2}\right)^2} \tag{3.97}$$

The solution to Eq. (3.97) in the time regime $\mathcal{O}(1) \ll t \ll \tau_{\rm eq}$ is, to leading order,

$$\delta\mu_2(t) \simeq \frac{\Delta^2}{8} \frac{1}{\log\frac{2t}{\sqrt{\pi}}} \tag{3.98}$$

and the behavior of μ_1 comes out to be

$$\mu_1 \simeq \frac{4JQ\tilde{K}}{\Delta^2(1 + DQ)} \mu_2^2 \frac{\delta\mu_2}{\bar{\mu}_2} \frac{2}{1 + 2\frac{\delta\mu_2}{\bar{\mu}_2}} \simeq \frac{8J\bar{Q}}{\Delta^2(1 + D\bar{Q})} T\delta\mu_2(t) \tag{3.99}$$

Notice that the above expression vanishes linearly when equilibrium is approached, since then $\delta\mu_2 \to 0$. This is at variance with the quadratic behavior at $T = 0$, cf. Eq. (3.92). In the present case, though, $d\mu_1/d\mu_2$ can still be considered small with respect to $\mathcal{O}(1)$ if T is low enough. We recall that we are actually assuming as a starting point that $\alpha^2 \sim 1/T$ is large, cf. Eq. (3.93), and this is precisely equivalent to assume that $\bar{\mu}_2 \sim T$ is small.

Arrhenius relaxation to equilibrium

For times even longer than the timescale of the aging regime, the system finally relaxes, exponentially fast, to equilibrium.

From the equations of motion studied above [see for instance Eq. (3.87)] we find that the relaxation time to equilibrium, i.e., the inverse of the acceptance ratio [cf. Eq. (3.93)] is

$$\tau_{eq} \propto e^{\alpha^2} \tag{3.100}$$

From the definition of α, Eq. (3.27), its asymptotic value at low temperature turns out to be

$$\alpha(T) = \sqrt{\frac{\Delta^2}{8\bar{\mu}_2(T)}} = \sqrt{\frac{A_s}{T}} \tag{3.101}$$

with

$$A_s \equiv \frac{\Delta^2 \tilde{K}_\infty}{8} = \frac{\Delta^2 K D}{8(D + J^2)} + O(T) \tag{3.102}$$

Accordingly, Eq. (3.100) is the Arrhenius law

$$\tau_{eq} \sim \exp\left(\frac{A_s}{T}\right) \tag{3.103}$$

We notice that, in general, in all quantities computed above, the zero temperature and the infinite time limits commute.

3.2.4 Dynamics of the *fragile* glass model

We now analyze what happens to the dynamics if the constraint Eq. (3.65) is switched on ($m_0 > 0$). First of all, it must be implemented in the dynamic rules in order to affect the entire evolution and not just its trivial static limit. We, therefore, let Δ^2 depend on the distance from the constraint, i.e., on the whole $\{x_i\}$ configuration before the MC step:

$$\Delta^2(t) \equiv 8[m_0 + \mu_2(t)] [\mu_2(t)]^{-\gamma} \tag{3.104}$$

where γ is larger than zero. We abbreviate

$$\Lambda(t) \equiv [\mu_2(t)]^{-\gamma} \tag{3.105}$$

The nearer the system goes to the constraint (i.e., the smaller the value of $\mu_2 = m_2 - m_1^2 - m_0$), the larger the variance becomes, thus implying almost always a refusal of the proposed updating. In this way, in the neighborhood of the constraint, the dynamics is slower and proceeds through very rare but typically very large moves, a mechanism that can be interpreted as *activated dynamics*.[5]

---

[5]This can be realized as well by taking $m_0 = 0$, yielding, eventually, a generalized Arrhenius relaxation.

When the constraint is reached, Λ becomes infinite and the system dynamics is stuck forever. The system no longer evolves toward equilibrium but is blocked in one single ergodic component of the configuration space. At large enough temperatures, the combination $\mu_2(t)$ will remain strictly positive throughout the whole relaxation process and the dynamics will be qualitatively identical to that of the strong glass. Contrary to the latter, however, the dynamics will get stuck forever out of equilibrium at some temperature T_0, Eq. (3.66), identifiable with the Kauzmann temperature.

In terms of Λ, the parameter α [Eq. (3.27)] can be rewritten as

$$\alpha = \sqrt{\frac{\Lambda}{1 + \frac{\mu_1^2}{m_0 + \mu_2}}} \simeq \sqrt{\Lambda} \tag{3.106}$$

for $\mu_1^2 \ll \mu_2 + m_0$. When all system parameters are fixed (aging setup) the equations of motion, in terms of μ_1 and μ_2, are still Eqs. (3.81)-(3.82).

Detailed balance

The question whether detailed balance is satisfied or not is nontrivial in this case. Indeed, it happens to be satisfied for this kind of dynamics only for large N. For exact detailed balance we should have

$$p(x|m_1, m_2) \exp(-\beta x) = p(-x|m_1, m_2) \tag{3.107}$$

but now, the probability distribution also depends on the harmonic oscillator configuration, through Δ^2, as defined in Eq. (3.104). When we perform the inverse move $\{x_i'\} \rightarrow \{x_i\}$, hence, the right-hand side of the detailed balance consists of $p(-x|m_1', m_2'; \Delta'^2) \neq p(-x|m_1, m_2; \Delta^2)$. Expanding this probability distribution in powers of $1/N$, however, we obtain

$$p(-x|m_1', m_2'; \Delta'^2) = p(-x|m_1, m_2; \Delta^2) + \mathcal{O}(\Delta^2/N) \tag{3.108}$$

Terms of $\mathcal{O}(\Delta^2/N)$ were already neglected in the approximation of the energy shift x done in Eq. (3.14). So, inasmuch as the whole approach is valid, i.e., for large N, detailed balance is also satisfied. It would be, instead, slightly violated in a finite N numerical simulation. Even though $\Delta^2 \propto \Lambda(t)$ grows as the system approaches equilibrium (it even diverges at the Kauzmann temperature), in our approach, we first perform the thermodynamic limit computing the dynamic equations and only eventually the limit $t \rightarrow \infty$.[6]

---

[6]This is what is done, e.g., in spherical p-spin models representing the mean-field approximation for structural glasses [Cugliandolo & Kurchan, 1993, 1994]. If we did the opposite, there would have been a region around the Kauzmann temperature where the detailed balance would have been violated and the dynamics would have been different from the one discussed here. However, to probe the latter order of limits is not our aim since we are interested in the slow relaxation that takes place in systems with a large number (Avogadro-like) of variables.

Aging dynamics

In the previous section the dynamics was performed within the same framework but at fixed Δ: the relaxation time is diverging at low temperature with an Arrhenius law, typical of *strong* glasses. Setting $m_0 = 0$ in Eq. (3.104) it would yield a strong glass system relaxing slowly to equilibrium with a generalized Arrhenius law $\tau_{eq} \sim \exp(A/T)^\gamma$, with a generic γ, and the slowing down would be enhanced with respect to the strong glass dynamics with a fixed Δ^2 because of the increase of the variance of the proposed oscillator move as equilibrium is approached.

Here we will, instead, develop a model representing a *fragile* glass with a Kauzmann transition at a finite temperature, keeping $m_0 > 0$. First of all, we look at the probability distribution of energy shifts in the MC update procedure. The variance Δ_x of Eq. (3.15) is now diverging as $\Lambda \to \infty$. Indeed, it is

$$\Delta_x = \Delta^2 \tilde{K}^2(m_0 + \mu_2 + \mu_1^2) = 8\Lambda\tilde{K}^2(m_0 + \mu_2)^2 + O\left(\Lambda\mu_1^2\right) \qquad (3.109)$$

In the time regime where Λ is large ($\mu_2 \ll 1$), x^2/Δ_x will be small. This allows us to approximate, in the aging regime of our interest, the Gaussian distribution of x as the exponential distribution

$$p(x|m_1, m_2) \simeq \frac{\exp\left(-\Lambda\right)}{\sqrt{2\pi\Delta_x}} \exp\left(\frac{x\,\bar{x}}{\Delta_x}\right)\left(1 - \frac{x^2}{2\Delta_x} + \frac{x^4}{8\Delta_x^2}\right) \qquad (3.110)$$

Looking at the asymptotic equation for μ_2 and noticing that the infinite time limit is $\bar{\mu}_2 = T/\tilde{K}_\infty(T) - m_0$, we introduce the temperature-like abbreviation

$$T^\star(t) \equiv \tilde{K}(t)\left[m_0 + \mu_2(t)\right] \qquad (3.111)$$

tending to T as $t \to \infty$, provided $T \geq T_0$.

To the leading order, the acceptance ratio, Eq. (3.28), is

$$A_0(t) \simeq \Upsilon(t) \equiv \frac{e^{-\Lambda}}{4T^\star\sqrt{\pi\Lambda}}\left\{\int_{-\infty}^0 dx\, e^{\frac{x}{2T^\star}} + \int_0^\infty dx\, e^{-\left(\beta - \frac{1}{2T^\star}\right)x}\right\}$$

$$= \frac{e^{-\Lambda}}{\sqrt{\pi\Lambda}}\frac{T^\star}{2T^\star - T} \qquad (3.112)$$

with

$$\Upsilon(t) \equiv \frac{e^{-\Lambda(t)}\left[1 - r(t)\right]}{\sqrt{\pi\Lambda(t)}} \qquad (3.113)$$

We further introduce another parameter, a sort of reduced temperature, that is small as times are large:

$$r \equiv \frac{T^\star - T}{2T^\star - T} \qquad (3.114)$$

In this notation the dynamic Eqs. (3.81)-(3.82) become

$$\dot{\mu}_1 = 4\Upsilon \left\{ JQ\tilde{K}(m_0 + \mu_2)r \left[1 - \frac{3(1 - 2r + 2r^2)}{\Lambda} + O\left(\frac{1}{\Lambda^2}\right) \right] \right. \tag{3.115}$$

$$\left. - \frac{\Lambda\mu_1(1 + QD)}{1 + \frac{\mu_1^2}{m_0 + \mu_2}} \left[1 - \frac{1 - 3r + 4r^2}{\Lambda} + O\left(\frac{1}{\Lambda^2}\right) \right] \right\}$$

$$\dot{\mu}_2 = -8\Upsilon \left\{ (m_0 + \mu_2)r \left(1 - \frac{3(1 - 2r + 2r^2)}{\Lambda} \right) \right. \tag{3.116}$$

$$\left. - \Lambda\mu_1^2 \left[1 - \frac{1 - 3r + 4r^2}{\Lambda} + O\left(\frac{1}{\Lambda^2}\right) \right] \right\}$$

The solutions to Eqs. (3.115) and (3.116) depend on the relative sizes of μ_1 and μ_2, and thus also on γ, as well as on r, which has a different behavior above T_0, where $T^\star$ tends to T in the infinite time limit, and below T_0, where $T^\star$ never equals the heat-bath temperature (we will see in Sec. 3.3 what is the physical meaning of $T^\star$).

The resolution of Eq. (3.116) can be carried out by neglecting the second term, proportional to $\Lambda\mu_1^2$, i.e., assuming $\mu_1^2 \ll \mu_2^\gamma$. If $\gamma < 1$, this is a stricter hypothesis than the one leading to Eq. (3.82) for the strong glass case ($\mu_1^2 \ll \mu_2$). We will actually see that this hypothesis is not satisfied as $\gamma < 1$ (see the end of the present section). The rest of Eq. (3.116) can be rewritten as a closed equation for Λ:

$$\dot{\Lambda} = \frac{8\gamma m_0}{\pi} e^{-\Lambda} \Lambda^{\frac{1}{2} + \frac{1}{\gamma}} r(\Lambda) \left[1 - r(\Lambda) \right] \tag{3.117}$$

where $r(\Lambda(t))$ is defined in Eq. (3.114) and where only the leading terms in Λ are kept. The solution we obtain will, then, be valid in the long time regime and for $T \gtrsim T_0$ or $T < T_0$ (in the latter case γ must be larger than one). Indeed, above T_0, $\Lambda = 1/(\bar{\mu}_2 + \delta\mu_2)^\gamma$. We can neglect $\bar{\mu}_2$ with respect to $\delta\mu_2(t)$ at temperatures very close to the Kauzmann temperature and for times that are not extremely long, so that we are far from thermalization and the dynamics still displays aging behavior. However, as is clear from Fig. 3.1, as soon as we go too far from T_0, we cannot neglect the asymptotic value $\bar{\mu}_2$ anymore.

The behavior of r discriminates between the aging regimes above and below T_0:

$$\begin{array}{llll} T \geq T_0, & \Lambda(t) \to \bar{\Lambda}(T) \sim \frac{1}{(T - T_0)^\gamma}, & r \simeq \mu_2 - \bar{\mu}_2 & \\ T < T_0, & \Lambda(t) \to \infty, & r \simeq r_\infty \sim T_0 - T & \end{array} \tag{3.118}$$

In Appendix 3.A we present the details of the resolution of Eq. (3.117). Here we directly present the results in terms of the distance from the config-

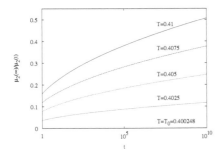 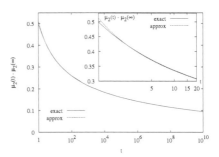

FIGURE 3.1

The ratio of $\bar{\mu}_2/\mu_2(t)$ at different temperatures, at and above the Kauzmann temperature, is plotted. Too far away from T_0 the contribution of $\bar{\mu}_2$ to $\mu_2(t)$ becomes relevant at shorter time decades. The case is plotted with $K = J = 1$, $H = L = 0.1$, $m_0 = 5$ and with a Vogel-Fulcher exponent $\gamma = 2$. Reprinted figure with permission from [Leuzzi & Nieuwenhuizen, 2001a]. Copyright (2001) by the American Physical Society.

FIGURE 3.2

The difference $\mu_2(t) - \bar{\mu}_2$ is plotted for $T = 0.41$, slightly above the Kauzmann temperature $T_0 = 4.00248$, for which $\bar{\mu}_2 = 0.09763$. The full curve represents the exact solution to Eq. (3.116), with initial condition $\Lambda(0) = 1$. The dashed curve is a plot of the approximated solution, Eq. (3.119). In the inset the initial behavior is shown. Reprinted figure with permission from [Leuzzi & Nieuwenhuizen, 2001a].

urational space constraint μ_2:

$$\mu_2(t) \simeq \frac{1}{[\log(t/t_0) + \omega \log\left(\log(t/t_0)\right)]^{1/\gamma}} + \left[\frac{T}{\tilde{K}_\infty(T)} - m_0\right] \theta(T - T_0) \quad (3.119)$$

The expressions for the parameters t_0 and ω depend on the temperature phase as reported in Eq. (3.A.8). $\theta(x)$ is the Heaviside function: $\theta(x \geq 0) = 1$ and $theta(x < 0) = 0$.

In Fig. 3.2 we show the exact solution, numerically computed, to Eq. (3.116) for a particular choice of the parameter values, $K = J = 1$, $H = L = 0.1$, $m_0 = 5$, $\gamma = 2$ and we compare it with the approximated solution yielded by Eq. (3.119). We can see that, already after one decade, the behaviors coincide.

The dynamics of $\mu_1(t)$ is less general. As $T < T_0$ its behavior is, actually, strongly model dependent. The ratio of Eqs. (3.115) and (3.116) yields the equation

$$\frac{d\mu_1}{d\mu_2} = \frac{\mu_1(1 + QD)(\Lambda + 2 - 3r + 2r^2) - JQT^\star r}{2r(m_0 + \mu_2) - \mu_1^2(\Lambda + 2 - 3r + 2r^2)} \quad (3.120)$$

With respect to the relative weights of μ_1 and μ_2, we can identify different regimes, where the solution displays different behaviors.

1. Aging regime above the Kauzmann temperature: $T \gtrsim T_0, \forall \gamma$.
   The leading term of the solution is given by the stationary solution of
   Eq. (3.120). We expand r for long times, $r = r_1 \delta \mu_2$ (the expression for
   r_1 is reported in Appendix 3.A, Eq. (3.A.9)), thus yielding the following
   long time behavior of μ_1:

   $$\mu_1(t) \simeq \frac{TJ\bar{Q}r_1}{1 + \bar{Q}D} \frac{\delta\mu_2(t)}{\Lambda} \tag{3.121}$$

2. Aging regime below the Kauzmann temperature, enhanced separation
   of timescales: $T < T_0, \gamma > 1$.
   In this and in the following cases $\delta\mu_2(t) = \mu_2(t)$. Also in this dynamic
   regime the adiabatic approximation can be carried out and the second
   term in the denominator of Eq. (3.120) is again negligible. The leading
   term of r in its expansion in powers of μ_2, r_∞, is of $\mathcal{O}(1)$ [cf. Eq.
   (3.A.11)]. Therefore we get

   $$\mu_1(t) = \frac{J\tilde{K}_\infty m_0 r_\infty \bar{Q}}{1 + \bar{Q}D} \frac{1}{\Lambda(t)} + \mathcal{O}\left(\frac{\mu_2}{\Lambda}\right) \tag{3.122}$$

3. Non-generic regimes.
   Below T_0 and for $\gamma = 1$ many different regimes occur, that are model
   dependent. Furthermore, in some of these regimes the assumption at the
   basis of the adiabatic approximation (μ_1 relaxing on faster timescales
   than μ_2) is not valid anymore. This also occurs in the regime for $\gamma < 1$.
   We will not need the details of these regimes for the test of the out-
   of-equilibrium thermodynamics formulated on the HOSS model in Sec.
   3.3. For completeness, however, we report them in Appendix 3.A.

The one time dependent variables $\mu_1(t)$ and $\mu_2(t)$ provide the dynamic
behavior of every observable in the long, preasymptotic, time regime, i.e.,
as we will explicitly see in Sec. 3.3.4, in the aging regime. When the time
increases further the dynamics eventually relaxes to equilibrium exponentially,
as $\exp(-t/\tau_{eq})$.

Vogel-Fulcher relaxation time to equilibrium

The relaxation time to equilibrium is the characteristic time on which the
system initially out of equilibrium (because, for instance, of a sudden quench
to low temperature) relaxes toward equilibrium. It can be defined, for instance, from the dynamical equations of the observables $m_{1,2}$ of the harmonic
oscillators $\dot{m}_i = -m_i/\tau_{eq}$, as the time at which the quantity of interest goes
to $1/e$ of its initial value. Equivalently, it is identified with the inverse of the
acceptance ratio in the asymptotic time limit, $\tau_{eq} \sim 1/\bar{A}_0$, cf. Eq. (3.112).
Hence, in any temperature regime the relaxation time turns out to have an

exponential behavior in Λ:

$$\tau_{\text{eq}} \sim e^{\Lambda} = \exp\left(\frac{1}{\mu_2}\right)^{\gamma} \tag{3.123}$$

Making use of Eq. (3.119) we find the behavior for the relaxation time versus temperature when $T > T_0$: $\mu_2(t)$ tends asymptotically to $\bar{\mu}_2(T) = T/\tilde{K}_{\infty}(T) - T_0/\tilde{K}_{\infty}(T_0)$, and, near enough to the Kauzmann temperature, we can linearize the latter as $\bar{\mu}_2 \simeq (T - T_0)\tilde{K}_{\infty}(T_0)$. We get, thus, the exponential law

$$\tau_{\text{eq}} \propto \exp\left(\frac{\tilde{K}_{\infty}(T_0)}{T - T_0}\right)^{\gamma} \tag{3.124}$$

This is a generalized VF law where γ can have any value and, in particular, the value $\gamma = 1$. In practice, the exponent is used in experiments to make the best Vogel-Fulcher-like fitting of the relaxation time [Angell, 1995; McKenna, 1989]. In our model γ is a constant; it has no prescribed value since we do not make any connection with a microscopic system. Its value has relevance on the speed of structural relaxation [see, e.g., Eq. (3.119)] and there are three qualitatively different dynamic regimes depending on its value: $\gamma > 1$, $\gamma = 1$ and $0 < \gamma < 1$. We have seen that, for $\gamma = 1$ (the original VF law) the situation is actually *non*-generic. We anticipate that the difference induced on the aging dynamics by different values of γ will imply that a consistent thermodynamic picture cannot be always constructed in terms of a single extra temperature-like parameter, not even for the very simple HOSS model. We will analyze in Sec. 3.3 what the limits and the reasons for this are.

Below T_0 no relaxation to equilibrium can occur and the system is stuck for ever out of equilibrium. However, also in this case it is possible to define a characteristic time describing the structural relaxation. We will come back to this in Sec. 3.4 after having applied the two temperature thermodynamic theory, presented in Secs. 2.3, 2.4 and 2.5, to the HOSS model.

3.2.5 Adam-Gibbs relation in the HOSS model

The configurational entropy S_c for the fragile HOSS glass model is ill-defined at zero temperature because we are using continuous variables, cf. Sec. 3.2.1. To cure such a violation of the Nernst principle we can, however, subtract the constant $S_c^{(0)} \equiv S_c(T_0)$ from the configurational entropy, thus obtaining

$$S_c = \frac{N}{2}\log\left(1 + \frac{\mu_2}{m_0}\right) \tag{3.125}$$

Expanding this regularized entropy in powers of μ_2 we find the leading order: $S_c \simeq \frac{N}{2m_0}\mu_2$. Using this result we also find, from Eq. (3.123), the following relation between the configurational entropy and the relaxation dynamics in the fragile HOSS:

$$\tau_{\mathrm{eq}} \propto \exp\left[\frac{B}{S_{\mathrm{c}}(T)}\right]^{\gamma} \qquad (3.126)$$

with $B \equiv N/(2m_0)$. Looking at Eq. (1.14) of Sec. 1.4, we recognize that the above relation is just the AG relation generalized to exponents γ also different from one, provided the identification $B = C/T$ is done, where C in the original work [Adam & Gibbs, 1965] was an extensive quantity proportional to the difference between the Gibbs free energy of the system and a reduced Gibbs free energy computed counting only those configurations including CRRs and, thus, contributing to the glass relaxation.

3.3   Out-of-equilibrium thermodynamics

The history of a system that is far from equilibrium can be expressed by means of a number of effective parameters, like the *effective temperature* or other *effective fields*, in order to recast the out-of-equilibrium dynamics into a thermodynamic approach. The aim is, indeed, to yield a formalism in which any two metastable states can be connected inside a phase diagram of a certain (reasonably small) number of thermodynamic parameters.

   The number of effective parameters needed to make such a translation is, in principle, equal to the number of independent observables considered. For a certain class of system, however, some partial thermalization takes place and the effective parameters pertaining to processes evolving on the same characteristic (long) timescale become equal to each other in time, at least on a wide time window. Examples of out-of-equilibrium regimes governed by a single effective temperature have been considered in Chapter 2 and more will be presented in Chapters 4, 6 and 7.

3.3.1   Quasi-static approach

Given the solution of the dynamics, i.e., the time dependence of the functions $m_1(t)$ and $m_2(t)$, a quasi-static approach can be followed by computing the partition function Z_e of all the macroscopically equivalent states (those having the same values for $m_{1,2}$) at a given time t. The measure on which this out-of-equilibrium partition function is evaluated is not the Gibbs measure. In order to generalize the equilibrium thermodynamics we introduced an effective temperature T_e and an effective field H_e, and substitute the equilibrium measure by a mimicking function

$$\omega(T, T_e, H_e) \equiv \exp\left[-\frac{1}{T_e}\mathcal{H}_{\mathrm{eff}}(\{x_i\}, T, H_e)\right] \qquad (3.127)$$

The Hamiltonian $\mathcal{H}_{\text{eff}}$ is the one derived in Eq. (3.54), where we have substituted the true external field H with the effective field H_e.

The quantities T_e and H_e are, at this step of the computation, simply fictitious parameters. However, as soon as we get the expression of the candidate *effective* thermodynamic potential $F_e \equiv -T_e \log Z_e$ as a function of macroscopic variables $m_{1,2}$ and effective parameters, we can determine T_e and H_e from the conditions of minimum F_e with respect to m_1 and m_2 and evaluate the resulting analytic expressions at the dynamic values $m_1 = m_1(t)$ and $m_2 = m_2(t)$. Counting all the macroscopically equivalent states at time t, at which the dynamic variables take values m_1 and m_2, we obtain

$$Z_e\left(m_1, m_2; T_e, H_e\right) \equiv \int \mathcal{D}x \, \exp\left[-\frac{1}{T_e}\mathcal{H}_{\text{eff}}(\{x_i\}, T, H_e)\right] \qquad (3.128)$$

$$\times \delta\left(Nm_1 - \sum_i x_i\right) \delta\left(Nm_2 - \sum_i x_i^2\right)$$

From this partition function we can build an effective thermodynamic potential as a function of T_e and H_e, besides T and H, where the effective parameters depend on time through the slowly varying $m_1(t)$ and $m_2(t)$, solutions of the dynamics. The parameters T_e and H_e are actually a way of describing the evolution in time of the system out of equilibrium in a thermodynamic language. The effective free energy takes the form

$$F_e(t) = U\left(m_1(t), m_2(t)\right) - T S_{\text{ep}}\left(m_1(t), m_2(t)\right) \qquad (3.129)$$
$$- T_e(t) S_{\text{c}}\left(m_1(t), m_2(t)\right) + [H - H_e(t)] N m_1(t)$$

with

$$T_e(t) = \tilde{K}\left(m_1(t), m_2(t)\right)\left[m_2(t) - m_1^2(t)\right] = T^\star(t) \qquad (3.130)$$
$$H_e(t) = H - \tilde{K}\left(m_1(t), m_2(t)\right)\mu_1(t) \qquad (3.131)$$

where the last term of F_e replaces the $-HNm_1$ occurring in U [see Eq. (3.55)] with $-H_e N m_1$.

The quantity

$$S_{\text{c}}(t) = \frac{N}{2} \log \frac{m_2(t) - m_1^2(t)}{m_0} \qquad (3.132)$$

is the configurational entropy (3.57) and the formal expressions for U and S_{ep} are given in Eqs. (3.55) and (3.56) (now m_1 and m_2 are time-dependent quantities). Furthermore, we recognize in Eq. (3.130) that the parameter $T^\star$, introduced in Eq. (3.111) in the study of the dynamics and entering in many essential expressions, turns out to be the effective temperature for the HOSS model.

The formalism so far discussed holds both for the strong and fragile dynamic cases and also for the HO model without fast processes (for which $J = L = 0$, $\tilde{K} = K$ and $\tilde{H} = H$). Indeed, the new thermodynamic-like parameters are specified only as the actual dynamic behaviors of $m_{1,2}(t)$ (equivalently $\mu_{1,2}(t)$) are inserted.

In the regimes where $\mu_1 \ll \mu_2$, the effective temperature $T_e(t)$ alone is enough for a complete thermodynamic description of the dominant, out-of-equilibrium, physical phenomena (i.e., $H_e \simeq H$, for $t_0 \ll t \ll \tau_{eq}$). The difference between the effective field and the external field is negligible with respect to $T_e - T$. These are the dynamic regimes occurring for long times in the HO model (Sec. 3.1), in the strong HOSS model (Sec. 3.2.3) and in the fragile HOSS model (Sec. 3.2.4) as $T \geq T_0$ and also for $T < T_0$ provided that $\gamma > 1$. In the cases where $\gamma \leq 1$, on the contrary, the effective field H_e is also needed to map the history of the aging system (μ_1 is no longer negligible, see Appendix 3.A). We will come back to the regime of validity of a two temperature thermodynamic (T and T_e) in Sec. 3.5 where we will tackle the Kovacs effect and its implementation on the HOSS model.

3.3.2   Effective temperature from generalized laws

Let us suppose that we are considering aging regimes where $H_e = H$. The effective free energy (3.129) is, then,

$$F_e = U - TS_{ep} - T_eS_c \qquad (3.133)$$

and its differential, using Eqs. (3.54), (3.55) and (3.56), turns out to be:

$$dF = -S_{ep}dT - S_cdT_e - MdH \qquad (3.134)$$

Above, we have combined the generalization of the second law of thermodynamics for the heat variations,

$$đQ = TdS_{ep} + T_edS_c \qquad (3.135)$$

with the first law of thermodynamics

$$dU = đQ + đW = TdS_{ep} + T_edS_c - MdH \qquad (3.136)$$

writing the variation in work done on the system as $đW = -MdH$ (we define $M = Nm_1$).

This is the same expression obtained in the two temperature picture of Sec. 2.4. At equilibrium, where $T_e = T$, this reduces to the usual expression for ideal reversible quasi-static transformations, $đQ = TdS_{tot}$, with $S_{tot} = S_{ep} + S_c$.

If we now add the possibility of an extra effective field, the candidate thermodynamic potential that we obtain as a starting point is

$$F_e = U - TS_{ep} - T_eS_c + (H - H_e)M \qquad (3.137)$$

If we generalize Eq. (3.135) as

$$đQ = TdS_{ep} + T_edS_c + (H_e - H)dM \qquad (3.138)$$

the induced internal energy variation turns out to be

$$dU = đQ + đW = TdS_{\text{ep}} + T_e dS_c + (H_e - H)dM - MdH \qquad (3.139)$$

and, hence,

$$dF = -S_{\text{ep}}dT - S_c dT_e - MdH_e \qquad (3.140)$$

Once again, at equilibrium, where $H_e = H$ and $T_e = T$, this reduces to the usual expression $đQ = TdS_{\text{tot}}$. In Eq. (3.138), however, there is an odd mixing of "mechanical" (observables conjugated to external fields imposed on the system) and thermal objects. This is, in a way, the price to pay to generalize the two temperature thermodynamics including an extra effective field, H_e, that is not a true field but a quantity representing the (thermal) history of the system.

Starting from Eq. (3.136) we can derive the effective temperature through a generalization of the relation $T = \partial U / \partial S$, valid at equilibrium for a system of internal energy U and entropy S, with the derivative taken at constant magnetization (or volume). For simplicity, we put $H_e = H$ in the derivation (eventually we will apply our result to the case $\mu_1 \ll \mu_2$ where this setting is correct). Out of equilibrium, together with the previous equation of state for equilibrium processes (where S has to be substituted by S_{ep}) the following generalization also holds:

$$T_e = \left. \frac{\partial U}{\partial S_c} \right|_{S_{\text{ep}}} \qquad (3.141)$$

A more feasible identity, where the variable to be kept constant during the transformation is the heat-bath temperature, rather than the entropy of the fast processes, can be reworked [Franz & Virasoro, 2000; Crisanti & Ritort, 2000c; Mézard & Parisi, 2001]. To this aim, let us introduce the function $\bar{F}$:

$$\bar{F} \equiv F_e + T_e S_c \qquad (3.142)$$

inducing $d\bar{F} = T_e dS_c - S_{\text{ep}} dT$.[7] Through this auxiliary potential function we can then rewrite the effective temperature as

$$T_e = \left. \frac{\partial \bar{F}}{\partial S_c} \right|_T \qquad (3.143)$$

This result is a firm prediction for systems that satisfy the assumption of a two temperature thermodynamics. Writing the right-hand side of Eq. (3.143) as $\dot{\bar{F}}/\dot{S_c}$ and using Eqs. (3.81, 3.82), one obtains

$$T_e(t) = \tilde{K}(m_1(t), m_2(t)) \left[m_2(t) - m_1^2(t) \right] + \mathcal{O}(\mu_1) \qquad (3.144)$$

that is the thermodynamic effective temperature derived in section 3.3.1 [Eq. (3.130)]. We notice that the error is of order μ_1, exactly the order of difference between H_e and H that we are neglecting.

---

[7]In Chapter 7, Secs. 7.2, 7.3 and 7.5 we will analyze the meaning of $\bar{F}$ in a more general context and its relationship with the total free energy.

3.3.3   Dynamic transition rate and effective temperature

In the dynamic scenario that we have been analyzing in Sec. 3.2, an effective temperature can be directly related to the transition rate from a configuration of harmonic oscillators $\{x_i\}$ of energy E to a configuration of energy E' such that

$$x = E' - E \tag{3.145}$$

We can ask ourselves whether just looking at the frequency by which relaxation steps occur we might be able, in principle without knowing any details of the system under probe, to find the effective temperature. At the basis of this connection there is the assumption that partial equilibration occurs on a certain hypersurface in the configurational space (e.g., the constant energy surface in the HO model or the one fixed by $\mu_2 = $ const in the HOSS model) before moving to lower energy configurations.

This procedure has been originally devised by Ritort [2004] (see also [Crisanti & Ritort, 2003; Crisanti & Ritort, 2004; Garriga & Ritort, 2005]) in the HO model. Here we will generalize it to the HOSS model, as well. The starting point for the HO model is a microcanonical argument, since we are dealing with systems that partially equilibrate on the energy hypersurface $E = $ const. Because the system evolves exclusively through entropic barriers, the ratio between the probability of transition from a configuration i to a configuration f and its inverse $f \rightarrow i$ is simply proportional to the ratio of the number of available configurations having energy $E' = E(f)$ to the one having energy $E = E(i)$:

$$\frac{p(x)}{p(-x)} \propto \frac{\omega(E')}{\omega(E)} \tag{3.146}$$

where the density ω of configurations at energy E is the exponential of the configurational entropy $S_c(E)$. Here we are referring to slow relaxing α modes, taking for granted that the thermalization of all fast processes has already taken place. This is why the contribution of the transitions is connected to the configurational entropy and no contribution from the S_{ep} is considered.

The rate of transition $\nu(x)$ from a configuration at energy E to one at energy E' is defined as the ratio of the probability $p(x)$ of proposing a move yielding an energy shift x to the characteristic time $\tau(E)$ that the system spends in the configuration at energy E:

$$\nu(x) = \frac{p(x)}{\tau(E)} \tag{3.147}$$

Assuming that, for energy changes $x = \mathcal{O}(1)$, $\tau(E) \simeq \tau(E')$ one finds

$$\frac{\nu(x)}{\nu(-x)} \simeq \frac{p(x)}{p(-x)} \simeq e^{S_c(E')-S_c(E)} \simeq e^{\frac{\partial S_c(E)}{\partial E}x} = e^{\beta_e x} \tag{3.148}$$

where we have used the effective temperature definition Eq. (3.141).[8] So the effective temperature also shows up in the ratio of the transition rate to its time-reversed one.

Partial equilibration over the constant energy hypersurface

According to the above picture, measuring the rate of transition of a process allows one to determine an independent estimate of the effective temperature. To begin we show the explicit derivation of T_e from the rate's ratio in the case of the HO model (Sec. 3.1) without an external field. The initial configuration $\{x_k^{(i)}\}$ is at energy $E = U = KM_2/2$ [$H = 0$, implying $m_1 = 0$ in Eq. (3.1)] and the energy shift in the MC update is given by $Ky_2/2 \equiv K(M_2' - M_2)/2$. The proposed x_i changes to a new configuration of energy U' are randomly distributed with a variance Δ^2. The accessible points at energy U' from $\{x_k^{(i)}\}$ thus satisfy the relation

$$\sum_{k=1}^{N} \left(x_k - x_k^{(i)} \right) = \Delta^2 \tag{3.149}$$

We draw in Fig. 3.3 a two dimensional projection of the two energy hyperspheres. The larger one has a radius $R = \sqrt{M_2}$, the smaller one, $R' = \sqrt{M_2'}$. The small dashed circle centered on $\{x_k^{(i)}\}$ intersects the $U'[\{x_k\}] = \text{const}$ surface in two points. In an N dimensional representation this corresponds to an $N - 2$ dimensional hypersphere. We call $R_\cap$ its radius. The probability of the proposed update is proportional to the surface of the intersecting region:

$$p(x) \propto R_\cap^{N-2} \tag{3.150}$$

From Fig. 3.3 one deduces the expression for $R_\cap$:

$$R_\cap^2 = \Delta^2 - \frac{\left(R^2 - R'^2 + \Delta^2 \right)^2}{4R^2} \tag{3.151}$$

Using this geometrical relation for the HO model one has [cf. Eqs. (3.15)-(3.17) with $p(y_1) = \delta(y_1)$]

$$R_\cap^2 = \Delta^2 - \frac{\left(y_2 - \Delta^2 \right)^2}{4Nm_2} = \Delta^2 \left[1 - \frac{(x - \bar{x})^2}{N\Delta_x} \right] \tag{3.152}$$

Consequently, the density of configurations turns out to be

$$\omega \propto R_\cap^{N-2} \sim \exp \left\{ -\frac{(x - \bar{x})^2}{2\Delta_x} \right\} \tag{3.153}$$

---

[8]In the HO model no fast processes are involved and, indeed, $x = \delta E = \delta U$. In the HOSS model, there would be an extra contribution from the entropy of equilibrated processes and $x = \delta E$ would rather be the variation of $U - TS_{ep} = F_e + T_e S_c$, cf. Eqs. (3.54), (3.69) and (3.142). In this case Eq. (3.143) can be used instead.

that is, indeed, proportional to the exchange probability $p(x)$ as we know it from Eqs. (3.15, 3.16, 3.17).

The expression for T_e is eventually determined by the recipe Eq. (3.148)

$$\frac{p(x)}{p(-x)} = \exp\left\{\frac{2\overline{x}}{\Delta_x}x\right\} \to T_e = \frac{\Delta_x}{2\overline{x}} = Km_2 \qquad (3.154)$$

This result coincides with the previous computations of T_e, as, e.g., in Eqs. (3.130, 3.141), where, for the model at issue, it is $m_1 = H = J = 0$ (and, therefore, $\tilde{K} \to K$).

Partial equilibration over the hypersurface $\mu_2 = $ const

To extend the previous approach to the presence of an external field H and, further, of fast processes, one has to observe that in these other cases the hypersurface over which the $\{x_i\}$ are partially equilibrated is not the one at constant energy, but the one identified by the condition $\sum_i x_i^2/N - (\sum_i x_i/N)^2 = $ const. Indeed, the slow observable carrying the structural relaxation is now the variable $\mu_2(\{x_i\})$.

In the HOSS model, as a relaxation MC step is performed,

$$N[\mu_2(\{x_k'\}) - \mu_2(\{x_k\})] = y_2 - 2m_1y_1 = \frac{2x}{\tilde{K}(\{x_i\})} + 2\mu_1(\{x_i\})y_1 \simeq \frac{2x}{\tilde{K}(\{x_k\})} \qquad (3.155)$$

where, in neglecting $\mathcal{O}(\mu_1)$ terms, we have exploited the knowledge of the dynamics for the strong glass case and for the regimes of our interest in the

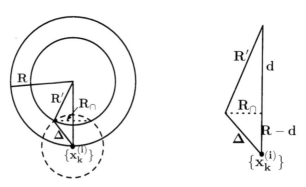

FIGURE 3.3

A pictorial representation in two dimension of the transition between two configurations of harmonic oscillator positions, $\{x_k\}$ in the N dimensional space of their degrees of freedom. The point $\{x_k^{(i)}\}$ stands for the initial configuration, at energy U. The locus of final configurations at energy U is given by the intersection between the dashed circle of radius Δ and the inner circle of radius R'. Eq. (3.151) is easily obtained looking at the triangle on the right-hand side of the figure.

fragile glass case [$T \gtrsim T_0$, $\forall \gamma$ and $T < T_0$ provided $\gamma > 1$, cf. Eqs. (3.92, 3.121, 3.122) in Secs. 3.2.3, 3.2.4].

The different choice in the slow relaxing observable changes the argument of the functions in Eqs. (3.146, 3.148): $U \to M_2 - M_1^2 = N\mu_2$. However, the fact that the μ_2 shift in a MC update is proportional to the energy shift x, makes Eq. (3.148) holding also when external fields (and/or *fast* spherical spins) are present. This is very important since it establishes a compatibility between the present approach and the introduction of perturbing fields, e.g., for studying the response functions, the two-time dynamics and analyzing the FDR out of equilibrium (Sec. 3.3.4).

We again refer to Fig. 3.3 where now the radii are $R = \sqrt{N(\mu_2 + m_0)}$, $R' = \sqrt{N(\mu_2' + m_0)}$ and

$$R_\cap^2 = \Delta^2 \left[1 - \frac{\left(N(\mu_2 - \mu_2') - \Delta^2 \right)^2}{4N(\mu_2 + m_0)\Delta^2} \right] \simeq \Delta^2 \left[1 - \frac{\left(2x/\tilde{K} + \Delta^2 \right)^2}{4N(\mu_2 + m_0)\Delta^2} \right] \quad (3.156)$$

implying a configuration density

$$w \propto R_\cap^{N-2} \sim \exp\left\{ -\frac{\left(x - \frac{\Delta^2 \tilde{K}}{2} \right)^2}{2\Delta^2 \tilde{K}^2 \mu_2} \right\} = \exp\left\{ -\frac{(x - \bar{x})^2}{2\Delta_x} \right\} \quad (3.157)$$

In the last equality we have used Eqs. (3.70, 3.16), consistently with the present approximation ($\mu_1^2 \ll \mu_2$).

The formula leading to the *transition rate* effective temperature is Eq. (3.148) that, together with Eq. (3.70), yields

$$T_e = \tilde{K}(\mu_2 + m_0) + O(\mu_1) \quad (3.158)$$

It coincides with the quasi-static derivation Eq. (3.130), and with the generalized Maxwell derivation, Eq. (3.143). Setting $m_0 = 0$ the T_e for the strong glass model is obtained. Setting further $\tilde{K} = K$, the one for the HO model with an external field is recovered, cf. Eq. (3.154). Therefore, also this phase space argument is consistent with the thermodynamic approach.

3.3.4 FDR and effective temperature

At equilibrium, temperature plays the role of a proportionality factor between thermodynamic fluctuations and response to external perturbations in the FDT. Out of equilibrium the fluctuation-dissipation equivalence is not granted by a theorem anymore, but a generalization can be devised, as explained in Sec. 2.8. We will look here at this relation in the fragile glass HOSS model, in the aging dynamics far from equilibrium.

First we define and compute the correlation and response functions that, unlike the one-time quantities $m_1(t)$ and $m_2(t)$, depend in a nontrivial way on two times when the system is out of equilibrium, thus showing directly the loss of time translation invariance (TTI) with respect to the case at equilibrium.

The correlation function between the thermodynamic fluctuation of a quantity $m_a(t)$ at time t and that of a quantity $m_b(t')$ at an earlier time t' are defined as

$$C_{ab}(t, t') \equiv N \langle \delta m_a(t) \delta m_b(t') \rangle, \quad \delta m_a \equiv m_a - \langle m_a \rangle, \quad a, b = 1, 2 \quad (3.159)$$

where $\langle .... \rangle$ is the average over the dynamic processes, i.e., the harmonic oscillators.

The response of an observable m_a at time t to a perturbation in a conjugate field H_b at some previous time t' takes the form

$$G_{ab}(t, t') \equiv \frac{\delta \langle m_a(t) \rangle}{\delta H_b(t')} \quad a, b = 1, 2 \quad (3.160)$$

In the models that we are considering, $H_1 = H$ and $H_2 = -K/2$.

We will probe the most general case treated in this chapter, i.e., we will use the formalism for the fragile HOSS model. Setting, then, $m_0 = 0$ will give back the strong model, while eliminating the fast-slow interaction ($J = 0$) will lead to the two time functions for the HO model in an external field (cf. Sec. 3.1).

Equations of motion of the correlation functions

The two-time behavior of the HO and HOSS models with parallel MC dynamics can also be studied analytically. Indeed, using the approach exposed for one time observables in Sec. 3.1.1 we can obtain the MC equations of motion for two-time quantities.[9]

The following equations hold for the equal time correlation functions in all models with MC dynamics for which the formal decomposition of probability distributions in two Gaussian distributions, cf. Eq. (3.15), is allowed:

$$\frac{d}{dt} C_{ab}(t, t) = \int dx W(\beta x) \left\{ \bar{y}_a(x) \bar{y}_b(x) + \Delta_y \left(-\frac{\mathcal{H}_1}{\mathcal{H}_2} \right)^{a+b-2} \right. \quad (3.161)$$

$$\left. + \sum_{c=1,2} \frac{\partial}{\partial m_c} [\bar{y}_a(x) C_{cb}(t, t) + \bar{y}_b(x) C_{ca}(t, t)] \right\} p(x|m_1, m_2) \quad a, b = 1, 2$$

The functions $\bar{y}_a(x)$ and the parameters Δ_y, H_1 and H_2 entering the above equation are those specifying the model. For the HOSS model they are

---

[9]The relevant manipulations to yield the equations for correlations and response functions can be found in Nieuwenhuizen [2000]; Leuzzi & Nieuwenhuizen [2001a].

[recalling Eqs. (3.14), (3.16), (3.17) and using (3.67) and (3.130)]:

$$\bar{y}_1(x) = \frac{\mu_1}{m_2 - m_1^2 + \mu_1^2} \frac{\bar{x} - x}{\tilde{K}} = \left(4\Lambda\mu_1 - \mu_1 \frac{x}{T_e}\right) + \mathcal{O}(\mu_1^3) \quad (3.162)$$

$$\bar{y}_2(x) = \frac{2}{\tilde{K}}\left(x + \tilde{H}\,\bar{y}_1(x)\right) \quad (3.163)$$

$$\bar{x} = \frac{\Delta^2 \tilde{K}}{2} = 4T^*\Lambda \quad (3.164)$$

$$\Delta_y = \frac{\Delta^2(m_2 - m_1^2)}{m_2 - m_1^2 + \mu_1^2} = 8(m_0 + \mu_2)\Lambda + \mathcal{O}(\Lambda\mu_1^2) \quad (3.165)$$

$$\mathcal{H}_1 = \tilde{H}; \qquad \mathcal{H}_2 = -\frac{\tilde{K}}{2}$$

Expanding the integrals appearing in Eq. (3.161) for large times (i.e., small μ_2 and $\mu_1 \sim \mu_2^{1+\gamma}$) one has the system of linear differential equations

$$\dot{C}_{11}(t,t) = c(t) + 2d_{y_1}^{(1)}(t)C_{11}(t,t) + 2d_{y_1}^{(2)}(t)C_{12}(t,t) \quad (3.166)$$

$$\dot{C}_{12}(t,t) = 2m_1(t)\,c(t) + d_{y_2}^{(1)}(t)C_{11}(t,t) \quad (3.167)$$
$$+ \left[d_{y_1}^{(1)}(t) + d_{y_2}^{(2)}(t)\right]C_{12}(t,t) + d_{y_1}^{(2)}(t)C_{22}(t,t)$$

$$\dot{C}_{22}(t,t) = 4m_1(t)^2 c(t) + o(t) + 2d_{y_2}^{(1)}(t)C_{12}(t,t) + 2d_{y_2}^{(2)}(t)C_{22}(t,t) \quad (3.168)$$

with

$$c(t) \equiv 8\left[m_0 + \mu_2(t)\right]\Lambda(t)\left(1 - \frac{1 - 2r + 4r^2}{\Lambda(t)}\right)\Upsilon(t) \quad (3.169)$$

$$o(t) \equiv 32\left[m_0 + \mu_2(t)\right]^2\Upsilon(t) \quad (3.170)$$

The functions $d_{y_k}^{(j)}$ are the derivative with respect to m_j of the MC integral of y_k. They are reported in Appendix 3.B, together with the relevant terms of their long time expansion both in the case (slightly) above the Kauzmann temperature T_0 and below it (in the latter case we only tackle the case for $\gamma > 1$ as we did for the one-time variables in Sec. 3.2.4). Indeed, due to the complicated form of the equations we are not able to find analytic solutions valid at any time. We are obliged to compute approximate solutions valid on given timescales and temperature intervals. We will present the solutions in the aging regime, for times that are long but not as long as the relaxation time to equilibrium, τ_{eq}. For high temperature the present approximation will be valid as far as $\bar{\mu}_2 \ll \delta\mu_2$, that is for T not too much larger than T_0, cf., e.g., Fig. 3.1. The regime where T is well above T_0 exposes little glassy behavior and will not be discussed. Below T_0, the approximation chosen will hold for the two-time variable behavior in the regime where $\mu_1 \ll \mu_2$, that is for $\gamma > 1$ [cf. Eq. (3.122)].

The study of the dynamics of correlation and response functions for times longer than τ_{eq}, when the system approaches equilibrium, can also be performed [Leuzzi & Nieuwenhuizen, 2001a] but is not of interest in the present context.

Dynamics in the aging regime for $T > T_0$

In the aging regime, for temperatures just above the Kauzmann temperature T_0, we can neglect $\bar{\mu}_2$ with respect to $\delta\mu_2(t)$.[10]

To find the solutions of Eqs. (3.166)-(3.168) we first perform an adiabatic expansion neglecting the time derivatives of the correlations functions. Indeed, to first order approximation, $\dot{C}_{ab}$ is proportional to $\dot{\mu}_2$: it is of $\mathcal{O}(\mu_2\Upsilon)$, negligible with respect to the right-hand side terms (of $\mathcal{O}(\Lambda\Upsilon)$). For the case $T \gtrsim T_0$, in the aging regime, $\bar{\mu}_2$ is negligible, $\mu_1 \sim r/\Lambda$ and $r \sim \mu_2(t)$, leading to

$$C_{ab}(t,t) \simeq \frac{[2m_1(t)]^{a+b-2}}{1 + DQ(t)}\{m_0 + \mu_2(t)\} + \underbrace{\mathcal{O}\left(1/\Lambda\right)}_{\text{observable dependent}} + \mathcal{O}(\mu_2^2) \quad (3.171)$$

Notice that the inclusion of the terms of order $\mathcal{O}\left(1/\Lambda\right)$ breaks the proportionality between the three equal time correlation functions. If $\gamma > 1$ (strictly) this is not an issue since the correction is of an order smaller than μ_2, but if $\gamma \leq 1$ the three expressions become much more involved than Eq. (3.171). We report them in Appendix 3.B for completeness. The main consequence is that the FDR will depend on the conjugated observable-field couple used. The correction terms breaking the proportionality are actually all proportional to some power of m_0, implying that the problem arises only for the fragile glass case. Here we will concentrate, however, on the case where $\mathcal{O}\left(1/\Lambda\right) \ll \mathcal{O}(\mu_2)$.

Once we have the equal time solutions, we can solve the equations for the two-time functions. Extending the parallel MC dynamics introduced in Sec. 3.1.1 to the two-time variables, we are able to derive the equations

$$\partial_t C_{ab}(t,t') = d_{y_a}^{(1)}(t)C_{1b}(t,t') + d_{y_a}^{(2)}(t)C_{2b}(t,t') \qquad a,b=1,2 \quad (3.172)$$

valid for the response functions $G_{ab}(t,t')$, as well. The coefficients $d_{y_a}^{(j)}$ are those appearing in Eqs. (3.166)-(3.168) and are reported in Appendix 3.B [Eqs. (3.B.18)-(3.B.21)].

Combining the correlation function equations, Eq. (3.172), and expanding down to order $\mathcal{O}(\Upsilon)$ the coefficients of the correlation functions, one obtains

$$\partial_t \left[C_{2b}(t,t') - 2m_1(t)C_{1b}(t,t')\right] \qquad\qquad (3.173)$$
$$\simeq \left[C_{2b}(t,t') - 2m_1(t)C_{1b}(t,t')\right]d_1(t) + 2\dot{m}_1(t)C_{1b}(t,t'), \qquad b=1,2$$

---

[10]This means, in particular, that in expressions (3.A.33)-(3.A.36) we have to put $\bar{\mu}_2$ equal to zero everywhere, including inside the constants $\bar{Q}$, $\bar{P}$ and $\tilde{K}_\infty$ [defined respectively in (3.79), (3.80) and (3.57)] and we can write $\delta\mu_2(t) = \mu_2(t)$.

Since

$$\dot{m}_1(t) = 4\Lambda\mu_1 \left[A_0(t) - 4T^*(t)\Lambda(t)A_1(t)\right] \sim \mathcal{O}(\mu_2\Upsilon) + \mathcal{O}(\mu_2^2\Lambda\Upsilon) \quad (3.174)$$

[cf. Eq. (3.30) generalized to the fragile HOSS model: $\alpha \to \sqrt{\Lambda}$, $K \to \tilde{K}$] the first term in the right-hand side of Eq. (3.173) can be neglected at the order of the current approximation. In the adiabatic approximation, the latter equation reduces to

$$C_{2b}(t,t') \simeq 2m_1(t)C_{1b}(t,t') \qquad b = 1,2 \quad (3.175)$$

and

$$\partial_t C_{1b}(t,t') \simeq \left[d_{y_1}^{(1)}(t) + 2m_1(t)d_{y_1}^{(2)}(t)\right] C_{1b}(t,t') \qquad b = 1,2 \quad (3.176)$$

yielding the aging behavior

$$C_{ab}(t,t') \simeq C_{ab}(t',t')\frac{h(t)}{h(t')} \quad (3.177)$$

In the above expression we have introduced the time sector function

$$h(t) \equiv \exp \int_{-\infty}^{t} d\tau \left[d_{y_1}^{(1)}(\tau) + 2m_1(\tau)d_{y_1}^{(2)}(\tau)\right] \quad (3.178)$$

$$\simeq \exp\left\{-4 \int_{-\infty}^{t} d\tau\,\Upsilon(\tau)\left[\Lambda(\tau) - 1\right]\left[1 + Q(\tau)D\right]\right\} \quad (3.179)$$

The two-time dependence is all included in the ratio of the time sector function computed at different times, $h(t)/h(t')$ (compare with Sec. 1.3).

Response functions

The equal time response functions can be recovered as the limit

$$\lim_{\Delta t \to 0} \frac{1}{\Delta t}\frac{\delta \langle m_a\,(t + \Delta t)\rangle}{\delta H_b(t)}\bigg|_{H_b=0} \quad (3.180)$$

of the two-time response function $G_{ab}(t + \Delta t, t)$. Neglecting $\mathcal{O}(\Upsilon^2)$ terms (called *switch terms* in [Nieuwenhuizen, 2000]), such a limit, for $a = b = 1$, becomes

$$G_{11}(t^+, t) = -\beta \int dy_1 dy_2\, W'(\beta x)\, y_1^2\, p(y_1, y_2|m_1, m_2) \quad (3.181)$$

$$= \frac{4\Upsilon\Lambda}{\tilde{K}} - \frac{2\Upsilon}{\tilde{K}} + \mathcal{O}(\mu_2\Upsilon)$$

where $W'(\beta x)$ is the derivative of the Metropolis transition probability $W(\beta x)$, Eq. (3.10), with respect to its argument:

$$W'(\beta x) = \begin{cases} -e^{-\beta x} & \text{if } x > 0 \\ 0 & \text{if } x \geq 0 \end{cases} \quad (3.182)$$

The function $p(x|m_1, m_2)$ is the Gaussian distribution of x conditioned by the values of m_1, m_2 at the time the update is proposed [see Eq. (3.15) for its definition] and Δ_y and $\bar{y}_1(x)$ are given in Eqs. (3.162), (3.165).

Analogue expressions are derived for G_{12} and G_{22}:

$$G_{12}(t^+, t) = -\beta \int dy_1 dy_2 \, W'(\beta x) \, y_1 \, y_2 \, p(y_1, y_2|m_1, m_2)$$

$$= \frac{8m_1}{\tilde{K}} \Upsilon \Lambda - \frac{4m_1 \Upsilon}{\tilde{K}} + \mathcal{O}\left(\mu_2 \Upsilon\right) \tag{3.183}$$

$$G_{22}(t^+, t) = -\beta \int dy_1 dy_2 \, W'(\beta x) \, y_2^2 \, p(y_1, y_2|m_1, m_2)$$

$$= \frac{16m_1^2}{\tilde{K}} \Upsilon \Lambda - \frac{8m_1^2 \Upsilon}{\tilde{K}} - 32\Upsilon m_0^2 + \mathcal{O}\left(\mu_2 \Upsilon\right) \tag{3.184}$$

Notice here that the $O(\Upsilon)$ term of G_{22} has an extra $-32m_0^2$, breaking the proportionality between the three response functions written above in the fragile glass case, exactly as it happens for the correlation functions. The existence of this term, anyway, has no consequences for the leading term of the FDR.

The equations describing the evolution in t of the response to a perturbation at t' have the same shape as those for the correlation functions (3.172). The solutions are then

$$G_{ab}(t, t') = G_{ab}(t', t') \frac{h(t)}{h(t')} \tag{3.185}$$

With these results we can generalize the FDT defining another effective temperature, T_e^{FD}, by means of the ratio between the derivative with respect to the initial time (also called the "waiting" time) t' of the correlation function C_{ab} and the corresponding response function G_{ab}:

$$T_{e;a,b}^{FD}(t, t') \equiv \frac{\partial_{t'} C_{ab}(t, t')}{G_{ab}(t, t')} \tag{3.186}$$

To compute it we need:

$$\partial_{t'} C_{ab}(t, t') = (\partial_{t'} C_{11}(t', t')) \frac{h(t)}{h(t')} - C_{ab}(t', t') \frac{h(t)}{h^2(t')} \partial_{t'} h(t') \tag{3.187}$$

$$= 4\Upsilon(t')[\Lambda(t') - 1][1 + Q(t')D]C_{ab}(t', t') \frac{h(t)}{h(t')} + \mathcal{O}(\mu_2 \Upsilon)$$

$$\simeq 4\left[2m_1(t')\right]^{a+b-2} [m_0 + \mu_2(t')] \Upsilon(t') [\Lambda(t') - 1] \frac{h(t)}{h(t')}$$

Eventually, we get the main result of this section:

$$T_e^{FD}(t, t') = T_e^{FD}(t') \simeq T^*(t') \left[1 + \mathcal{O}\left(\frac{1}{\Lambda(t')}\right) + \mathcal{O}\left(\mu_2(t')^2\right)\right], \quad \forall \, (a, b) \tag{3.188}$$

where T^* was first introduced in Sec. 3.2.4, Eq. (3.111) and, later on, in Sec. 3.3, it was identified with the effective temperature [Eqs. (3.130, 3.141, 3.158)]. We recall that $\Lambda^{-1} = \mu_2^\gamma$.

Looking back to Eq. (3.188) we see that T_e^{FD} coincides, in the timescale of our interest, with the effective temperature T_e that we got by previous approaches, only if $1/\Lambda$ is negligible with respect to μ_2, as we assumed already in Eq. 3.171. This is true, indeed, if the exponent γ of the generalized VF law (3.124), is greater than one. Otherwise, as we stressed since the beginning of the computation, the last correction is no longer sub-leading: already for $\gamma = 1$, $T_e^{FD} \to T_e$ exclusively in the infinite time limit, when they both tend to T, i.e., for timescales longer than those of the considered aging regime. As already discussed in Sec. 3.2.2, where we presented the results of the dynamics of the one-time observables, the value of the exponent γ discriminates between different regimes. For $\gamma > 1$ an out of equilibrium thermodynamics can be built in terms of a single additional effective parameter (the effective temperature T_e). For $\gamma \le 1$, T_e alone does not give consistent results in the generalization of the equilibrium properties to the non equilibrium case and in order to cure this inconsistency, more effective parameters are needed. This discrepancy was already clear from section 3.2.4, for the regimes below T_0 where the one-time variables had different behaviors depending on the value of γ being greater, equal to or lesser than 1. For $T > T_0$ such a difference was apparently absent but it emerges here at the level of two-time variable dynamics.

Low temperature case: $T < T_0$, $\gamma > 1$

Our approach allows us to study the regime even below the Kauzmann temperature T_0. In the latter case, though, we have qualitatively different behaviors depending on the value of γ, i.e., on the relative weight of μ_1 and μ_2. We describe here the case $\gamma > 1$, where $\mu_1 \ll \mu_2$ [cf. Eq. (3.122)] and where we can still hope to find a unique effective temperature linking out-of-equilibrium dynamics with thermodynamic properties of the model in the same way as it does above T_0. For $\gamma > 1$, according to the results shown in Sec. 3.3.1 and 3.2.2, it is, indeed, not necessary to introduce any effective thermodynamic parameter other than the effective temperature, and the analysis can be carried out in a way similar to the one above T_0 (Sec. 3.3.4).[11]

The equations of motion for the equal time correlation functions are identical to the Eqs. (3.166)-(3.168). What changes are sub-leading terms in the time-dependent coefficients $d_{y_a}^{(j)}$. Solutions to these equations are obtained, as before, in the adiabatic approximation and expanding all the functions in

---

[11]In expanding the time-dependent coefficients of C_{ab} in the equations of motion [$d_{y_{1,2}}^{(1,2)}$, in Eqs. (3.B.18)-(3.B.21)] we now have to take into account that r never vanishes, while the asymptotic value of $\mu_2(t)$, denoted by $\bar{\mu}_2$, is zero. For the technical discussion about which terms are now leading and sub-leading, we refer to Appendix 3.B.

powers of $\mu_2(t)$. We notice that, because of the presence of the Kauzmann transition, we lose the static limit.

Below T_0, the two-time correlation functions turn out to be, at the leading order:

$$C_{ab}(t,t') \simeq \frac{[2m_1(t')]^{a+b-2}}{1+DQ(t')}\left[m_0 + \mu_2(t') + \mathcal{O}(\mu_1(t'))\right]\frac{h(t)}{h(t')}$$
$$a,b = 1,2 \qquad (3.189)$$

where $h(t)$ is still given by Eq. (3.178) at order $\Lambda\Upsilon$. Also in this case, the leading orders of different correlation functions are proportional to each other (it is $\gamma > 1$). The breaking of proportionality occurs only in the correction terms. For the response functions, from Eqs. (3.181)-(3.184) we get

$$G_{11}(t,t^+) \simeq \frac{4\Upsilon\Lambda}{\tilde{K}} - \frac{2\Upsilon(1-2r)^2}{\tilde{K}} + \frac{8\Upsilon(\Lambda\mu_1)^2}{T_e} \qquad (3.190)$$

$$G_{12}(t,t^+) \simeq 2m_1G_{11}(t,t^+) + \frac{16r\Lambda\mu_1\Upsilon}{\tilde{K}} \qquad (3.191)$$

$$G_{22}(t,t^+) \simeq 4m_1^2G_{11}(t,t^+) + \frac{64m_1r\Lambda\mu_1\Upsilon}{\tilde{K}} + \frac{8m_0(1-2r)^2\Upsilon}{\tilde{K}} \qquad (3.192)$$

where this time the contributions $\Lambda\mu_1$ and $(\Lambda\mu_1)^2$ are both of the order Υ and we have to take them both into account. Notice that the Gs differ by terms of order $O(\Upsilon)$ (the first sub-leading order).

The two-time behavior of the response functions is as in Eq. (3.185), provided only terms of order $\Lambda\Upsilon$ are considered in the exponent of the time sector function $h(t)$.

The last thing that we need, before computing T_e^{FD}, is the derivative

$$\partial_{t'}C_{ab}(t,t') = \left[\partial_{t'}C_{11}(t',t') - C_{11}(t,t')\frac{\partial_{t'}h(t')}{h(t')}\right]\frac{\tilde{h}(t)}{\tilde{h}(t')} \qquad (3.193)$$

$$\simeq 4\Upsilon(t')\left[m_0 + \mu_2(t')\right]\Lambda(t')\frac{\tilde{h}(t')}{\tilde{h}(t)} \qquad (3.194)$$

It follows that

$$T_e^{\mathrm{FD}}(t,t') \equiv \frac{\partial_{t'}C_{ab}(t,t')}{G_{ab}(t,t')} \simeq T^\star(t')\left[1 + \mathcal{O}\left(\frac{1}{\Lambda}\right) + \mathcal{O}(\mu_2^{1+\gamma})\right] = T_e^{\mathrm{FD}}(t')$$
$$\forall\,(a,b) \qquad (3.195)$$

Now $\mathcal{O}(1/\Lambda) = \mathcal{O}(\mu_1) \ll \mathcal{O}(\mu_2)$, because $\gamma > 1$: in the long time regime $T_e^{\mathrm{FD}}(t)$ coincides with $T_e(t)$ for any (a,b) couple of conjugated variables. The differences remain at the level of sub-leading terms, therefore beyond the aging regime.

3.3.5 Heat flow of α processes

From Eqs. (3.56, 3.59, 3.130 and 3.131) the complete expression for the rate of change of the heat flow turns out to be:[12]

$$\dot{\mathcal{Q}} = \frac{N\tilde{K}}{2} \frac{1}{1 + QD - \tilde{K}JQ\mu_1} \left\{ \dot{\mu}_1 \left[2\mu_1 + \frac{TQ\tilde{K}}{J} \left(w + \frac{T}{2} \right) \right] \right. \tag{3.196}$$

$$\left. + \dot{\mu}_2 \left[1 + DQ - \frac{TQ\tilde{K}^2}{D} \left(w + \frac{T}{2} \right) \right] \right\}$$

The rate of the heat flowing out of the system is $-\dot{\mathcal{Q}}$. Referring to the aging regimes described in section 3.2.3 for the strong glass model and in section 3.2.4 for the fragile one, the quantity $\dot{\mathcal{Q}}$ is proportional to $\dot{\mu}_2$ (that is negative). In the non-generic dynamic regimes for the fragile glass model reported in Appendix 3.A, the heat flow is $-\dot{\mathcal{Q}} \propto -(\dot{\mu}_1 + \dot{\mu}_2)$ in the cases 3a and 3b $(T < T_0, \gamma = 1, \epsilon \geq 1)$ and $-\dot{\mathcal{Q}} \propto -\dot{\mu}_1$ in the cases 3c, 3d, 3e $(T < T_0, \gamma = 1, \epsilon \leq 1)$ and 4 $(T < T_0, \gamma < 1)$.

In every dynamic aging regime examined, $\dot{\mu}_1$ and $\dot{\mu}_2$ are negative and this implies that the heat flow of the out-of-equilibrium system is positive in its approach to equilibrium, no matter what the values of the parameters of the model are. This is just what one expects: during relaxation heat is dumped in the bath, in order to allow the system to go to states of lower energy.

3.3.6 Effective temperature from a fluctuation formula

From the expression of $m_1(t; T)$ as function of H we can compute the quantity $\chi^{\text{fluct}} \equiv \left. \frac{\partial m_1}{\partial H} \right|_{T,t}$ that is the contribution to susceptibility in a cooling-heating setup caused by a change in the field H at fixed time (also called *fluctuation susceptibility*). In Sec. 2.7, cf. Eq. (2.93), we saw that in a cooling experiment the whole susceptibility can, indeed, be written as

$$\chi_{ab} \equiv \left. \frac{\partial m_a}{\partial H_b} \right|_T = \left. \frac{\partial m_a}{\partial H_b} \right|_{T,T_e} + \left. \frac{\partial m_a}{\partial T_e} \right|_{T,H_b} \left. \frac{\partial T_e}{\partial H_b} \right|_T =$$

$$= \left. \frac{\partial m_a}{\partial H_b} \right|_{T,t} - \left. \frac{\partial m_a}{\partial T_e} \right|_{T,H_b} \left. \frac{\partial T_e}{\partial H_b} \right|_{T,t} + \left. \frac{\partial m_a}{\partial T_e} \right|_{T,H_b} \left. \frac{\partial T_e}{\partial H_b} \right|_T$$

$$\equiv \chi_{ab}^{\text{fluct}} + \chi_{ab}^{\text{loss}} + \chi_{ab}^{\text{conf}} \tag{3.197}$$

Here we are considering an aging situation, so only the first term is relevant. Assuming separation of timescales, we propose the following form for

---

[12]We use $\mathcal{Q}$ in order to avoid confusion with the abbreviation Q, Eq. (3.79), widely used for the HOSS model, that is not a heat.

$\chi_{ab}^{\text{fluct}}(t,t)$

$$\chi_{ab}^{\text{fluct}}(t,t) = \left.\frac{\partial m_a}{\partial H_b}\right|_{T,t} = N\frac{\langle \delta m_a(t)\delta m_b(t)\rangle_{\text{fast}}}{T} + N\frac{\langle \delta m_a(t)\delta m_b(t)\rangle_{\text{slow}}}{T_e^{\text{fl}}(t)}$$

(3.198)

where $\langle \ldots \rangle_{\text{fast/slow}}$ is the average, respectively, over fast and slow processes. The fast ones are governed by the heat-bath temperature, the slow ones by some effective temperature T_e^{fl} depending on the timescale t. Through $\chi_{ab}^{\text{fluct}}(t)$ one can look at the connection between the fluctuation effective temperature T_e^{fl} and the other effective temperatures so far defined. To work it out we start, e.g., from:

$$\chi_{11}^{\text{fluct}}(t) \equiv \left.\frac{\partial m_1}{\partial H}\right|_{T,t} = N\frac{\langle \delta m_1(t)\delta m_1(t)\rangle}{T_e^{\text{fl}}} = \frac{C_{11}(t,t)}{T_e^{\text{fl}}}$$

(3.199)

Using the following expression for m_1, obtained from Eq. (3.62),

$$m_1(t;T,H) = -\frac{L}{J} + \frac{D}{J\tilde{K}\left(m_1(t;T,H),m_2(t;T,H);T\right)}$$

(3.200)

the fluctuation susceptibility χ_{11}^{fl} turns out to be:

$$\left.\frac{\partial m_1}{\partial H}\right|_{T,t} = \frac{1}{\tilde{K}(1+QD)} + \mathcal{O}\left(\mu_1\right)$$

(3.201)

Here we are neglecting terms like $\partial \mu_1/\partial H$ and $\partial \mu_2/\partial H$, of order μ_1 or higher (we deal, hence, with the regimes $[T > T_0, \forall \gamma]$ and $[T < T_0, \gamma > 1]$ where $\mu_1 \ll \mu_2$). Taking Eq. (3.171), we see that the leading term of C_{11} can be written as

$$C_{11}(t,t) = \frac{m_0 + \mu_2}{1 + QD} + \mathcal{O}\left(\mu_2^\gamma\right)$$

(3.202)

and this leads to

$$T_e^{\text{fl}} = \tilde{K}(m_0 + \mu_2) + \mathcal{O}(\mu_2^\gamma)$$

(3.203)

thus coinciding with previous definitions of the effective temperature, see Eqs. (3.130, 3.141, 3.158, 3.188) at the order of our interest, i.e., $\mathcal{O}(\mu_2)$. At higher orders and in other regimes (i.e., with $\gamma \leq 1$) there will be nonuniversalities. If $\gamma \leq 1$ the terms of $\mathcal{O}(\mu_2^\gamma)$ become dominant with respect to $\mathcal{O}(\mu_2)$, leading to the same situation that we had for FDR, namely, the off-equilibrium thermodynamic description cannot be implemented by means of a unique effective parameter.

3.4   Below the Kauzmann transition

As we anticipated in Sec. 3.2.1 for the fragile model version, a thermodynamic phase transition to a glass occurs at $T_K \equiv T_0 = \tilde{K}_\infty(T_0)\, m_0$. To see how the

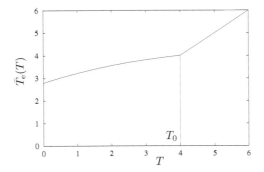

FIGURE 3.4

In the static regime, the effective temperature is shown as a function of the heat-bath temperature. At high temperatures they coincide but below the Kauzmann point T_0, $\bar{T}_e$ never reaches T, not even in the infinite time limit: the system remains forever out of equilibrium. Values of the constants are $K = J = 1$, $H = L = 0.1$, $m_0 = 5$.

transition takes place we first look at the asymptotic behavior of the effective temperature. When $T \geq T_K$ and $t \to \infty$, T_e approaches the heat-bath temperature T. When $T < T_K$, instead, T_e never reaches such a temperature. It rather goes toward some limiting value $\bar{T}_e(T)$ that we can compute from Eq. (3.130), rewritten for clarity in the explicit form

$$\bar{T}_e^4 - (2Km_0 + T)\bar{T}_e^3 + m_0 \left[(Km_0 + T)K + \frac{D^2 - J^4}{J^2} \right] \bar{T}_e^2 \quad (3.204)$$

$$- 2 \left(\frac{D}{J} \right)^2 Km_0^2 \, \bar{T}_e + \left(\frac{D}{J} \right)^2 K^2 m_0^3 = 0$$

i.e., a quartic equation for the effective temperature in the infinite time limit.[13] The same equation evaluated at $\bar{T}_e = T = T_K$ yields the value of the Kauzmann temperature T_0 as a function of the parameters of the model. In Fig. 3.4 we plot $\bar{T}_e$ versus T for a specific choice of parameter values.

From Eq. (3.130), or Fig. 3.4, we observe that $d\bar{T}_e/dT|_{T_K^-} < 1$, whereas, coming from above the Kauzmann temperature, one has $d\bar{T}_e/dT|_{T_K^+} = 1$. The derivative of $\bar{T}_e(T)$ shows, thus, a discontinuity at $T = T_0$. Any thermodynamic function, like U and $M \equiv Nm_1$, will depend on the heat-bath temperature both explicitly and through this effective temperature. For the

---

[13]The present equation is equivalent to Eq. (3.84) for $\tilde{K}(T, t)$ in the infinite time limit, since $\bar{T}_e(T)$ is just $\tilde{K}_\infty(T)m_0$.

specific heat we will have, for instance:

$$C \equiv \frac{1}{N}\frac{dU}{dT}\bigg|_H = \frac{1}{N}\frac{\partial U}{\partial T}\bigg|_H + \frac{1}{N}\frac{\partial U}{\partial \bar{T}_e}\bigg|_{H,T}\frac{dT_e}{dT}\bigg|_H \qquad (3.205)$$

This is of the same form as $C = c_1 + c_2\,(\partial T_e/\partial T)_p$ assumed originally by Tool [1946] (see also Sec. 2.2) for the study of caloric behavior in the glass formation region.

The discontinuity in $d\bar{T}_e/dT\big|_H$ causes a discontinuity in the specific heat and also in the quantity $-\partial M/\partial T\big|_H{}^{14}$ because both of these quantities contain terms proportional to $\partial \bar{T}_e/\partial T\big|_H$. Different from the glassy regime above T_0, this result holds even for infinite time, since no relaxation to the equilibrium, globally stable, "crystal" state takes place.

3.4.1 Instantaneous relaxation time

In Sec. 3.2.4, we find a VF form for the relaxation time above T_K when both $t, t_w \to \infty$. Far from equilibrium, in the aging regime, we might as well define a time dependent τ_α, giving the characteristic timescale on which the α processes are taking place.

For $T \gtrsim T_K$, in the aging regime, $\bar{\mu}_2$, the static part of μ_2, is negligible with respect to the dynamic part $\delta\mu_2$ so that for the effective temperature we have the following expansion:

$$T_e(t) \simeq T + \tilde{K}_\infty \frac{1 + D\bar{Q} + \bar{P}}{1 + \bar{Q}D}\delta\mu_2(t) \qquad (3.206)$$

$$\simeq T_0\left(1 + \frac{1}{m_0}\frac{1 + D\bar{Q} + \bar{P}}{1 + \bar{Q}D}\delta\mu_2(t)\right) + \mathcal{O}(T - T_0)$$

where, in the last expression, $\bar{Q}$ and $\bar{P}$, cf. Eqs. (3.79), (3.80), are computed at T_0 and $\tilde{K}_\infty(T_0) = T_0/m_0$.

With the above formula we determine the expression

$$\tau_\alpha(t) \propto \exp\left(\frac{1}{\delta\mu_2(t)}\right) \simeq \exp\left(\frac{A(T)}{T_e(t) - T}\right)^\gamma \simeq \exp\left(\frac{A(T_0)}{T_e(t) - T_0}\right)^\gamma,$$

$$A(T) \equiv \tilde{K}_\infty(T)\left(1 + \frac{\bar{P}(T)}{1 + \bar{Q}(T)D}\right) \qquad (3.207)$$

We can do the same below the Kauzmann temperature, provided that the VF exponent is $\gamma > 1$. For $T < T_0$ the relaxation time always diverges for $t \to \infty$. However, an instantaneous relaxation time can be considered and

---

[14]This quantity has been termed "magnetizability" by Nieuwenhuizen [1998c, 2000]. It is the analogue of a thermal expansivity of the model for which the external field is the pressure and M is the volume.

expressed in terms of the effective temperature using the first order expansion of T_e in μ_2:

$$T_e(t) = \bar{T}_e + \tilde{K}_\infty \left(1 + \frac{\bar{P}}{1 + \bar{Q}D} \right) \mu_2(t). \tag{3.208}$$

We find, from Eq. (3.123)

$$\tau_\alpha(t) = \tau_\alpha(T, T_e(t)) \propto \exp\left(\frac{A(T)}{T_e(t) - \bar{T}_e(T)} \right)^\gamma \tag{3.209}$$

where $A(T)$ is given in Eq. (3.207) and $\bar{T}_e(T) = \tilde{K}_\infty(T)m_0$. The aging behavior just above and well below T_0 are, thus, intimately related. Eq. (3.209) resembles a VF law where the heat-bath temperature has been substituted by a time-dependent effective temperature $T_e(t)$ and the Kauzmann temperature by its asymptotic value $\bar{T}_e$. Such a relation could hold very well in more general systems exhibiting universal behavior in the aging regime.

When $\gamma \leq 1$, in those regimes where μ_1 cannot be neglected with respect to μ_2, there is no simple expression for τ_α, because a unique effective thermodynamic parameter is not enough to describe the out-of-equilibrium dynamics of the system and those are the nonuniversal regimes.

3.5 Kovacs effect: limits of two temperature thermodynamics

We use the HOSS model for fragile glass studied up to now to implement a particular experimental protocol in order to get some insight into the memory effect initially observed by Kovacs [1963], see Sec. 1.7. We show that, in spite of its simplicity, this model captures the phenomenology of the Kovacs effect: it makes it possible to implement the protocol not only with temperature shifts but with magnetic field shifts as well, and it allows us to obtain analytical expressions for the evolution of the thermodynamic observables (in specific time regimes). The HOSS model property, of displaying both fast and slow processes, turns out to be necessary for the memory effect to occur.

The system is prepared at a temperature T_i (step **1**) and quenched to a region of temperature close to the T_K, i.e., $T_l \gtrsim T_K$ (step **2**). After a time t_a, during which a given observable reaches the value it would have at equilibrium at a certain temperature $T_f > T_l$, the system is heated up to T_f (step **3**), cf. Fig. 1.7. Solving Eqs. (3.30)-(3.31), or equivalently Eqs. (3.81)-(3.82), we determine the evolution of the system both in step 2 and 3 of the protocol. In step 2, the time t_a at which $m_1(t_a; T_l) = \bar{m}_1(T_f)$, is such that:

$$m_1(t_a^+; T_f) = \bar{m}_1(T_f) \tag{3.210}$$
$$m_2(t_a^+; T_f) = m_2(t_a; T_l)$$

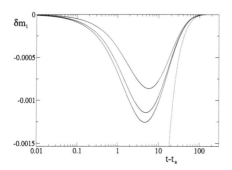

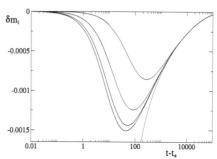

FIGURE 3.5

Fragile glass with $\gamma = 1$. The Kovacs protocol is implemented with a quench from temperature $T_i = 10$ to T_l, and a final jump to the intermediate temperature $T_f = 4.3$ at $t = t_a$. The full curves, bottom to top, refer to $T_l = 4.005, 4.05, 4.15$. The dashed curve refers to $T_l = T_f$ (simple aging with no final temperature shift). Reprinted figure with permission from [Aquino et al., 2006a]. Copyright (2006) by the American Physical Society.

FIGURE 3.6

Fragile glass with $\gamma = 2$. The Kovacs protocol is implemented with a quench from temperature $T_i = 10$ to T_l, and a final jump to $T_f = 4.3$. The full curves, bottom to top, refer to $T_l = 4.005, 4.05, 4.15, 4.25$. The dashed line refers to $T_l = T_f$ (simple aging). The hump occurs on longer timescales with respect to to the case $\gamma = 1$. Reprinted figure with permission from [Aquino et al., 2006a].

The evolution of the fractional "magnetization"

$$\delta m_1(t) = \frac{m_1(t) - \bar{m}_1(T_f)}{\bar{m}_1(T_f)} \tag{3.211}$$

after step 3 $(t > t_a)$ for different values of T_l is reported in Figs. 3.5 and 3.6 respectively for $\gamma = 1$ and $\gamma = 2$. In both the figures, the values for the parameters of the model are: $J = K = 1$, $L = H = 0.1$, $m_0 = 5$. For such parameter values, the Kauzmann temperature turns out to be $T_K = 4.00248$ (see also Fig. 3.2).

Since the equilibrium value of m_1 decreases with increasing temperature (as opposed to what happens to the volume in the original experiment), we observe a reversed "Kovacs hump." The curves keep the same properties typical of the Kovacs effect, the minima occur at a time that decreases and have a depth that increases with increasing magnitude $T_f - T_l$ of the final switch of temperature. As expected, since increasing γ corresponds to further slowing the dynamics, the effect is displayed on a longer timescale in the case of $\gamma = 2$ as compared to $\gamma = 1$.

The sign of the Kovacs hump

We rewrite Eq. (3.30) exploiting the expressions for the acceptance rate $A_0(t)$ and the average energy shift $A_1(t)$ [Eqs. (3.34), (3.36)]:

$$\dot{m}_1 = \mu_1 \beta \frac{\Delta^2 \tilde{K}}{2}\, b(\alpha) \tag{3.212}$$

where $b(\alpha)$ is defined in Eq. (3.35) and $\alpha^2 = \Delta^2/(8(m_0 + \mu_2 + \mu_1^2))$. Actually, since in the last step of the protocol $m_1(t = t_a) = \bar{m}_1(T_{\mathrm{f}})$ and $b(\alpha)$ is always positive, from Eq. (3.212) one soon realizes that the hump for this model can be either positive or negative, depending on the sign of the term $\mu_1 = \tilde{H}/\tilde{K} - m_1$ at $t = t_a^+$. As we know from its definition, this term is zero when both $m_1 = \bar{m}_1(T_{\mathrm{f}})$ and $m_2 = \bar{m}_2(T_{\mathrm{f}})$, so one would expect $m_2(t = t_a^+) = \bar{m}_2(T_{\mathrm{f}})$ to be the border value determining the positivity or negativity of the hump. Since $\tilde{H}(\bar{m}_1(T_{\mathrm{f}}), m_2(t))$ decreases with increasing m_2 while $\tilde{K}(\bar{m}_1(T_{\mathrm{f}}), m_2(t))$ increases, it follows that the condition for a positive hump is:

$$m_2(t = t_a^+) < \bar{m}_2(T_{\mathrm{f}}) \tag{3.213}$$

For shifts of temperature in a wide range close to the Kauzmann temperature T_K, where the dynamics is slower and the effect is expected to show up significantly on a long timescale, the condition (3.213) is never fulfilled and, hence, a negative hump is expected.

Kovacs protocol at constant temperature with magnetic field shifts

Interchanging the roles of T and H, the Kovacs protocol can be implemented at constant temperature, by changing the magnetic field H. Indeed, from Eqs. (3.63), (3.66), one notices that the value of the transition temperature T_K depends on H as well. Therefore, the protocol must be implemented in the following way. The temperature is kept fixed at T_i. This is a Kauzmann temperature for a specific value of the field $H = H_K$, i.e., it is the solution of the equation [cf, Eq. (3.66)]

$$T_i = m_0 \tilde{K}_\infty(T_i, H_K) \tag{3.214}$$

The temperature T_K decreases with decreasing H. So, if we work at $T = T_i$ with magnetic fields $H < H_K$, we are sure to implement every step of the protocol keeping the system always above the H-driven Kauzmann transition. The three steps are schematically depicted in Fig. 3.7. We start with the system equilibrated at $T = T_i$ and $H = H_i \ll H_K$, and, at time $t = 0$, we shift instantaneously the field to a larger value H_l, such that $H_i < H_l \lesssim H_K$. Then we let the system age for a time t_a such that $m_1(t_a; H_l) = \bar{m}_1(H_f)$. At this time the field is shifted to H_f (with $H_f < H_l < H_K$). The subsequent evolution of the fraction magnetization $\delta m_1(t)$ is shown in Fig. 3.8. Again the curves show all the typical properties of the Kovacs hump, with a very slow relaxation back to equilibrium due to the fact that H_f has been chosen very close to H_K.

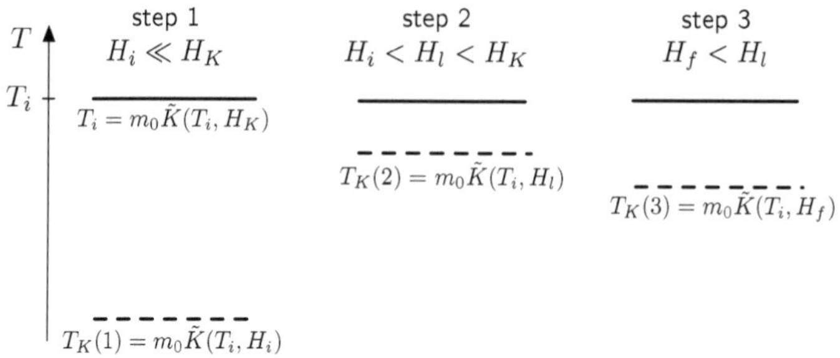

FIGURE 3.7

A pictorial sketch of the Kovacs protocol implemented at fixed temperature T_i varying the external field and, therefore, the relative Kauzmann transition temperature according to $T_K(H) = m_0 \tilde{K}_\infty(T_i, H)$. In the first step T_K is much less than T_i because $H_i \ll H_K$, the field value for which T_i is the Kauzmann temperature. At step 2, increasing the field makes T_K larger, slightly below T_i, so that the system turns out to be just above the Kauzmann transition. Then, in step 3, by lowering $H \rightarrow H_f$, the Kauzmann temperature decreases.

3.5.1 Analytical solution in the long-time regime

Through a numerical solution of the dynamics, we have, thus, seen that the HOSS model reproduces the phenomenology of the Kovacs effect, showing the same qualitative properties of the Kovacs hump as obtained in experiments (see, for example, [Kovacs, 1963; Josserand *et al.*, 2000]) or in other models [Berthier & Bouchaud, 2002; Buhot, 2003; Bertin *et al.*, 2003; Mossa & Sciortino, 2004; Cugliandolo *et al.*, 2004].

The peculiarity of the present model is that, by carefully choosing the working conditions in which the protocol is implemented, it further provides us with an analytical solution for the evolution of the variable of interest. We will make use of μ_1 and μ_2, whose equations of motion are Eqs. (3.81)-(3.82).

We will choose to implement steps 2 and 3 of the protocol in a range of temperatures very close to the Kauzmann temperature T_K. As exhaustively discussed in Sec. 3.2.4, cf. Eq. (3.119), in the long time regime the variable $\mu_2(t)$ decays logarithmically to its equilibrium value, that is small for $T \sim T_K$. So, if t_a is large enough, the following equation is shown to be valid [from Eq. (3.120)], assuming $\Lambda \mu_1^2 \ll \mathcal{O}(1)$ and $r \ll \mathcal{O}(\Lambda)$:

$$\frac{d\mu_1}{d\mu_2} = \frac{\mu_1(1 + QD)(\bar{\mu}_2 + \delta\mu_2)^{-\gamma}}{2r(m_0 + \bar{\mu}_2 + \delta\mu_2)} - \frac{J\,Q\,\tilde{K}}{2} \tag{3.215}$$

where now the variable $\delta\mu_2(t) = \mu_2(t) - \bar{\mu}_2$ is used and barred variables always refer to the equilibrium condition. Of course, choosing T_l close to T_K and waiting a long time t_a so that the system approaches equilibrium, allows

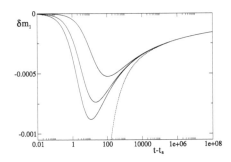 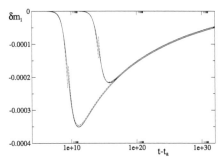

FIGURE 3.8

The Kovacs protocol at constant $T_i = 4.2$, with a sequence of field shifts. $H_K = 2.24787$. The initial field is $H_i = 0.1$, the field is then switched to H_l and is left to evolve (step (2)). At $t = t_a$, $m_1(t_a, H_l) = \bar{m}_1(H_f)$ and the field is switched to $H_f = 2.17$ (step (3)). The curves, bottom to top, refer to $H_l = 2.22$, 2.20, 2.18, the dashed line refers to the case $H_l = H_f$ ($\gamma = 1$, $J = K = 1$ and $L = 0.1$). Reprinted figure with permission from [Aquino *et al.*, 2006a].

FIGURE 3.9

Numerical solution (continuous lines) compared to the analytical solution for short times (dot-dashed) and for intermediate-long times (dashed), in the aging regime. $T_i = 10$, $T_f = 4.018$. The left curve is realized with $T_l = 4.005$, the right one with $T_l = 4.008$. For long times the system loses memory of the intermediate temperature. Reprinted figure with permission from [Aquino *et al.*, 2006a].

only small temperature shifts for the final step of the protocol, meaning also that T_f will be close to T_K. All the coefficients which appear in Eq. (3.215), can be assumed constant and equal to their equilibrium values at T_f with a very good approximation. The equation can then be easily integrated:

$$
\mu_1(\delta\mu_2) = \exp\left[-\frac{1+\bar{Q}D}{\gamma(\delta\mu_2)^\gamma}\;{}_2F_1\left(\gamma,\gamma,\gamma+1,-\frac{\bar{\mu}_2}{\delta\mu_2}\right)\right] \tag{3.216}
$$

$$
\times\left\{\mu_1^+\exp\left[\frac{1+\bar{Q}D}{\gamma(\delta\mu_2^+)^\gamma}\;{}_2F_1\left(\gamma,\gamma,\gamma+1,-\frac{\bar{\mu}_2}{\delta\mu_2^+}\right)\right]\right.
$$

$$
\left.-\frac{J\bar{Q}\tilde{K}_\infty}{2}\int_{\delta\mu_2^+}^{\delta\mu_2}dz\exp\left[\frac{1+\bar{Q}D}{\gamma z^\gamma}\;{}_2F_1\left(\gamma,\gamma,\gamma+1,-\frac{\bar{\mu}_2}{z}\right)\right]\right\}
$$

where the superscript $^+$ indicates that μ_1 is evaluated at $t = t_a^+$ and ${}_2F_1$ is the hyper-geometric function (see Appendix 3.A.3). This expression simplifies in cases where $\gamma = 1$, $3/2$, 2. Here, we limit ourselves to the case $\gamma = 1$ which corresponds to ordinary VF relaxation law. The other solutions and relative coefficients are reported in Appendix 3.A.3. In the case $\gamma = 1$, the solution can be written as

$$\mu_1(t) = \left(\frac{\delta\mu_2(t)}{\delta\mu_2(t) + \bar{\mu}_2}\right)^{\frac{(1+\bar{Q}D)}{\bar{\mu}_2}} \left[\mu_1^+ \left(\frac{\delta\mu_2^+ + \bar{\mu}_2}{\delta\mu_2^+}\right)^{\frac{1+\bar{Q}D}{\bar{\mu}_2}}\right. \tag{3.217}$$

$$\left. -\frac{J\bar{Q}\tilde{K}_\infty}{2}\int_{\delta\mu_2^+}^{\delta\mu_2(t)} dz \left(\frac{z}{z+\bar{\mu}_2}\right)^{-\frac{1+\bar{Q}D}{\bar{\mu}_2}}\right]$$

where:

$$\int_a^b dz \left(\frac{z}{z+\eta}\right)^\alpha = \left.\frac{x^{\alpha+1} \, {}_2F_1(1-\alpha, -\alpha, 2-\alpha, -\frac{x}{\eta})}{\eta^\alpha(1+\alpha)}\right|_{x=a}^{x=b}$$

One can then expand $m_1(t)$ in terms of μ_1 and $\delta\mu_2$, and obtain the following expression for the Kovacs curves:

$$\delta m_1(t) = \frac{\tilde{K}_\infty(\bar{w}+T/2)}{\tilde{K}_\infty(\bar{w}+T/2)+D}\left\{\frac{\mu_1(t)-\mu_1^+}{\tilde{m}_1(T_f)} + 2(\delta\mu_2(t)-\delta\mu_2^+)\right\} \tag{3.218}$$

where the coefficients are approximately constant in the regime chosen and can be evaluated at equilibrium.

For small $t-t_a$, a linear approximation for the variable $\delta\mu_2$, with a slope given by Eq. (3.82) evaluated at $t = t_a^+$, turns out to be very good. Inserting this expression in Eq. (3.217) to get $\mu_1(t)$ and then in Eq.(3.218), a good approximation of the first part of the hump for small and intermediate $t-t_a$ is obtained, as shown in Fig. 3.9.

When $t-t_a$ is very large, we can use Eq. (3.217) and the pre-asymptotic approximation for $\mu_2(t)$ [Eq. 3.119]

$$\mu_2(t) = \left(\log\frac{t}{t_0} + \frac{1}{2}\log\log\frac{t}{t_0}\right)^{-1/\gamma}$$

Inserting this expression in Eq. (3.217) to obtain $\mu_1(t)$ and then in Eq. (3.218), a good approximation for the hump and the tail of the Kovacs curves is obtained. In Fig. 3.9 we show the agreement between the analytical expression so obtained and the numerical solution.

3.5.2   Effective temperature and effective field

The out-of-equilibrium state of the system can be expressed by a number of effective parameters which is, in general, equal to the number of independent observables considered. In the HOSS model, given the solution of the dynamics, a quasi-static approach was followed to generalize the equilibrium thermodynamics (see Sec. 3.3.1) by computing the partition function of all the macroscopic equivalent states at a given time t and providing the effective

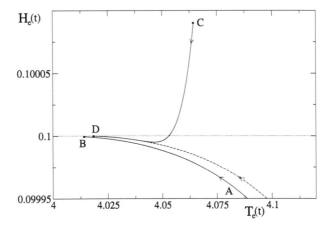

FIGURE 3.10

Effective field vs. effective temperature in the Kovacs protocol. The continuous line AB refers to the last part of step 2 of the protocol, i.e., aging at $T_l = 4.005$ after a quench from $T_i = 10$ (we do not show the full line starting at $T_e = 9.13$ and $H_e = 0.0826$, outside our picture). The continuous line CD represents the evolution of the system in step (3) of the protocol, after an instantaneous switch of the bath temperature from $T_i = 4.005$ to $T_f = 4.018$ (resulting in a jump from point B to C). The non-monotonicity of the curve CD, i.e., of the evolution of H_e after the jump, is the signature of the Kovacs effect. The dashed line represents simple aging at T_f after a quench from T_i ($H = 0.1$). Reprinted figure with permission from [Aquino et al., 2006a].

temperature $T_e(t) = \tilde{K}(m_1(t), m_2(t))[m_2(t) - m_1^2(t)]$, and the effective field $H_e(t) = H - \tilde{K}(m_1(t), m_2(t))\mu_1(t)$ [Eqs. (3.130, 3.131)]. We plot H_e as a function of T_e in a Kovacs' setup, in Fig. 3.10.

We see that in step 2 of the protocol (lower continuous line), equivalent to a simple aging experiment, the effective magnetic field relaxes monotonically to the value H. In step 3 (upper continuous line CD), after the final jump of the bath temperature, represented in the figure by the jump from point B to point C, the effective magnetic field goes through a non-monotonic evolution before relaxing to the equilibrium value H. This is where the Kovacs effect occurs. A conclusion that can be drawn is that a thermodynamic-like picture in terms of the effective temperature alone is not possible in the Kovacs setup unless at the cost of neglecting effects of the order of magnitude of the Kovacs effect itself! So, while in an aging experiment in the long time regime $H_e(t) - H$ is very small compared to $T_e(t) - T$ (so that one can consider $H_e = H$ and use only T_e as effective parameter), in the Kovacs protocol, it is in the non-monotonic evolution of the effective field that the memory effect manifests itself.

An additional effective field is, then, needed to recover a thermodynamic-like picture of the system inclusive of the Kovacs effect. The dashed line in the figure represents a simple aging experiment at $T = T_{\rm f}$, cf. the dashed lines in Figs. 3.5, 3.6 and 3.8. In this case a thermodynamic-like picture with only an effective temperature is feasible, assuming $H_{\rm e} = H$. From Fig. 3.10 one can argue that such a picture would be possible also in step 3 of the protocol (curve CD) since the two curves, for $T_{\rm e}$ close enough to $T_{\rm f}T_{\rm f}$, coincide. But this happens when $\delta m_1(t)$ is beyond the hump and the signature of the memory effect, the non-monotonic evolution, is lost. This analysis confirms the results obtained by Mossa & Sciortino [2004] in a molecular model liquid, where the impossibility of a thermodynamic-like picture with only the effective temperature was based on a potential energy landscape (PEL) analysis (see Chapter 6 for an introduction to the PEL method). In the latter model, in the last step of the Kovacs protocol, the system happens to explore regions of the PEL never explored in equilibrium, and, therefore, a simple mapping to an equilibrium condition at a different temperature (the effective temperature by definition) is not possible.

3.6 Measuring effective temperature in HO models

We introduced the problem of the direct thermologic measurement of the effective temperature in Sec. 2.9. One important and not well understood issue is whether effective temperatures relative to two apart off-equilibrium aging glass systems tend to become equal as they are put in contact, as it would happen at equilibrium according to the zeroth law of thermodynamics. The point is, then, to see whether such a law can be generalized out of equilibrium where only one or none of the two systems is actually at equilibrium with the heat-bath.

Since we are off equilibrium, another question to address would be on which timescale such partial thermalization would take place. Indeed, it cannot be taken for granted that two different aging systems, each one with its own effective temperature, will interact in such a way to yield an equal $T_{\rm e}$ and the reason might be the same preventing the two systems to equilibrate: the hypothesized extremely low thermal conductivity (cf. Sec. 2.9).

Garriga & Ritort [2001b,a] face this problem making use of the HO model of Sec. 3.1 and explicitly computing the heat flow between two off-equilibrium systems. They consider a system A composed by N harmonic oscillators $\mathbf{x}_A = \{x_{i,A}\}$ and a system B composed by N harmonic oscillators $\mathbf{x}_B = \{x_{i,B}\}$. The two systems are connected by means of a weak, local interaction. The Hamiltonian of the whole set is:

$$\mathcal{H} = \frac{K_A}{2} M_{2,A} + \frac{K_B}{2} M_{2,B} - \epsilon\, Q_{AB} \tag{3.219}$$

where

$$M_2^{(A)} = Nm_2^{(A)} \equiv \sum_{i=1}^{N} |\mathbf{x}_A|^2 \qquad (3.220)$$

$$M_2^{(B)} = Nm_2^{(B)} \equiv \sum_{i=1}^{N} |\mathbf{x}_B|^2 \qquad (3.221)$$

$$Q_{AB} = Nq_{AB} \equiv \sum_{i=1}^{N} \mathbf{x}_A \cdot \mathbf{x}_B \qquad (3.222)$$

The dynamics imposed is, as usual, the parallel Monte Carlo dynamics introduced in Sec. 3.1.1, where, however, the distributions of the updates of the positions of the oscillators in A and in B have different widths (the variances are $\Delta_A \neq \Delta_B$). The interesting case occurs when the dynamics is implemented separately on the two subsystems, that is, when the two sets of variables are sequentially updated and the Metropolis algorithm, Eq. (3.10), only depends on the energy shifts relative to one set at a time:

$$\delta E_{A,B} = x^{(A,B)} = \frac{K_{A,B}}{2} y_2^{(A,B)} - \epsilon \, \delta Q_{AB} \qquad (3.223)$$

The only coupling between A and B is, then, through the Hamiltonian coupling.

The equations of motion both for the one-time and the two-time variables are those considered throughout this chapter. In the present case there are no fast variables ($J = L = 0$), the external field H is zero and one works at a zero heat-bath temperature. For small ϵ, in the aging regime, the one-time dynamics evolves, thus, as

$$m_2^{(A,B)} \simeq \frac{K_A K_B \Delta_{A,B}^2}{8(K_A K_B - \epsilon^2)} \frac{1}{\log t} \qquad (3.224)$$

$$q_{AB} \simeq \epsilon \frac{K_A \Delta_A^2 + K_B \Delta_B^2}{16(K_A K_B - \epsilon^2)} \frac{1}{\log t} \qquad (3.225)$$

The two-time variables are the correlation functions $C_A(t, t_w)$ and $C_B(t, t_w)$, cf. Eq. (3.159) with $a = b = 1$, involving the fluctuations between variables belonging to the same subsystem, and $C_{A|B}(t, t_w)$ and $C_{B|A}(t, t_w)$ describing the correlation between the fluctuations in B at time t_w and in A at time t, and vice versa.

Analogously, four response functions can be defined. Two of them, $G_A(t, t_w)$ and $G_B(t, t_w)$,[15] cf. Eq (3.160) with $a = b = 1$, are the responses to small, external perturbations acting on the same subsystem, whereas the other two,

---

[15] $C(t, t_w) \equiv N \langle \overline{\delta m_1(t) \delta m_1(t_w)} \rangle$, where $\delta m_1(t) = m_1(t) - \bar{m}_1$ and $m_1 = 1/N \sum_i x_i$. If the external fields are zero, as in Eq. (3.219), $\bar{m}_1 = 0$.

$G_{A|B}(t, t_w)$ and $G_{B|A}(t, t_w)$ describe the response at time t of the system A (respectively B) to a linear perturbation of system B (resp. A) at time t_w:

$$G_{A|B}(t, t_w) \equiv \frac{\delta \langle m_1^{(A)} \rangle}{\delta H_B(t_w)} \bigg|_{H_B=0} \tag{3.226}$$

The equations governing their evolution are similar to Eq. (3.161) for the equal time correlations and to Eq. (3.172) for the two-time correlation and response functions. Now, however, only fluctuations and perturbations in m_1 are considered and no fast spherical spins are present (i.e., $\tilde{K} \to K_{A,B}$ and $\tilde{H} \to H_{A,B}$, independent of the oscillator variables). Furthermore, the unperturbed values of the external fields are zero. The two-time behavior of such quantities is necessary to define the FDR, that is interpretable as an effective temperature (cf. Secs. 2.8 and 3.3.4).

The formal resolution in the aging regime is carried out by Garriga & Ritort [2001b] yielding, for the FDRs at small coupling ϵ:

$$T_{e,A}^{FD}(t_w) \equiv \frac{\partial_{t_w} C_A(t, t_w)}{G_A(t, t_w)} \simeq K_A m_{2,A}(t_w) + \mathcal{O}(\epsilon^2) \sim \frac{1}{\log t_w} \tag{3.227}$$

$$T_{e,B}^{FD}(t_w) \equiv \frac{\partial_{t_w} C_B(t, t_w)}{G_B(t, t_w)} \simeq K_B m_{2,B}(t_w) + \mathcal{O}(\epsilon^2) \sim \frac{1}{\log t_w} \tag{3.228}$$

Both expressions display the same logarithmic decay to zero (the heat-bath temperature). However, in the aging regime, their ratio never equals one, unless both systems have the same spring constant and the same MC updates distribution. Indeed

$$\frac{T_{e,A}^{FD}}{T_{e,B}^{FD}} = \frac{K_A \Delta_A^2}{K_B \Delta_B^2} \tag{3.229}$$

This is easy to understand because, if the spring constant of A is larger (at equal Δ), the proposed energy shift in a MC collective move will, typically, be larger and the transition probability will, hence, be smaller and the relaxation relatively slower (i.e., the effective temperature will be larger). Similarly, fixing Ks, if the distribution of the updates of system A is wider, the proposed move will be, typically, larger and, therefore, less probable. Again $T_{e,A}^{FD} > T_{e,B}^{FD}$.

This implies that no "effective thermalization" holds on the timescales of the aging regime $(t, t_w \gg \mathcal{O}(1)$ and $|t - t_w| \ll t_w)$ and the zeroth law of thermodynamics is satisfied only when equilibrium is reached, when both effective temperatures relax to the heat-bath temperature.

3.6.1 Heat flux between off-equilibrium systems

In order to better comprehend this, one has to observe the heat transfer between the two subsystems. According to the proposals of Cugliandolo *et al.*

[1997] and Exartier & Peliti [2000], Garriga & Ritort [2001a] define the heat flux between the two systems A and B as the balance of the supplied powers:

$$J(t, t_w) \equiv \epsilon \frac{\partial}{\partial t} \left[C_{A|B}(t, t_w) - C_{B|A}(t, t_w) \right]$$

$$= \frac{\epsilon}{N} \left[\dot{\mathbf{x}}_A(t) \cdot \mathbf{x}_B(t_w) - \dot{\mathbf{x}}_B(t) \cdot \mathbf{x}_A(t_w) \right] \qquad (3.230)$$

Let us, now, consider A as the thermometer and B the out-of-equilibrium system whose effective temperature should be measured. In order for A to be a good thermometer, the power it supplies to the system B should be negligible. In Eq. (3.230) this amounts to considering $\partial_t C_{A|B} \ll \partial_t C_{B|A}$. We said, in Sec. 2.9, that one possible explanation for the impossibility of devising a thermometer able to measure the effective temperature of the slow modes might be that its response time should be comparable with the typical timescales on which those modes evolve in the aging regime. A normal thermometer, indeed, with a fast response, interacts with the fast degrees of freedom equilibrated at the heat-bath temperature and can, therefore, only measure the latter. However, the condition $\partial_t C_{A|B} \ll \partial_t C_{B|A}$ can only be satisfied if $\dot{\mathbf{x}}_A \ll \dot{\mathbf{x}}_B$, that is if the relaxation time of the thermometer is smaller than the one of the system under probe (notice that β processes are not considered in the present analysis).

If, at time $s \sim \mathcal{O}(1)$, we couple a generic thermometer A to a generic glassy system B and we measure the heat transfer occurring between the waiting time t_w and the time t, Eq. (3.230), for both t and t_w much larger than s and next to each other, its analytic expression will be

$$J(t \simeq t_w, t_w) = \epsilon^2 \partial_u C_B(t, u)\big|_{u=t_w} C_A(t_w, t_w)(\beta_A - \beta_B(t_w)) \qquad (3.231)$$

$$-\epsilon^2 \int_s^t du \ \{ G_A(t, u) \left[+\beta_B(u) T_A \partial_u C_B(t, u) + \partial_t C_B(t, u) \right]$$

$$+ C_A(t, u) \left[\beta_A \partial_u^2 C_B(t, u) + \beta_B(u) \partial_u \partial_t C_B(t, u) \right] \}$$

where $\beta_B(u) = 1/T_{e,B}^{FD}(u)$ and $\beta_A = 1/T_A$ is the inverse temperature of the thermometer. It turns out that only if the characteristic response time of the thermometer (supposed at equilibrium) is much less than the time t at which the measurement is completed, the thermometer temperature can eventually become equal to the effective temperature.[16]

The heat flux is related to the thermal conductivity κ_T by the Fourier law, a linear relationship involving the temperature gradient in an off-equilibrium, though stationary, situation [de Groot & Mazur, 1962]

$$J = -\kappa_T \nabla T = L_0 \nabla \beta \qquad (3.232)$$

---

[16]Even though it is not clear whether this can happen for a t that is not infinitely long, for which also the system B would reach equilibrium and, therefore, we would find ourselves describing a standard equilibrium measurement.

where L_O is the Onsager coefficient. We stress, however, that the aging regime is not stationary and the results derived by using the above formula must, therefore, be carefully verified. Certainly, a sort of stationarity sets in if, in the aging timescale, one, exploits the idea of an effective temperature $T_e(t)$ that evolves so slowly to be considered constant in a given time interval (some decades) around t.

If both A and B are HO systems, as in Eq. (3.219), the heat flux relative to the above-described situation is

$$J(t \simeq t_w, t_w) \sim \epsilon^2 \frac{1}{t(\log t)^2}(\beta_A - \beta_B(t)) \sim -\epsilon^2 \frac{1}{t}(T_A - T_B(t)) \qquad (3.233)$$

Being in the aging regime, we have seen that $\beta_B = 1/T_{e,B}^{FD} \sim \log t_w$, cf. Eq. (3.228), implying $\nabla\beta \simeq \beta_A - \beta_B(t_w) \sim \log t_w$. This leads to $J \sim 1/(t_w \log t_w)$. The heat transferred between the thermometer A and the off-equilibrium system B in a certain interval of time, Δt, in the aging regime is, therefore,

$$\mathcal{Q} = \int_{t_w}^{t_w + \Delta t} dt' J(t') \propto \log \frac{\log(t_w + \Delta t)}{\log t_w} \qquad (3.234)$$

If $\Delta t \ll t_w$, then $\mathcal{Q} \sim \Delta t/(t_w \log t_w)$ and the total transferred heat cannot lead to $T_A = T_{e,B}^{FD}(t)$, since to achieve this the heat exchanged should decay as $\mathcal{Q} \sim 1/\log t_w$. To transfer enough heat to compensate the difference between the thermometer and the system, that is to measure the effective temperature, Δt must, then, be of the same order of t_w, that is the age of the system: the thermal conductivity decays too fast to allow for a partial "off-equilibrium" thermalization and the zeroth law of thermodynamics only holds very near to equilibrium.

3.7   Mode-dependent effective temperature

We have seen in Sec. 2.8 that the effective temperature indirectly measured by means of the FDR depends on the frequency of the AC measurement. In Sec. 7.5 we will further explore the possibility that the viscous liquid is heterogeneous in space, that is, composed by local regions belonging to different homogeneous glassy states (by state, meaning a minimum of the free energy landscape), a framework in which an effective temperature, however defined, might become dependent on the space, yielding a distribution of effective temperatures rather than a single valued thermodynamic parameter. It becomes important, then, to generalize the class of HO models studied in this chapter to the case in which more modes are present and probe how the effective temperature depends on the modes distribution.

Garriga & Ritort [2005] generalized the Hamiltonian of the HO model, Eq. (3.1), substituting the Hooke constant by a disordered parameter with a given distribution:

$$\mathcal{H} = \frac{K}{2} \sum_{i=1}^{N} \epsilon_i x_i^2 \qquad (3.235)$$

where the ϵ_i are all positive and distributed according to the function $g(\epsilon) = 1/N \sum_i \delta(\epsilon_i - \epsilon)$ that, for $N \to \infty$, is a continuous function.

The usual MC dynamics is applied (Sec. 3.1.1) and, for what concerns the novelty of the presence of the quenched randomness, the technical approach is analogous to the one developed to the backgammon model that will be exhaustively presented in the next chapter, in Sec. 4.3 and Appendices 4.A-4.D. Hence, here, we will only concentrate on the physical results in the case of the harmonic oscillators.

Thermodynamics is trivial and the free energy per oscillator reads

$$\beta f = -\frac{1}{2} \log \left(\frac{2\pi}{\beta K} \right) + \frac{1}{2} \int_0^\infty d\epsilon\, g(\epsilon)\, \log \epsilon \qquad (3.236)$$

The second term is zero in the ordered case $(g(\epsilon) = \delta(\epsilon - 1))$, yielding back Eq. (3.3).

In the dynamics, the probability distribution of the MC energy shift x, is still the Gaussian of Eq. (3.15) (with $H = 0$), but average and variance, Eq. (3.16), are generalized as

$$\bar{x} = \frac{\Delta^2 K}{2} \bar{\epsilon}, \qquad \Delta_x = \Delta^2 K^2 [m_2]_{\epsilon^2} \qquad (3.237)$$

where

$$\bar{\epsilon} \equiv \frac{1}{N} \sum_{i=1}^{N} \epsilon_i, \qquad [m_2]_{\epsilon^\ell} \equiv \left\langle \frac{1}{N} \sum_{i=1}^{N} \epsilon_i^\ell x_i^2 \right\rangle \qquad (3.238)$$

The brackets average is carried out over different initial conditions and dynamical histories of the system.

The equations of motion at zero temperature consist in the hierarchy

$$\frac{dU}{dt} = A_1(t) = \frac{\bar{x}}{2} \left[\operatorname{erfc}(\alpha) + \left(1 - \frac{\beta \bar{x}}{2\alpha^2} \right) b(\alpha) \right] \qquad (3.239)$$

$$\frac{d[m_2]_{\epsilon^\ell}}{dt} = \Delta^2 \left(\bar{\epsilon}^\ell - \bar{\epsilon} \frac{[m_2]_{\epsilon^{\ell+1}}}{[m_2]_{\epsilon^2}} \right) A_0(t) + \frac{[m_2]_{\epsilon^{\ell+1}}}{[m_2]_{\epsilon^2}} A_1(t) \qquad (3.240)$$

where A_0, A_1 and b are defined in Eqs. (3.34), (3.35) and (3.36), respectively, and $\alpha^2 = \bar{x}^2/(2\Delta_x) = \Delta^2 \bar{\epsilon}/(8[m_2]_{\epsilon^2})$. In the adiabatic approximation, $[m_2]_{\epsilon^2}$ can be linearly related to the energy, assuming that the system has equilibrated over the hypersurface of constant energy (cf., e.g., Sec. 3.2.3, Eqs. (3.91)-(3.92) for the relation between μ_1 and μ_2 or the microcanonic argument of Sec. 3.3.3) and the equations of motion can be closed and solved analytically. In particular, the acceptance ratio A_0, Eq. (3.34), does not depend on the disorder and the energy long time behavior only depends on the average $\bar{\epsilon}$:

$$A_0(t) \simeq \frac{1}{4t \log(2t/\sqrt{\pi})} \tag{3.241}$$

$$U(t) \simeq \frac{K\Delta^2\bar{\epsilon}}{16} \frac{1}{\log(2t/\sqrt{\pi}) + 1/2 \log[\log(2t/\sqrt{\pi})]} \tag{3.242}$$

3.7.1   Quasi-static effective temperature

The effective temperature defined in the quasi-static approach (cf. Sec. 3.3.1 for the HOSS model) reads, in this case,

$$T_e(t) = \frac{K}{\bar{\epsilon}}[m_2]_{\epsilon^2}(t) \tag{3.243}$$

returning Eq. (3.130), with $m_1 = 0$ and $\tilde{K} = K$, in the ordered case ($\epsilon = \bar{\epsilon} = 1$).

To probe the contributions related to different oscillation modes one introduces a mode-dependent energy density, describing the energy contribution of those oscillators of spring constant $K\epsilon$:

$$w(\epsilon, t) \equiv \frac{K}{2} \left\langle \frac{1}{N} \sum_{i=1}^{N} \delta(\epsilon_i - \epsilon) x_i^2(t) \right\rangle \tag{3.244}$$

whose equation of motion reads

$$\dot{w}(\epsilon, t) = \frac{2}{K} \frac{\epsilon w(\epsilon, t)}{[m_2]_{\epsilon^2}} A_1(t) + \Delta^2 \left[\frac{K}{2} g(\epsilon) - \frac{\bar{\epsilon} \, \epsilon w(\epsilon, t)}{[m_2]_{\epsilon^2}}\right] A_0(t) \tag{3.245}$$

Studying its dynamics it is possible to identify a threshold energy $\epsilon^\star$ discriminating between two qualitatively different dynamic regimes:

$$\epsilon^\star w(\epsilon^\star, t) = \frac{K}{2\bar{\epsilon}} g(\epsilon^\star)[m_2]_{\epsilon^2}(t) \tag{3.246}$$

All modes with $\epsilon > \epsilon^\star$ are equilibrated at the effective temperature yielded by Eq. (3.243), whereas modes with $\epsilon < \epsilon^\star$ are, instead, off equilibrium. In Fig. 3.11 we display the behavior of $w(\epsilon, t)$ for a system with a binary distribution of the disorder, $g(\epsilon) = 1/2[\delta(\epsilon - \epsilon_1) + \delta(\epsilon - \epsilon_2)]$ and we compare it with the expression

$$w(\epsilon, t) = \frac{K}{2} \frac{g(\epsilon)}{\epsilon \, \bar{\epsilon}} [m_2]_{\epsilon^2}(t) \tag{3.247}$$

valid when $\epsilon > \epsilon^\star$.

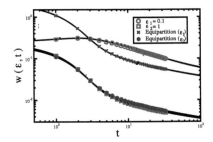

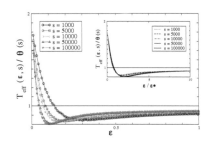

FIGURE 3.11

$w(\epsilon, t)$ at $T = 0$ for a binary distribution of the random energies. The numerical resolution of Eq. (3.245) and the approximated adiabatic expression ("equipartition"), Eq. (3.247), are compared. Symbols correspond to MC simulations for a system with $N = 5000$. Reprinted figure with permission from [Garriga & Ritort, 2005]. Copyright (2005) by American Physical Society.

FIGURE 3.12

Fluctuation-dissipation ratio vs. ϵ at different waiting times and at zero temperature. The distribution $g(\epsilon)$ of the quenched disorder is uniform between 0.01 and 1. In the inset the scaling of T_e^{FD} with the threshold energy is displayed. Reprinted figure with permission from [Garriga & Ritort, 2005].

3.7.2 Mode-dependent fluctuation-dissipation ratio

Let us define the two-time dependent correlation and response functions:

$$C(\epsilon, t, t_{\mathrm{w}}) = \left\langle \frac{1}{N} \sum_{i=1}^{N} \delta(\epsilon_i - \epsilon) x_i(t) x_i(t_{\mathrm{w}}) \right\rangle \qquad (3.248)$$

$$G(\epsilon, t, t_{\mathrm{w}}) = \left. \frac{\delta m_1(\epsilon, t)}{\delta H(t_{\mathrm{w}})} \right|_{H(t_{\mathrm{w}}) \to 0}; \qquad t > t_{\mathrm{w}} \qquad (3.249)$$

where

$$m_1(\epsilon, t) \equiv \left\langle \sum_{i=1}^{N} \delta(\epsilon_i - \epsilon) x_i(t) \right\rangle \qquad (3.250)$$

and enters in the Hamiltonian, Eq. (3.235), adding the term $-HNm_1$. Solving the equal time and the two-time equations of motion, cf., respectively, Eqs.

(3.161) and (3.172) with $J = L = 0$ and $\tilde{K} \to K\epsilon$, one finds

$$C(\epsilon; t, t') \simeq C(\epsilon; t', t') \frac{h(t)}{h(t')} \qquad (3.251)$$

$$C(\epsilon; t', t') \simeq \frac{2}{K} w(\epsilon, t') \qquad (3.252)$$

$$G(\epsilon; t, t') \simeq G(\epsilon; t', t') \frac{h(t)}{h(t')} \qquad (3.253)$$

$$G(\epsilon; t', t') \simeq \left(\frac{K\Delta^2 \bar{\epsilon}}{2} A_0(t') - A_1(\bar{\epsilon}, t') \right) \frac{g(\epsilon)}{K^2 [m_2]_{\epsilon^2}(t')} \qquad (3.254)$$

where the time sector function is now the ϵ-dependent

$$h(t) = \exp \left\{ -\frac{\epsilon}{K} \int_{-\infty}^{t} d\tau \left[\frac{K\Delta^2 \bar{\epsilon}}{2} A_0(\tau) - A_1(\bar{\epsilon}, \tau) \right] \frac{1}{[m_2]_{\epsilon^2}(\tau)} \right\} \qquad (3.255)$$

The mode dependent FDR eventually turns out to be

$$T_{\mathrm{e}}^{\mathrm{FD}}(\epsilon; t') \equiv \frac{\partial_{t'} C(\epsilon; t, t')}{G(\epsilon; t, t')} = \frac{1}{g(\epsilon)} \left[2\epsilon w(\epsilon, t') + 2K \frac{[m_2]_{\epsilon^2}}{(K/2\Delta^2 \bar{\epsilon} A_0(t') - A_1(\bar{\epsilon}, t'))} \right] \qquad (3.256)$$

For $\epsilon > \epsilon^*$ this reduces to

$$T_{\mathrm{e}}^{\mathrm{FD}}(\epsilon, t) \simeq \frac{K}{\bar{\epsilon}} [m_2]_{\epsilon^2} = T_{\mathrm{e}}(t) \qquad (3.257)$$

that is the quasi-static expression obtained in the adiabatic approximation, Eq. (3.243), independent of the mode frequency.

For $\epsilon < \epsilon^*$, instead,

$$T_{\mathrm{e}}^{\mathrm{FD}}(\epsilon, t) = 2T_{\mathrm{e}} - 2\frac{\epsilon \, w(\epsilon, t)}{g(\epsilon)} \qquad (3.258)$$

This implies, in particular, that those modes with larger timescales are at higher temperature with respect to modes with frequencies above the threshold. This can be observed in Fig. 3.12, where for oscillators with $\epsilon < \epsilon^*$, $T_{\mathrm{e}}^{\mathrm{FD}}(\epsilon, t)/T_{\mathrm{e}}(t) > 1$. In the $\epsilon \to 0$ limit, the FDR is twice the quasi-static effective temperature and this corresponds to the effective temperature of a diffusive process [Cugliandolo & Kurchan, 1994]. Processes with $\epsilon < \epsilon^*$ can, thus, be interpreted as diffusive, off-equilibrium, processes.

3.7.3   Transition rate effective temperature

The effective temperature can be estimated from the transition probability as well, as explained in Sec. 3.3.3. Looking at the energy shifts in the MC dynamics, Eq. (3.148) holds also in the present model, yielding a T_{e} coinciding with Eq. (3.243).

The same approach can be implemented considering, instead of the average energy shift, the mode-dependent energy shift

$$\delta E_\epsilon \equiv \left\langle \epsilon \frac{K}{2} \frac{1}{N} \sum_{i=1}^{N} \delta(\epsilon_i - \epsilon) x_i^2 \right\rangle = \epsilon w(\epsilon, t) \qquad (3.259)$$

yielding an effective temperature equal to Eq. (3.256), that reduces to Eq. (3.257), provided $\epsilon > \epsilon^*$ (partial thermalization regime).

3.A   HOSS equations of motion for one-time variables

First of all we recall some integrals and functions that will be useful not only for the HOSS model but throughout the book. The first ones are the erf function, the error function defined, as

$$\text{erf}(\alpha) \equiv \frac{2}{\sqrt{\pi}} \int_{-\infty}^{\alpha} dz \, e^{-z^2} \tag{3.A.1}$$

and its complementary erfc, defined as

$$\text{erfc}(\alpha) \equiv \frac{2}{\sqrt{\pi}} \int_{\alpha}^{\infty} dz \, e^{-z^2} = 1 - \text{erf}(\alpha) \tag{3.A.2}$$

3.A.1   Strong glass

The asymptotic expansion of erfc for large α is

$$\text{erfc}(\alpha) = \frac{e^{-\alpha^2}}{\alpha\sqrt{\pi}} \left[1 + \sum_{k=1}^{\infty} (-1)^k \frac{2k!}{k!(2\alpha)^{2k}} \right] \simeq \frac{e^{-\alpha^2}}{\alpha\sqrt{\pi}} \left(1 - \frac{1}{2\alpha^2} \right) \tag{3.A.3}$$

where, in the last part, we reported the terms that are relevant for expressing the slow relaxing dynamics of our concern, and that yield the following equation of motion [equivalent to Eq. 3.87]:

$$\dot{\alpha} = \frac{e^{-\alpha^2}}{\sqrt{\pi}} \tag{3.A.4}$$

whose solution is [cf. Eq. 3.88]

$$-i \, \text{erf}(i\alpha(t)) = \frac{2t}{\pi} + \text{const} \tag{3.A.5}$$

Using Eq. (3.A.3) for large α this can be approximately inverted as

$$\alpha(t) \simeq \sqrt{\log \frac{2t}{\sqrt{\pi}} + \frac{1}{2} \log \log \frac{2t}{\sqrt{\pi}}} \tag{3.A.6}$$

leading to Eq. (3.89) for the strong glass case.

3.A.2   Fragile glass

Dynamics of $\mu_2(t)$

Now we look at the equations for the fragile glass. Neglecting higher order terms in Eq. (3.82) the approximated equation for large Λ, i.e., large times, is:

$$\dot{\Lambda} = \frac{\Lambda^\omega}{t_0} e^{-\Lambda} \tag{3.A.7}$$

with

$$
\begin{aligned}
t_0 &= \frac{\sqrt{\pi}}{8(m_0 + \bar{\mu}_2)\gamma r_1}, & \omega &= \tfrac{1}{2}, & T &\geq T_0, \\
t_0 &= \frac{\sqrt{\pi}}{8 m_0 \gamma r_\infty (1 - r_\infty)}, & \omega &= \frac{2+\gamma}{2\gamma}, & T &< T_0.
\end{aligned}
\tag{3.A.8}
$$

This holds for $T \gtrsim T_0$ and for any $T < T_0$. The parameters r_1 and r_∞ come from the expansion of r, defined in Eq. (3.114), for small μ_2, respectively, above and below T_0.

- Above T_0, the first order expansion is

$$
r \simeq \frac{1}{T(1 + \bar{Q}D)} \left[\tilde{K}(1 + \bar{Q}D + \bar{P}) \delta\mu_2(t) - \mu_1(t) \frac{2\bar{P}D}{J} \right]
\tag{3.A.9}
$$

where $\bar{P}$ and $\bar{Q}$ were introduced, together with P and Q, in Eqs. (3.79) and (3.80). Even though we have expanded in μ_1 as well, this contribution turns out to be negligible for the present level of approximation (see also [Leuzzi & Nieuwenhuizen, 2001a; Leuzzi, 2002]). For future convenience we, then, define the abbreviation

$$
r_1 \equiv \frac{1}{m_0 + \bar{\mu}_2(T)} \frac{1 + \bar{Q}D + \bar{P}}{1 + \bar{Q}D}
\tag{3.A.10}
$$

To study the case above T_0, we use the variable $\delta\mu_2(t) = \mu_2(t) - \bar{\mu}_2$. Here we work at T not too far from T_0 and we concentrate on the aging regime rather than on the eventual Debye relaxation to equilibrium, i.e., we consider $\bar{\mu}_2(T) \ll \delta\mu_2$, approximating $\delta\mu_2(t) \simeq \mu_2(t) = \Lambda(t)^{-1/\gamma}$ in the aging regime. In this range of temperature $T^\star(t) - T \sim \mu_2(t)$.

- Below T_0, the qualitative behavior of $\mu_2(t)$ (in this case the $\bar{\mu}_2$ part is zero) is the same, but T is never reached. This implies that r tends to some asymptotic constant

$$
r_\infty = \frac{m_0 - T/\tilde{K}_\infty(T)}{2m_0 - T/\tilde{K}_\infty(T)}
\tag{3.A.11}
$$

The generic implicit solution of Eq. (3.A.7) is

$$
\Gamma\left(1 - \omega, -\Lambda(t)\right) \left[-\Lambda(t)\right]^\omega [\Lambda(t)]^{-\omega} = \frac{t}{t_0} + \text{const}
\tag{3.A.12}
$$

where $\Gamma(\kappa, x)$ is the Euler incomplete gamma function

$$
\Gamma(\kappa, a) = \int_a^\infty dx \; x^{\kappa-1} e^{-x}
\tag{3.A.13}
$$

In the case above T_0, i.e., where $\omega = 1/2$, Eq. (3.A.12) takes the special form $-i \, \text{erf}(i\Lambda) = t/t_0$, identical to the one for strong glasses (indeed, it is $\Lambda \sim \alpha^2$ as far as $\mu_1^2 < m_0 + \mu_2$).

Eq. (3.A.7) represents the dynamics for large times and low temperatures, i.e., large Λ. Consistently we expand the solution (3.A.12) as

$$e^{\Lambda}\Lambda^{-\omega} \simeq \frac{t}{t_0} \tag{3.A.14}$$

Inverting it one obtains:

$$\Lambda(t) \simeq \log(t/t_0) + \omega \log\log(t/t_0) \tag{3.A.15}$$

and, eventually Eq. (3.119).[17]

Dynamics of $\mu_1(t)$

For the dynamics of μ_1 we observe that many different regimes arise as the temperature drops below T_0. Here, we report them all for the sake of completeness, even though those of our interest are only the first two in the list. Indeed, our effort to map the aging dynamics of a system into a unique effective thermodynamic parameter turns out to be successful for the HOSS model only in those dynamic regimes where $\mu_1 \ll \mu_2$ (see the approach of Sec. 3.3.1 to understand why).

The starting equation is Eq. (3.120), i.e., the ratio of equations (3.115) and (3.116):

$$\frac{d\mu_1}{d\mu_2} = \frac{\mu_1(1+QD)(\Lambda+2-3r+2r^2) - JQT^\star r}{2r(m_0+\mu_2) - \mu_1^2(\Lambda+2-3r+2r^2)} \tag{3.A.16}$$

Depending on the temperature and the value of the exponent γ the above equation can be differently approximated in the regime of long times.

1. Aging regime above the Kauzmann temperature: $T > T_0$, $\forall \gamma$. Neglecting the term of $\mathcal{O}(\mu_1^2\Lambda)$ in the denominator, according to the order of the approximation already used in devising the equation of motion for μ_2, the solution is

$$\mu_1(t) = \frac{TJ\bar{Q}r_1}{1+\bar{Q}D}\frac{\delta\mu_2(t)}{\Lambda} + \mathcal{O}(\delta\mu_2^2) + \mathcal{O}(\delta\mu_2^{2\gamma+1}) \tag{3.A.17}$$

where r_1 is defined in Eq. (3.A.10). The parameter $\Lambda = (\bar{\mu}_2 + \delta\mu_2)^{-\gamma}$ can be further expanded in powers of $\delta\mu_2/\bar{\mu}_2$ provided that the temperature is not too close to T_0. As $T \to T_0$, implying $\bar{\mu}_2 \to 0$, one has instead to consider that $\delta\mu_2/\bar{\mu}_2 \ll 1$ only as the equilibration occurs (i.e., even beyond the aging relaxation) and $\mu_1 \sim (\delta\mu_2)^{1+\gamma}$.[18]

---

[17]Compare with $\lambda(t)$ in Sec. 4.1.2, in the framework of aging urn models.
[18]This is the same behavior as the $T = 0$ dynamics of the strong glass case at $T = 0$. There it was, furthermore, $\gamma = 1$ and therefore $\mu_1 \sim \mu_2^2$, cf. Eq. (3.92).

2. Aging regime below the Kauzmann temperature, enhanced separation of timescales: $T < T_0$, $\gamma > 1$:

$$\mu_1(t) = \frac{J\bar{T}^\star r_\infty \bar{Q}}{1 + \bar{Q}D} \frac{1}{\Lambda(t)} + \mathcal{O}(\mu_2^{1+\gamma}) \qquad (3.A.18)$$

with

$$\bar{T}^\star \equiv \lim_{t\to\infty} \tilde{K}(m_1(t), m_2(t))[m_0 + \mu_2(t)] = \tilde{K}_\infty m_0 \qquad (3.A.19)$$

$$\tilde{K}_\infty \equiv \lim_{t\to\infty} \tilde{K}(m_1(t), m_2(t)) \qquad (3.A.20)$$

$$r_\infty = \frac{\bar{T}^\star - T}{2\bar{T}^\star - T} \qquad (3.A.21)$$

3. Model-dependent aging regimes below the Kauzmann temperature: $T < T_0$, $\gamma = 1$. In this case the adiabatic expansion is no longer consistent. We have to solve equation (3.A.16) taking $d\mu_1/d\mu_2$ into account. Since $\gamma = 1$ one has $\mu_2 = 1/\Lambda$. To the leading order the equation takes the form

$$\frac{d\mu_1}{d\mu_2} = \frac{\mu_1 \Lambda (1 + \bar{Q}D) - J\bar{Q}\bar{T}^\star r_\infty}{2r_\infty m_0} + \mathcal{O}(\mu_1) + \mathcal{O}(\mu_2) + \mathcal{O}(\Lambda\mu_1^2) \qquad (3.A.22)$$

Defining the quantity $\epsilon \equiv \frac{(1+\bar{Q}D)}{2r_\infty m_0}$ we identify another five, model-dependent, sub-regimes in the case of $\gamma = 1$.

(a) $\epsilon > 1$. The solution is

$$\mu_1(t) \simeq \frac{J\bar{Q}\bar{T}^\star r_\infty}{2r_\infty m_0(\epsilon - 1)} \mu_2(t) - c_1 \frac{1}{\epsilon - 1} \mu_2^\epsilon(t) \qquad (3.A.23)$$

The exponent ϵ is always positive, at least in cooling, because $\bar{T}^\star > T$, making r_∞ and $\bar{Q}$ positive. c_1 is also positive because it is the exponential of the integration constant (the value of which depends on the initial conditions). Since $\epsilon > 1$, the second term in the right-hand side can be neglected and $\mu_1 \sim \mu_2$.

(b) $\epsilon = 1$. We find

$$\mu_1(t) \simeq -\frac{J\bar{T}^\star r_\infty \bar{Q}}{1 + \bar{Q}D} \frac{\log \mu_2(t)}{\Lambda(t)} + c_2 \mu_2(T) \qquad (3.A.24)$$

where c_2 is the integration constant and can take any value. In the long time dynamics the logarithm term will take over and, independently of the initial conditions, $\mu_1 > \mu_2$ and will be positive.

(c) $1/2 < \epsilon < 1$. The second term in Eqs. (3.A.22) is leading and the solution is

$$\mu_1(t) \simeq c_1 \frac{1}{1 - \epsilon} \mu_2^\epsilon(t) \qquad (3.A.25)$$

c_1 is a positive constant and $\mu_1 \gg \mu_2$ and positive.

(d) $\epsilon = 1/2$. When $\epsilon \le 1/2$ the second term in the denominator $(O(\mu_1^2))$, always neglected up to now, has to be taken into account. In this case the leading term in the denominator goes to zero and $J\bar{T}^\star r_\infty \bar{Q}$ can be neglected with respect to $\mu_1 \Lambda (1 + \bar{Q}D)$ in the numerator. We can, thus, solve the equation

$$\frac{d\mu_2}{d\mu_1} = \frac{2r_\infty m_0 - 2\mu_1^2 \Lambda}{\mu_1 \Lambda(1 + \bar{Q}D)} \qquad (3.A.26)$$

For $\epsilon = 1/2$ we obtain

$$\mu_2(t) = -\frac{2}{1 + \bar{Q}D} \, \mu_1^2(t) \log \mu_1(t) + c_2 \, \mu_1^2(t) \qquad (3.A.27)$$

which is not analytically invertible. It is clear anyway that in this sub-regime $\mu_1 \gg \mu_2$. c_2 can take any value.

(e) $\epsilon < 1/2$. The solution is

$$\mu_1(t) = \sqrt{\frac{m_0 \, r_\infty (1 - 2\epsilon)}{\Lambda(t)}} \qquad (3.A.28)$$

$$\times \left(1 + \frac{c_1}{2} \left(\frac{1 + \bar{Q}D}{2} \right)^{1/\epsilon} \left(\frac{1 - 2\epsilon}{\epsilon} \right)^{1/\epsilon - 1} \mu_2(t)^{1/2\epsilon - 1} \right)$$

where $c_1 > 0$. Again it holds that $\mu_1 \gg \mu_2$.

4. Aging regime below the Kauzmann temperature: reduced separation of timescales: $T < T_0$, $\gamma < 1$. Considering also the term of $O(\mu_1^2 \Lambda)$ in the denominator of Eq.(3.A.16), the solution is now

$$\mu_1(t) = \frac{1}{\sqrt{\Lambda(t)}} \sqrt{r_\infty m_0} \left[1 - \frac{1 + \bar{Q}D}{2m_0 r_\infty \gamma} \mu_2(t)\Lambda(t) \right] \qquad (3.A.29)$$

In this low temperature regime $\mu_1 \gg \mu_2$ once again.

For $\gamma = 1$, $\epsilon \le 1/2$ and for $\gamma < 1$ the solution to equation (3.A.22) involves only the absolute value of μ_1, thus giving two possible choices for the sign of the function $\mu_1(\mu_2)$. In order to guarantee continuity of μ_1 at the parameter values at which the dynamics changes regime, we require μ_1 to have the same sign in two contiguous regimes. That means that in Eqs. (3.A.27), (3.A.28), and (3.A.29) we chose the plus sign. This choice will also bring an always positive heat flux out of the relaxing glass (see Sec. 3.3.5).

Long time expansions

We collect here the expansions to first order in $\delta\mu_2(t)$ of the quantities m_1, r [defined in Eq. (3.114)], $\tilde{K}$ [Eq. (3.57)], P [Eq. (3.80)] and Q [Eq. (3.79)].

$$m_1(t) = \bar{m}_1 - \frac{\bar{P}D}{TJ(1+\bar{Q}D)}\delta\mu_2(t) \tag{3.A.30}$$

$$r(t) = \frac{1}{m_0 + \bar{\mu}_2}\left(1 + \frac{\bar{P}}{1+\bar{Q}D}\right)\delta\mu_2(t) \tag{3.A.31}$$

$$\tilde{K}(t) = \tilde{K}_\infty + \frac{\tilde{K}_\infty \bar{P}}{(1+\bar{Q}D)(m_0 + \bar{\mu}_2)}\delta\mu_2(t) \tag{3.A.32}$$

$$P(t) = \bar{P} + P_1\delta\mu_2(t) \tag{3.A.33}$$

$$Q(t) = \bar{Q} + Q_1\delta\mu_2(t) \tag{3.A.34}$$

with

$$P_1 \equiv \frac{\bar{P}}{(1+\bar{Q}D)(m_0+\bar{\mu}_2)}\left[\frac{J^2(m_0+\bar{\mu}_2)(3\bar{w}+T/2)}{2\bar{w}^2(\bar{w}+T/2)} + 1 + \bar{Q}D - \bar{P}\right] \tag{3.A.35}$$

$$Q_1 \equiv \frac{\bar{Q}}{(1+\bar{Q}D)(m_0+\bar{\mu}_2)}\left[\frac{J^2(m_0+\bar{\mu}_2)(3\bar{w}+T/2)}{2\bar{w}^2(\bar{w}+T/2)} - 3\bar{P}\right] \tag{3.A.36}$$

and, from Eqs. (3.52, 3.76),

$$\bar{w} \equiv \sqrt{J^2\bar{m}_2 + 2JL\bar{m}_1 + L^2 + \frac{T^2}{4}} = \sqrt{J^2\frac{T}{\tilde{K}_\infty}\theta(T-T_0) + \left(\frac{D}{\tilde{K}_\infty}\right)^2 + \frac{T^2}{4}} \tag{3.A.37}$$

As $T \le T_0$, $\delta\mu_2 \to \mu_2$.

3.A.3 Analytic expressions for the Kovacs effect

In Eq. (3.216) we have used the hypergeometric function, defined as:

$$_2F_1(a,b,c,z) = \frac{\Gamma(c)}{\Gamma(a)\Gamma(b)}\sum_{n=0}^{\infty}\frac{\Gamma(a+n)\Gamma(b+n)}{\Gamma(c+n)}\frac{z^n}{n!}$$

$$A_Q = 1 + Q(\bar{m}_1(T_f), \bar{m}_2(T_f))D = 1 + \bar{Q}(T_f)D$$

Introducing the abbreviation

$$B_Q^\gamma = \exp\left[A_Q\frac{{}_2F_1(\gamma,\gamma,\gamma+1,-\frac{\bar{\mu}_2}{\delta\mu_2^+})}{\gamma(\delta\mu_2^+)^\gamma}\right]$$

we present here the μ_1 behavior in the non-monotonic Kovacs effect, that is the solution to Eq.(3.215), in two specific model cases: $\gamma = 3/2, 2$. For $\gamma = \frac{3}{2}$

one has:

$$\mu_1(\delta\mu_2) = \left(\frac{1 - \sqrt{1 + \frac{\delta\mu_2}{\bar{\mu}_2}}}{1 + \sqrt{1 + \frac{\delta\mu_2}{\bar{\mu}_2}}}\right)^{\frac{(1+\bar{Q}D)}{\bar{\mu}_2^{3/2}}} e^{\frac{2(1+\bar{Q}D)}{\bar{\mu}_2\sqrt{\bar{\mu}_2+\delta\mu_2}}} \left(\mu_1^+ B_Q^{\frac{3}{2}}\right.$$

$$\left. - \frac{J\bar{Q}\tilde{K}_\infty}{2} \int_{\delta\mu_2^+}^{\delta\mu_2} dz \left(\frac{1 + \sqrt{1 + \frac{z}{\bar{\mu}_2}}}{1 - \sqrt{1 + \frac{z}{\bar{\mu}_2}}}\right)^{\frac{1+\bar{Q}D}{\bar{\mu}_2^{3/2}}} e^{-\frac{2(1+\bar{Q}D)}{\bar{\mu}_2\sqrt{\bar{\mu}_2+z}}}\right)$$

and for $\gamma = 2$:

$$\mu_1(\delta\mu_2) = \left(\frac{\delta\mu_2}{\delta\mu_2 + \bar{\mu}_2}\right)^{\frac{1+\bar{Q}D}{\bar{\mu}_2^2}} e^{\frac{1+\bar{Q}D}{\bar{\mu}_2^2}(1+\frac{\delta\mu_2}{\bar{\mu}_2})} \left(\mu_1^+ B_Q^2\right.$$

$$\left. - \frac{J\bar{Q}\tilde{K}_\infty}{2} \int_{\delta\mu_2^+}^{\delta\mu_2} dz \left(\frac{z}{z + \bar{\mu}_2}\right)^{-\frac{1+\bar{Q}D}{\bar{\mu}_2^2}} e^{-\frac{1+\bar{Q}D}{\bar{\mu}_2^2}(1+\frac{z}{\bar{\mu}_2})}\right)$$

3.B Monte Carlo integrals in one- and two-time dynamics

First we will analyze the case above the Kauzmann temperature. In this case the expansion shown in Eqs. (3.B.14)-(3.B.17) in powers of $\mu_2(t)$ becomes both an expansion in $\delta\mu_2(t)$ and in $\bar{\mu}_2$ [or, equivalently, in $1/\bar{\Lambda} = (\bar{\mu}_2/B)^\gamma$]. Indeed, we are interested in studying what happens for long times (but not as long as the relaxation time to equilibrium: $t_0 \ll t \ll \tau_{\rm eq}$) and near the Kauzmann temperature T_0, i.e., for small values of $\delta\mu_2(t)$ and even smaller values of $\bar{\mu}_2$ (or large values of $\bar{\Lambda}$).

The following exact relations hold:

$$\partial_{m_1}\mu_1 = -1 - LQ\tilde{K}, \qquad\qquad \partial_{m_2}\mu_1 = -\frac{J}{2}Q\tilde{K} \qquad (3.B.1)$$

We stress that $\partial_{m_1}\mu_1$ and $\partial_{m_2}\mu_1$ are still functions of μ_1 and μ_2, through $\tilde{K}$ and Q [see Eqs. (3.57) and (3.79)], and they can, thus, be expanded in powers of $\delta\mu_2$. They will appear very often in two specific combinations that we report here for convenience:

$$\partial_{m_1}\mu_1 + 2m_1\partial_{m_2}\mu_1 = -(1 + QD) \qquad (3.B.2)$$

$$\partial_{m_1}\mu_1 - 2\frac{L}{J}\partial_{m_2}\mu_1 = -1 \qquad (3.B.3)$$

In the following formulas, the derivatives of μ_1, as well as μ_1 itself, have to be considered as general, regular functions of μ_2. For the terms containing $\Lambda\mu_1$, from Eq. (3.121) we see that

$$\Lambda\mu_1 = \frac{J\bar{Q}Tr_1}{1+\bar{Q}D}\delta\mu_2 + \mathcal{O}\left(\delta\mu_2^2\right) \tag{3.B.4}$$

in the dynamic regime above T_0 (cf. Sec. 3.2.4, dynamics of μ_1, case 1), whereas in the second regime, $T < T_0$, $\gamma > 1$, it holds

$$\Lambda\mu_1 = \frac{J\bar{Q}\tilde{K}_\infty m_0}{1+\bar{Q}D} + \mathcal{O}\left(\mu_2(t)\right) \tag{3.B.5}$$

that is, of $\mathcal{O}(1)$.

We recall that x, defined in Eq. (3.14), is the energy difference between the current configuration of the system and the one proposed for the updating. The variable r [defined in Eq. (3.114)] is the distance of the effective temperature T_e from the heat-bath temperature (that is also the equilibrium value of T_e in the dynamic regime above the Kauzmann temperature). First we define the abbreviation:

$$\Upsilon \equiv \frac{e^{-\Lambda}(1-r)}{\sqrt{\pi\Lambda}} \tag{3.B.6}$$

that is the leading term of the acceptance ratio of the Monte Carlo dynamics:

$$A_0 \equiv \int dx\, W(\beta x)\, p(x|m_1, m_2) = \tag{3.B.7}$$

$$\Upsilon\left[1 - \frac{1 - 2r + 4r^2}{\Lambda} + \frac{3}{4\Lambda^2}\left(1 - 4r + 16r^2 - 24r^3 + 16r^4\right) + \mathcal{O}\left(\frac{1}{\Lambda^3}\right)\right]$$

We, then, give the behavior of the time derivative of the energy

$$A_1 \equiv \int dx\, W(\beta x)\, x\, p(x|m_1, m_2) = -4rT^\star\Upsilon\left[1 - \frac{3(1 - 2r + 2r^2)}{\Lambda}\right. \tag{3.B.8}$$

$$\left. + \frac{15}{4\Lambda^2}\left(3 - 12r + 28r^2 - 32r^3 + 16r^4\right) + \mathcal{O}\left(\frac{1}{\Lambda^2}\right)\right]$$

and the integro-differential equation for m_1 [obtained from Eq. (3.30) by substituting $K \to \tilde{K}$ and $H \to \tilde{H}$]:

$$m_1 = \int dx\, W(\beta x)\, \bar{y}_1(x)\, p(x|m_1, m_2) \tag{3.B.9}$$

$$= 4\mu_1\Upsilon\left[\Lambda - (1 - 3r + 4r^2) + \mathcal{O}\left(\frac{1}{\Lambda}\right)\right]$$

3.B.1   Coefficients of the two-time variables equations

In section 3.3.4 we compute the correlation and the response functions. In order to find their time dependence we need the following derivatives with respect to m_1 and m_2, taken as independent variables:

$$\frac{\partial T^\star}{\partial m_1} = 2\tilde{K}\left(P\frac{L}{J} - m_1\right), \qquad \frac{\partial T^\star}{\partial m_2} = \tilde{K}\left(P + 1\right) \qquad (3.B.10)$$

$$\frac{\partial r}{\partial m_1} = 2\frac{1 - 3r + 2r^2}{m_0 + \mu_2}\left(P\frac{L}{J} - m_1\right), \qquad \frac{\partial r}{\partial m_2} = \frac{1 - 3r + 2r^2}{m_0 + \mu_2}\left(P + 1\right) \qquad (3.B.11)$$

and

$$\frac{\partial \Upsilon}{\partial m_1} = -\Upsilon\left[\frac{m_1 \gamma}{\mu_2}\left(2\Lambda + 1\right) + 2\frac{1 - 2r}{m_0 + \mu_2}\left(P\frac{L}{J} - m_1\right)\right] \qquad (3.B.12)$$

$$\frac{\partial \Upsilon}{\partial m_2} = \Upsilon\left[\frac{\gamma}{2\mu_2}\left(2\Lambda + 1\right) - \frac{1 - 2r}{m_0 + \mu_2}\left(P + 1\right)\right] \qquad (3.B.13)$$

We present, then, the expansion of the coefficients of Eqs. (3.166)-(3.168) for the dynamics of the two-time observables. Intermediate objects that we need are:

$$d_0^{(1)} \equiv \frac{\partial}{\partial m_1}\int dx\; W(\beta x)p(x|m_1, m_2) \simeq -2m_1 d_0 - \frac{2L}{J}\frac{P(1 - 2r)}{m_0 + \mu_2}\Upsilon$$
$$\qquad (3.B.14)$$

$$d_0^{(2)} \equiv \frac{\partial}{\partial m_2}\int dx\; W(\beta x)p(x|m_1, m_2) \simeq d_0 - \frac{P(1 - 2r)}{m_0 + \mu_2}\Upsilon \qquad (3.B.15)$$

$$d_0 = \gamma\frac{\Upsilon}{\mu_2}\left[\Lambda - \frac{1 - 4r + 8r^2}{2} + \frac{3}{4\Lambda}\left(13 - 56r + 136r^2 - 160r^3 + 80r^4\right)\right]$$
$$- \frac{1 - 2r}{m_0 + \mu_2}\Upsilon$$

We notice that the above approximation for long time is valid both in the aging regimes for $T \gtrsim T_0$ (as far as $\bar{\mu}_2 \ll \delta\mu_2$, long times, but not infinite) and for $T < T_0$. The sub-leading corrections depend on γ being larger or smaller than one. The first two terms in d_0, instead, remain the most important two, whatever the values of T and γ. They are of order Λ/μ_2 and $1/\mu_2$, respectively. When $T \gtrsim T_0$ $\mathcal{O}(r) = \mathcal{O}(\mu_2)$, whereas for $T < T_0$ $\mathcal{O}(r) = \mathcal{O}(1)$:

$$d_1^{(1)} \equiv \partial_{m_1}\int dx\; W(\beta x)\; x\; p(x|m_1, m_2) \qquad (3.B.16)$$

$$\simeq -2m_1 d_1 - 8\frac{L}{J}\tilde{K}P(1 - 3r + 2r^2)\Upsilon + \begin{cases} \mathcal{O}\left(\Upsilon/\Lambda\right) & \text{if } T \gtrsim T_0 \\ \mathcal{O}\left(\Upsilon/(\Lambda\mu_2)\right) & \text{if } T < T_0,\; \gamma > 1 \end{cases}$$

$$d_1^{(2)} \equiv \partial_{m_2}\int dx\; W(\beta x)\; x\; p(x|m_1, m_2) \qquad (3.B.17)$$

$$\simeq d_1 - 4\tilde{K}P(1 - 3r + 2r^2)\Upsilon + \begin{cases} \mathcal{O}\left(\Upsilon/\Lambda\right) & \text{if } T \gtrsim T_0 \\ \mathcal{O}\left(\Upsilon/(\Lambda\mu_2)\right) & \text{if } T < T_0,\; \gamma > 1 \end{cases}$$

$$d_1 \equiv \Upsilon \left\{ -4\gamma T^\star \frac{\Lambda r}{\mu_2} + 2\gamma T^\star \frac{r}{\mu_2} \left(5 - 12r + 12r^2\right) - 4\tilde{K} \left(1 - 3r + 2r^2\right) \right\}$$

We stress that, as $T < T_0$, in order to use the above expressions one has to assume $\gamma > 1$ (see the discussion in Sec. 3.3.4).

Eventually, we come to the coefficients of the correlation and response functions in Eqs. (3.172):

$$d_{y_1}^{(1)} \equiv \partial_{m_1} \int dx \, W(\beta x) \, \overline{y}_1(x) \, p(x|m_1, m_2) \tag{3.B.18}$$

$$\simeq -2m_1 \Upsilon \left[2\gamma \frac{\Lambda^2 \mu_1}{\mu_2} + 4\Lambda \partial_{m_1}\mu_1 - 2\gamma(3 - 6r + 8r^2)\frac{\Lambda\mu_1}{\mu_2} \right.$$
$$\left. -4\partial_{m_1}\mu_1(1 - 3r + 4r^2) - 8\frac{\Lambda\mu_1}{m_0 + \mu_2} \right] - 16\frac{L}{J} P \frac{\Lambda\mu_1}{m_0 + \mu_2} \Upsilon + \mathcal{O}(\Upsilon\mu_2)$$

$$d_{y_1}^{(2)} \equiv \partial_{m_2} \int dx \, W(\beta x) \, \overline{y}_1(x) \, p(x|m_1, m_2) \tag{3.B.19}$$

$$\simeq \Upsilon \left[2\gamma \frac{\Lambda^2 \mu_1}{\mu_2} + 4\Lambda \partial_{m_2}\mu_1 - 2\gamma(3 - 6r + 8r^2)\frac{\Lambda\mu_1}{\mu_2} \right.$$
$$\left. -4\partial_{m_2}\mu_1(1 - 3r + 4r^2) - 8\frac{\Lambda\mu_1}{m_0 + \mu_2} \right] - 8P\frac{\Lambda\mu_1}{m_0 + \mu_2} \Upsilon + \mathcal{O}(\Upsilon\mu_2)$$

$$d_{y_2}^{(1)} \equiv \partial_{m_1} \int dx \, W(\beta x) \, \overline{y}_2(x) \, p(x|m_1, m_2) \tag{3.B.20}$$

$$\simeq 2m_1 d_{y_1}^{(1)} + \frac{2}{\tilde{K}} d_1^{(1)} + 16\Upsilon r \frac{L}{J} \frac{P}{1 + QD}$$

$$d_{y_2}^{(2)} \equiv \partial_{m_2} \int dx \, W(\beta x) \, \overline{y}_2(x) \, p(x|m_1, m_2) \tag{3.B.21}$$

$$\simeq 2m_1 d_{y_1}^{(2)} + \frac{2}{\tilde{K}} d_1^{(2)} + 8\Upsilon r \frac{P}{1 + QD} \tag{3.B.22}$$

All partial derivatives with respect to m_1 have been initially computed keeping m_2 fixed and vice versa (before considering time explicitly). Once time is inserted one knows the dynamic behavior of the coefficients and can expand them in powers of $\mu_2(t)$, as shown in Eqs. (3.B.14)-(3.B.17) We break the expansion at $\mathcal{O}\left(\Upsilon/(\Lambda\mu_2)\right)$ that is more than sufficiently refined to derive the dynamics of the correlation and response functions in all the regimes of our interest (regimes 1 and 2, according to the list in Appendix 3.A).

4

Aging urn models

Urn models consist, in their generic definition, of one or more sets of balls (particles, pawns, ...) and a number of urns (boxes, cells, states, ...) where the balls can be placed or taken from, according to a given extraction law regulating the evolution of the model.

They are a classical issue in probability theory [Feller, 1993; van Kampen, 1981], and have been used, as well, to build relational databases in learning theory [Gardy & Louchard, 1995; Boucheron & Gardy, 1997; Drmota et al., 2001], to determine the efficacy of vaccines [Hernandez-Suarez & Castillo-Chavez, 2000], to study epidemic spreading [Daley & Gani, 2001; Gani, 2004], population genetics [Hoppe, 1987] and to represent evolutionary processes [Schreiber, 2001; Benaïm et al., 2004], just to mention a few applications. The urn models have, furthermore, played a very important role in formulating fundamental concepts of statistical mechanics such as the approach to equilibrium and fluctuations out of equilibrium [Kac & Logan, 1987].

The prototype of the dynamic urn models in statistical mechanics and one of the most intensively studied ones has been the Ehrenfest model, otherwise called the "dog-flea" model, introduced with the aim of (critically) analyzing the H-theorem of Boltzmann [Ehrenfest & Ehrenfest, 1907]. In this model a given number N of fleas are randomly distributed over two dogs. The dogs stay near enough to each other so that the fleas can freely jump from one dog to the other one. In a probabilistic language, we have N distinguishable balls distributed between two urns. The dynamic rule of the model is elementary: at each time step a randomly chosen flea is "called" and changes dog. All fleas are equivalent. Even though extremely simple, the model has been a stimulus for many decades in physics and mathematics, even after it was exactly solved by Kac [1947], Siegert [1949] and Hess [1954] (see also [Kac, 1959; Emch & Liu, 2002]). To solve the problem means finding the evolution law for the number of fleas on each one of the two dogs. Since no interaction, nor energetic cost, nor constraint is involved, the occupation numbers at equilibrium are determined according to the requirement of maximum entropy. The dog-flea model has been generalized in various ways in the course of its century-long life. According to the classification of Godrèche & Luck [2001], we can identify a whole class of models whose common origin is the dog-flea model: the "Ehrenfest class" of dynamic urn models.

More generally, one can define a dynamic urn model specifying

1. the components (urns and balls)

2. their statistics

3. the cost function

4. the dynamic algorithm

5. the geometry

The base of the behavior of dynamic urn models fundamentally resides in the way the starting and the ending point of each dynamic step are chosen (the statistics, point 2.). In the case just considered, the *ball-to-box* choice of the statistics, actually *defines* the Ehrenfest class of models. Further specifications are the transition probability of *ball-to-box* moves and the energy function, but the discriminating ingredient is the statistics. We will see in the specific case of the backgammon model how this choice implies a Maxwell-Boltzmann statistics for the occupation probabilities at equilibrium (Sec. 4.1).

A qualitatively different class of models can, for instance, be defined by a *box-to-box* choice for the single move: we choose a box, we take any ball at random from that box, we put it in another box, randomly chosen. The class of models produced this way is called the *monkey* class [Godrèche & Luck, 2001]. The equilibrium statistics computed for monkey models turns out to be Bose-Einstein, even though nothing quantum is involved in the *box-to-box* update.

Example models belonging to this class are the "*B*-model" of Godrèche & Mézard [1995] and the zeta urn model [Bialas *et al.*, 1997]. The *B*-model is identical to the backgammon model but for the statistics, that is, however, crucial. The zeta urn model has a Hamiltonian $\mathcal{H} = \sum_{i=1}^{N} \log n_i + 1$, where n_i are the occupation numbers. Apart from nontrivial dynamic properties displaying aging and coarsening off-equilibrium behavior [Drouffe *et al.*, 1998; Godrèche & Luck, 2001], it also owns a static transition at finite temperature to a condensed phase. The monkey models are, however, not good models for glassy materials and we will not consider them any further.

Concentrating on Ehrenfest models, an important property that can be encoded, introducing disorder, is the existence of *collective modes*, that is, modes connected to the slowest processes evolving in a glassy system and carrying on the structural α relaxation.

Take a liquid well above the glass temperature, where correlations decay exponentially with time. One may consider the resultant behavior of the liquid as the superposition of different and independent harmonic modes. Each one of these energy modes corresponds to a normal mode of a system and describes a collective oscillation of the atoms around their local minimum. This is the harmonic approximation, known to work quite well in liquids (cf. Sec. 6.1.5). Nevertheless, already as the temperature undergoes the dynamic glass temperature (Sec. 1.1.1), other collective modes, different from the standard

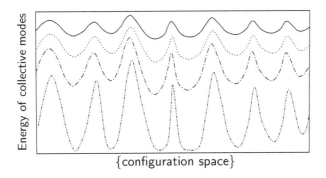

FIGURE 4.1

An over-simplified one dimensional projection of an energy landscape of collective modes: high energy collective modes are separated by low barriers while low energy collective modes are separated by high barriers.

vibrational ones, become important. The nature of these modes is quite different from the usual harmonic normal modes because they do not represent oscillations around a given configuration within a metastable well, but transitions among different wells.

As the characteristic energy of the collective modes depletes, the typical barrier separating these modes increases, leading to the opposite behavior with respect to the harmonic modes and to super-activation effects: while high energy collective modes are separated by low barriers, low energy collective modes are separated by high barriers and relax more slowly. A simple schematic representation of this scenario in a one dimensional configurational space is shown in Fig. 4.1.

Below the dynamic transition, relaxation dynamics proceeds by activation over the barriers characterizing the cooperativeness of the molecules over largely extended regions (see Sec. 1.1.1). In the model that we will discuss in Sec. 4.3, the size of these cooperative rearranging regions (CRR) cannot be inferred, but they can be related to the collective modes. As the temperature decreases, indeed, collective modes at low energy can be thought of as the representation of collective rearrangements of large regions, requiring higher activation energy and taking place at a lower rate.

We will tackle this problem at the end of the chapter by means of a generalization of the backgammon model where each urn has a distinct, randomly distributed, weight (or energy) [Leuzzi & Ritort, 2002]. Because of the different random energy values that the holes acquire, an analysis of the behavior of a material presenting several modes can be carried out in the situation in which those modes at larger energy (in absolute value) have thermalized while those at low energy relax too slowly to reach equilibrium on a given timescale (in the first order approximation of uncorrelated modes). This is qualitatively similar to the experimental results found, e.g., by Bellon *et al.* [2001] for the relaxation at different frequencies, Sec. 2.8.

4.1 The backgammon model

In the context of aging systems, we will be interested in a multistate Potts model [Potts, 1952; Wu, 1982] without energy barriers and without phase transitions that describes the typical dynamics encountered in glasses: the backgammon model, introduced by Ritort [1995]. The statics is simple and the dynamics displays a very slow relaxation at low temperature exclusively because of "entropic barriers." The dynamics of this model is equivalent to an urn model with M urns and N distinguishable balls, which, in turn, can be mapped into a biased random walk [Godrèche & Luck, 1996]. It is a generalization of the original Ehrenfest model in two ways: (i) the states are M instead of 2 and (ii) an energy function is introduced, so that the system can also be studied in temperature.

In this dynamic urn model, N balls are initially uniformly distributed among M boxes (point 1 in the scheme in the chapter introduction). At each time step a randomly chosen ball is moved to another nonempty randomly chosen urn (*ball-to-box* statistics, point 2). An energy function is defined as the number of occupied urns (point 3, cf. Eq. (4.1)) and a move of a ball into a new box is allowed if the energy does not increase (point 4).

According to the dynamic prescription it is clear that, once a box becomes empty, it remains as such forever and, thus, the number of nonempty boxes decreases until all balls are placed in one box. It is also easy to realize that the number of nonempty boxes decreases with a slower and slower rate as time goes by. Indeed, emptying a box with a few balls and placing them exclusively into boxes that are already occupied is an event that becomes more and more unlikely as the filled-in boxes available in the system decrease. The relaxation is hindered by the decrease of configurations available to decrease the energy, that is, by entropic barriers. This intuition is at the basis of the development of this model for slowly relaxing systems and this is the reason why, in the present chapter, we will concentrate on it to get as much information as possible on its glassy behavior. High entropic barriers means that there are only a few directions in phase space along which the system can evolve as the energy is decreased during the dynamics. As we will see, these unusual barriers are the source of typical features of glasses, like aging, time-dependent hysteresis effects and an Arrhenius-like relaxation time.

The backgammon model can be, as well, mapped into an asymmetric random walk with an absorbing site at the origin [Godrèche & Mézard, 1995]. This analogy can be of help in understanding the source of entropic barriers and the fact that the model evolves through rare events (Sec. 4.1.5 will be dedicated to this analysis).

The geometrical arrangement of the boxes (point 5.) plays no role in the backgammon model: any particle in any box can, indeed, move to any other box with the same probability. This amounts to considering the system in its mean-field approximation. The mean-field approximation in glass models induces a complete arrest in the dynamics because of the divergence of *energy* barriers (see, e.g., the spherical *p*-spin model in Sec. 7.2). In the mean-field case, indeed, the fluctuations leading to the formation of other glass phases (cf. Sec. 7.5) are neglected and the activated processes yielding the structural relaxation are inhibited. Nevertheless, the effect of *entropic* barriers is not much influenced by the geometry and the range of the interaction. The results obtained for the backgammon model are, indeed, a clear sign that entropic barriers alone can be the source of glassy behavior at low temperatures.

In this section we are going to schematically recall some of the features of this model and to connect them, when possible, to the idea of effective temperature in the context of slowly relaxing (glass-like) systems. For a complete study of the backgammon model the interested reader can refer to Ritort [1995]; Franz & Ritort [1995, 1996]; Godrèche & Luck [1996]; Franz & Ritort [1997]; Lipowsky [1997]; Arora *et al.* [1999]; Godrèche & Luck [1999, 2001] and also to the Proceedings of the Barcelona meeting on Dynamically Facilitated Models [Workshop, 2002] and to the reviews of Ritort & Sollich [2003] and Crisanti & Ritort [2003], respectively about facilitated models and the fluctuation-dissipation ratio for systems out of equilibrium.

The model is composed by N distinguishable particles that can occupy M different states, labeled by r. The energy of a given configuration is equal to the number of unoccupied states. The Hamiltonian of the model is

$$\mathcal{H} = -\sum_{r=1}^{M} \delta_{n_r,0} \tag{4.1}$$

where $\delta_{n,0}$ is a Kronecker delta, $n_r = 0,\ldots,N$ is the occupation number of state r and the number of particles is

$$\sum_{r=1}^{M} n_r = N \tag{4.2}$$

4.1.1 Equilibrium thermodynamics

The thermodynamic properties are easily computed, both using the canonical or the grand canonical ensemble. The system does not undergo any phase transition. The canonical partition function is

$$\mathcal{Z}_C = \sum_{\{n_r\}} \binom{N}{n_1 \ldots n_M} \delta\left(N - \sum_{r=1}^{M} n_r\right) e^{-\beta\mathcal{H}} \tag{4.3}$$

$$= \sum_{\{n_r\}} \frac{N!}{\prod_{r=1}^{M} n_r!} \delta\left(N - \sum_{r=1}^{M} n_r\right) \exp\left\{\beta \sum_r \delta_{n_r,0}\right\}$$

The factor $N!/\prod_r n_r!$ (with $0! = 1$) in the partition function $\mathcal{Z}_C$ is introduced to account for distinguishing particles. This factor leads to an overextensive entropy, $S \sim N \log N$ that can be cured eliminating from $\mathcal{Z}_C$ the overcounting term $N!$ in the numerator.

Using the integral representation for the delta function, one has

$$\mathcal{Z}_C = \int_0^{2\pi i} \frac{d\lambda}{2\pi i} \exp\left\{-\lambda N + \log\left[\sum_{\{n_r\}} \prod_{r=1}^{M} \frac{\exp\left(\lambda n_r + \beta \delta_{n_r,0}\right)}{n_r!}\right]\right\} \quad (4.4)$$

The partition function can be computed for large N and M in the saddle point approximation taking the maximum of the integrand. The maximum condition yields the self-consistency equation for $\lambda^\star$

$$\sum_{n=0}^{N} n \frac{\exp\left[n\lambda^\star + \beta \delta_{n,0}\right]}{n!} = \sum_{n=0}^{N} \frac{\exp\left[n\lambda^\star + \beta \delta_{n,0}\right]}{n!}$$

$$\longrightarrow \qquad z e^z = \rho\left(e^\beta + e^z - 1\right) \quad (4.5)$$

where we have defined the quantities $z \equiv \exp \lambda^\star$ and $\rho \equiv N/M$ (number of particles per state) and we used the fact that

$$\sum_{\{n_r\}} \prod_{r=1}^{M} g(n_r) = \left[\sum_{n=0}^{N} g(n)\right]^M$$

In the context of the grand canonical ensemble z is a fugacity, as we will see in a while.

The partition function for large N, M is, thus,

$$\mathcal{Z}_C \simeq \sum_{\{n_r\}} \prod_{r=1}^{M} \frac{\exp\left[n\lambda^\star + \beta \delta_{n,0}\right]}{z^{1/\rho} \, n!} \simeq \left(\frac{e^z z^{1-1/\rho}}{\rho}\right)^M \quad (4.6)$$

and the probability distribution of having different occupation numbers factorizes. Imposing normalization,

$$\sum_{\{n_r\}} \prod_{r=1}^{M} P_{n_r}(z, \beta) = \sum_{n=0}^{N} P_n(z, \beta) = 1 \quad (4.7)$$

one finds the distributions

$$P_n(z, \beta) = \rho \frac{z^{n-1} e^{\beta \delta_{n,0}}}{n! e^z}; \qquad n = 0, \ldots, N \quad (4.8)$$

In terms of these probability distributions, the energy per state and the z parameter come out to be

$$u = \frac{U}{M} = -P_0 = -\rho \frac{e^\beta}{z e^z}; \qquad z = -\log P_1 \quad (4.9)$$

The free energy per particle is

$$\frac{F}{N} = -\frac{T}{\rho} \left[z + \left(1 - \frac{1}{\rho} \right) \log z - \log \rho \right]$$ (4.10)

We notice that, for low temperature, the entropy of the system is ill-defined, as it happened for the HO models, cf. Sec. 3.1, diverging logarithmically with T. This pathology derives from the fact that the statistics of the model is Maxwell-Boltzmann [Kim *et al.*, 1996] (cf. Sec. 4.1.2).[1]

The same results can be obtained in the grand canonical ensemble, without making use of the saddle point approximation. In the grand canonical ensemble, the partition function is

$$\mathcal{Z}_{GC} = \sum_{N=0}^{\infty} \zeta^N \mathcal{Z}_C = \left[\sum_{n=0}^{\infty} \frac{e^{\beta \delta_{n,0}} \zeta^n}{n!} \right]^M = \left(e^\beta + e^\zeta - 1 \right)^M$$ (4.11)

where $\zeta \equiv \exp(\beta\mu)$ is the fugacity (μ is the chemical potential). This contains the sum over all possible occupation numbers of the relative weights of having n particles in a state. The grand potential per particle is, thus,

$$\frac{\Omega}{N} = -\frac{1}{\rho\beta} \log \left(e^\beta + e^\zeta - 1 \right)$$ (4.12)

The canonical and the grand canonical thermodynamic potentials are Legendre transforms of each other with respect to the number of particles and the chemical potential: $\Omega = F - N\mu$. This allows for the identification of the previously defined z and the fugacity ζ. The self-consistency Eq. (4.5) is obtained in this case from the condition

$$\frac{\partial\Omega}{\partial\mu} = -N \quad \text{or,} \quad \text{equivalently,} \quad \frac{\partial F}{\partial\mu} = 0$$ (4.13)

i.e., by fixing the number of particles equal to N. In the following, we will set the number of states equal to the number of particles ($M = N$, $\rho = 1$).

At low temperature, approximated relations can be used, reducing Eq. (4.5) to $ze^z \simeq e^\beta$ and the internal energy per state to $u \simeq -1 + 1/z \simeq -1 + T - T^2 \log T$. In Fig. 4.2 we plot the exact expression for u, together with these approximations for low T. They practically coincide. On the same interval also the linear approximation in T is shown, that, because of the logarithmic corrections, only holds near zero temperature.

---

[1]The entropy behaves correctly, instead, in models belonging to the monkey class, displaying a Bose-Einstein statistics [Godrèche & Mézard, 1995; Godrèche & Luck, 2001, 2005]. In these models, though, the relaxation to equilibrium is faster even at zero temperature and the effect of the presence of entropic barriers is smaller.

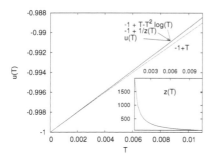

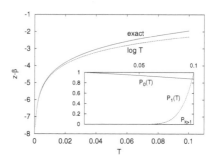

FIGURE 4.2

Energy versus temperature at equilibrium for the backgammon model at low T, in different approximations. The three curves on top, superimposing, are, respectively, the exact solution of Eqs. (4.5, 4.9) and the functions $-1 + 1/z(T)$ and $-1 + T + T^2 \log T$. The bottom curve is the linear approximation in T. Inset: fugacity vs. T: as $T \to 0$ z diverges.

FIGURE 4.3

Difference between fugacity and inverse temperature as $T \to 0$. Together with the exact curve, we also plot $z - \beta \sim - \log \beta$, approximation obtained for large z and β in Eq. (4.5). Inset: P_0 and P_1 versus T. All probabilities for states with more than one particle are practically zero in the T-range represented [cf. Eq. (4.8)].

4.1.2 Dynamics

As in any other urn model, the dynamics is defined by the statistics procedure by which balls are extracted from the boxes. The dynamics specifying the backgammon model is the following.

1. A ball is chosen at random among N with a uniform probability, this implies that a departure box d containing n_d particles is extracted with probability n_d/N out of the N available states

2. An arrival box a is chosen with a uniform distribution (independent of n_a)

3. The move is accepted according to the Monte Carlo algorithm (see Sec. 3.1.1), i.e., with probability 1 if the energy U decreases or is unchanged and with probability $W(\Delta U) = e^{-\beta \Delta U} = e^{-\beta}$ if the energy is increased by that move.

Note that the departure box satisfies $n_d \geq 1$ and it is chosen with probability n_d/N (*ball-to-box*). This dynamics corresponds to Maxwell statistics where particle are distinguishable and differs from the one corresponding to Bose statistics [Godrèche & Mézard, 1995; Kim *et al.*, 1996; Prados & Sanchez-Rey, 1997; Godrèche & Luck, 1996, 1999] where particles are indistinguishable and departure boxes are chosen with uniform probability $1/N$ (*box-to-box*).

Franz & Ritort [1996] derived the mean-field equations for the MC dynamics of the backgammon model. They form a set of hierarchical equations. For a generic occupation number n the evolution of P_n is described by an infinite hierarchy of equations:

$$\frac{dP_n(t)}{dt} = (n+1)\left[P_{n+1} - P_n\right] + P_{n-1} \tag{4.14}$$

$$+ P_0\left(e^{-\beta} - 1\right)\left[\delta_{n,1} - \delta_{n,0} - nP_n + (n+1)P_{n+1}\right]$$

$$n = 0, \ldots, N$$

with $P_{-1} = 0$ and the initial condition $P_n(0) = \delta_{n,1}$. Defining the shortcuts

$$\lambda \equiv \frac{1}{1 + P_0(e^{-\beta} - 1)}; \qquad \nu \equiv e^{-\beta} + P_1(1 - e^{-\beta}) \tag{4.15}$$

this set can be formally written as a Markovian system of the form [Godrèche & Luck, 1999]:

$$\frac{dP_n(t)}{dt} = \sum_{m=0}^{N} \mathcal{M}_{nm}[P_0(t), P_1(t)]P_m(t) \tag{4.16}$$

$$= \begin{cases} \frac{n+1}{\lambda(t)}P_{n+1}(t) - \frac{n+\lambda(t)}{\lambda(t)}P_n(t) + P_{n-1}(t), & n \geq 2 \\ \frac{2}{\lambda(t)}P_2(t) - 2P_1(t) + \nu(t)P_0(t), & n = 1 \\ P_1(t) - \mu(t)P_0, & n = 0 \end{cases}$$

where $\sum_{n\geq0}\mathcal{M}_{nm}[P_0(t), P_1(t)] = 0$. We stress, however, that the dynamics is actually non-Markovian because the λ and ν terms are functions of P_0 and P_1 in their turn, making the set of equations nonlinear, cf. Eqs. (4.14) or (4.16).

The derivation of the above equations is postponed to the later section 4.3 and to Appendix 4.A where they appear as a specific case of a generalized model. The solution for the distribution at equilibrium ($dP_n/dt = 0$) is provided by Eq. (4.8).

In particular, for $n = 0$, one obtains,

$$\frac{\partial P_0}{\partial t} = P_1(1 - P_0) - P_0 e^{-\beta}(1 - P_1) \tag{4.17}$$

yielding the energy relaxation $u(t) = -P_0(t)$, once $P_1(t)$ is known. Generally, it is not possible to exactly solve even the above equation without solving the whole hierarchy of equations, see Eq. (4.14). Not even at zero temperature, where the equation simplifies in

$$\frac{\partial P_0}{\partial t} = P_1(1 - P_0) \tag{4.18}$$

For long times it is, however, possible to yield an approximate analytical solution, within the context of different approximations, all relying on the assumption that the relaxation to 1 of $P_0(t)$ is "slow."

We now shortly report the resolution provided by Godrèche & Luck [1996] and, in the next section, we will analyze another - somewhat simpler - approach: an adiabatic approximation connected to the introduction of an effective temperature.

Let us introduce a generating function of the probability distributions:

$$\mathcal{G}(x,t) \equiv \sum_{n=0}^{N} x^n P_n(t) \tag{4.19}$$

The first term of the polynomial series is $\mathcal{G}(0,t) = P_0(t)$ and the closure condition on the probabilities reads now $\mathcal{G}(1,t) = 1$. The function $\mathcal{G}(x,t)$ satisfies the partial differential equation [cf. Eq. (4.14)]:

$$\frac{\partial \mathcal{G}(x,t)}{\partial t} = (x-1)\left[\mathcal{G}(x,t) - \frac{1}{\lambda}\frac{\partial \mathcal{G}(x,t)}{\partial x} - \frac{\lambda-1}{\lambda}\right] \tag{4.20}$$

with initial condition $\mathcal{G}(x,0) = \sum_n x^n \delta_{n,1} = x$, where we used the parameter λ defined in Eq. (4.15). The equation can be formally solved by the method of characteristics, i.e., defining the auxiliary variable

$$y(x,t) \equiv (1-x)\exp\left\{-\int_0^t \frac{du}{\lambda(u)}\right\} = (1-x)e^{-\tau(t)}; \tag{4.21}$$

$$\tau(t) \equiv \int_0^t \frac{du}{\lambda(u)}$$

The differential equation for $\hat{\mathcal{G}}(y,t) = \mathcal{G}(x(y,t),t)$ is somewhat simpler than Eq. (4.20):

$$\frac{\partial \hat{\mathcal{G}}(y,t)}{\partial t} = -ye^{\tau(t)}\left[\hat{\mathcal{G}}(y,t) - P_0(t)(1-e^{-\beta})\right] \tag{4.22}$$

and its implicit solution reads

$$\hat{\mathcal{G}}(y,t) = (1-y)\exp\left\{-y\int_0^t du\ e^{\tau(u)}\right\} \tag{4.23}$$

$$+y\int_0^t du\ P_0(u)(1-e^{-\beta})\exp\left\{\tau(u) - y\int_u^t dv\ e^{\tau(v)}\right\}$$

If computed at $x = 0$, i.e., $y = e^{-\tau(t)}$, it reduces to the nonlinear integral equation

$$P_0(t) = \left[1 - e^{-\tau(t)}\right]\exp\left\{-\int_0^t du\ e^{\tau(u)-\tau(t)}\right\} \tag{4.24}$$

$$+\int_0^t du\ P_0(u)(1-e^{-\beta})\frac{d}{du}\exp\left\{-\int_u^t dv e^{\tau(v)-\tau(t)}\right\}$$

The above equation can be solved only numerically [Franz & Ritort, 1996], but, at least for very low temperatures, a very good approximated solution can be computed analytically. We move, therefore, to consider the zero temperature case, where Eq. (4.24) simplifies: integrating by part and using Eq. (4.18) we find

$$1 - e^{-\tau(t)} = \int_0^t du \, P_1(u) \, [1 - P_0(u)] \exp \left\{ \int_0^u dv \, e^{\tau(v) - \tau(t)} \right\} \qquad (4.25)$$

For completeness, we report that the generating functional method was first applied to the problem in the random walk formalism [Godrèche & Mézard, 1995] that we will consider later, see Sec. 4.1.5.

Eq. (4.25) cannot be analytically solved unless for long times. In the latter case it can be approximated under the assumption that $P_0(t)$ evolves "slowly enough" with respect to the evolution of all other quantities involved (this will be checked *a posteriori* as the solution is computed, cf. Eq. (4.34)) [Godrèche & Luck, 1996]. Neglecting $e^{-\tau(t)}$ with respect to one, splitting the right-hand side in two parts and using the zero temperature differential Eq. (4.18), Eq. (4.25) becomes:

$$1 \simeq \int_0^t du \, P_1(u) \, [1 - P_0(u)] \qquad (4.26)$$

$$+ \int_0^t du \, P_1(u) \, [1 - P_0(u)] \left[\exp \left\{ \int_0^u dv \, e^{\tau(v) - \tau(t)} \right\} - 1 \right]$$

$$= P_0(t) + \int_0^t du \, P_1(u) \, [1 - P_0(u)]$$

$$\times \left[\exp \left\{ \int_0^t dv \, e^{\tau(v) - \tau(t)} - \int_u^t dv \, e^{\tau(v) - \tau(t)} \right\} - 1 \right]$$

The integral in the right-hand side is dominated by values of $u \sim t$. We, therefore, use the following approximations:

$$\tau(t) - \tau(v) = \int_v^t ds \, [1 - P_0(s)] \sim \epsilon[1 - P_0(t)] \qquad (4.27)$$

$$\int_u^t dv \, e^{\tau(v) - \tau(t)} \sim \frac{1 - e^{\epsilon[1 - P_0(t)]}}{1 - P_0(t)} \qquad (4.28)$$

where

$$\epsilon \equiv t - v; \qquad \epsilon \ll t \qquad (4.29)$$

We now use the abbreviation $\lambda(t) \equiv (1 - P_0(t))^{-1}$, cf. Eq. (4.15), and we change the integration variable from ϵ to $\eta \equiv e^{\epsilon/\lambda(t)}$. Inserting Eqs. (4.27) and (4.28) into Eq. (4.26), we obtain

$$1 \simeq 1 - \frac{1}{\lambda(t)} + P_1(t) \int_0^1 \frac{d\eta}{\eta} \, [\exp\{\lambda\eta\} - 1] \qquad (4.30)$$

that, together with Eq. (4.18) (rewritten as $d\lambda/dt = \lambda P_1$), leads to

$$\frac{dt}{d\lambda} \sim I(\lambda) = \int_0^1 \frac{d\eta}{\eta} \left(e^{\eta\lambda} - 1\right) = \sum_{n\geq 1} \frac{\lambda^n}{n\,n!} = \text{Ei}(\lambda) - \log\lambda - \text{C} \sim \frac{e^\lambda}{\lambda} \sum_{n\geq 0} \frac{n!}{\lambda^n}$$
(4.31)

where Ei is the exponential-integral function[2] and $\text{C} = 0.577215\ldots$, the Euler constant. The initial condition $P_0(0) = 0$ reads now $t(\lambda = 1) = 0$. Integrating, one has

$$t(\lambda) \sim \int_0^1 \frac{d\eta}{\eta^2} \left(e^{\eta\lambda} - 1 - \lambda\eta\right) = \sum_{n\geq 1} \frac{\lambda^{n+1}}{n\,(n+1)!} \sim \frac{e^\lambda}{\lambda} \sum_{n\geq 0} \frac{(n+1)!}{\lambda^n} \quad (4.32)$$

The error performed in the above equation is of the order $d\lambda/dt \sim e^{-\lambda}$, exponentially small. As a consequence, the asymptotic series in Eq. (4.32) can be considered valid at all orders in $1/\lambda = 1 - P_0$. At the first sub-leading order, one has a dynamic relaxation qualitatively identical to the one found for the harmonic oscillator models considered in the previous chapter [cf., e.g., Eqs. (3.89), (3.119) and Appendix 3.A]:

$$t \simeq \frac{e^\lambda}{\lambda} \left(1 + \frac{1}{\lambda}\right) \quad (4.33)$$

implying

$$P_0(t) \simeq 1 - \frac{1}{\log t + \log\log t} \quad (4.34)$$

In Fig. 4.4 we reproduce the long time behavior of $\lambda(t)$, comparing the analytical approximated long time prediction with the numerical resolution of the differential equations, see Eq. (4.14).

We conclude by mentioning that the presence of exponentially small corrections (non-perturbative in the control parameter λ) is usually a by-product of *adiabatic approximations*, such as the one that will be implemented in the next section.

4.1.3   Adiabatic approximation and effective temperature

To solve Eq. (4.17), without solving the whole hierarchy, Eq. (4.14), for any n, yet an alternative approach can be followed, assuming *separation of timescales* between processes involving empty states and processes involving occupied states. The relaxation dynamics is, thus, led by the processes of emptying a box, whereas all exchanges of particles among boxes containing already at least one particle occur on a shorter timescale and are practically at equilibrium on the constant energy (or P_0) surface. This is based on the fact

---

[2]The integral definition of this function is, for positive argument x, $\text{Ei}(x) \equiv - \lim_{\epsilon\to 0^+} \left[\int_{-x}^{-\epsilon} dt\, e^{-t}/t + \int_\epsilon^\infty dt\, e^{-t}/t\right]$.

that moves between nonempty states are energy and entropy costless, whereas emptying completely a box becomes more and more difficult as the number of filled-in boxes decreases. Indeed, as the dynamics goes on, all the balls will be gathered in a few, very crowded boxes and to empty one of them, many coordinated moves must occur at the same step in the evolution dynamics. This *adiabatic approximation* is somewhat cruder than the one adopted in the previous section to find the long time behavior of the probability distributions of the occupancies of the urns, but it leads, nevertheless, to the same results at the leading order (the difference will be of the order $1/\log t$ in the evaluation of $\lambda = 1 - P_0 \simeq \log t + \log \log t$). The basic point in both approaches, actually, is the same (even though with different and differently refined implementations): that the relaxation of P_0 is slower than the one of any other distribution.

In order to implement the adiabatic approximation one can introduce an effective temperature $T^*(t)$ [and an effective fugacity $z^*(t) = z(T^*(t))$, related to T^* by Eq. (4.5)] and express all the distributions $P_{n\geq 1}(t)$ as if they were at equilibrium at heat-bath temperature $T^*(t)$ [cf. Eq. (4.8)], i.e.,

$$P_{n\geq 1}(z^*(t)) = \frac{[z^*(t)]^{n-1}}{n!e^{z^*(t)}} \tag{4.35}$$

This time-dependent effective temperature is *defined* by

$$u(t) = -P_0(\beta^*(t), z^*(t)) = -\frac{e^{\beta^*}}{z^*e^{z^*}} \tag{4.36}$$

together with the condition

$$z^*(t)e^{z^*(t)} = e^{\beta^*(t)} + e^{z^*(t)} - 1 \tag{4.37}$$

If local equilibrium is attained on the hypersurface of constant energy, we can relate P_1 to P_0 using Eq. (4.35). The simplest way of dealing with Eq. (4.17) is to write both P_1 and P_0 in terms of the time-dependent fugacity z^* writing down a dynamical equation for z^*. Plugging Eqs. (4.35)-(4.37) into Eq. (4.17) and solving for $z^*(t)$, the equation of motion is

$$\frac{\partial z^*}{\partial t} = \frac{z^*[\exp(z^*) - 1]}{\exp(z^*) - z^* - 1}\left(e^{-z^*} - z^*e^{-\beta}\right) \tag{4.38}$$

In the long time limit, at low temperature, z^* diverges (see inset of Fig. 4.2 for the equilibrium behavior) and the leading asymptotic behavior is given by the solution of

$$\frac{\partial z^*}{\partial t} \simeq z^*\exp(-z^*) - e^{-\beta}z^*(z^* - 1) \tag{4.39}$$

that, in its implicit form, reads:

$$t = \text{const} + \int_1^{z^*(t)} \frac{du}{u}\, \frac{1}{e^{-u} - (u-1)e^{-\beta}} \tag{4.40}$$

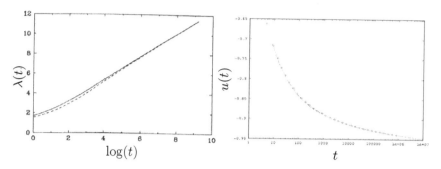

FIGURE 4.4

Inverse density of filled-in boxes λ as a logarithmic function of time. The full curve represents the numerical solution of the exact differential Eq. (4.14), the dashed curve represents the approximated analytical solution for long times, Eq. (4.33). Reprinted figure with permission from [Godrèche & Luck, 1996]. Copyright (1996) by the Institute of Physics Publishing.

FIGURE 4.5

Energy vs. t at $T = 0$. Three curves are compared: (i) numerical resolution of Eq. (4.14) (random initial conditions at $t = 0$), (ii) numerical resolution of adiabatic Eq. (4.38) (random initial conditions), (iii) Monte Carlo simulations for $N = 10^5$ particles/states. Reprinted figure with permission from [Franz & Ritort, 1996]. Copyright (1996) by Springer.

At zero temperature this simply yields $t \simeq \exp(z^\star)/z^\star$. Inverting it, one finds, for long times,

$$z^\star(t) \simeq \log t + \log \log t \tag{4.41}$$

$$u(t) \simeq -1 + \frac{1}{\log t + \log \log t} \tag{4.42}$$

identical to Eq. (4.34).

We stress that these are approximated solutions, valid for long times, of the already approximated adiabatic Eqs. (4.35), (4.37). However, one can observe that the right-hand side of Eq. (4.39) coincides with the first term in the $1/\lambda$ expansion, Eq. (4.31), and the solution (4.42) coincides with Eq. (4.34).[3] The inverse effective temperature can be computed as

$$\beta^\star(t) \simeq z^\star + \log z^\star \simeq \log t + \log \log t + \log \left(\log t + \log \log t - 1 \right) \tag{4.43}$$

The validity of the approximation can be appreciated comparing the results with the one obtained by exactly solving the adiabatic Eq. (4.38) or with the

---

[3] Even though the effective fugacity $z^\star$ and the inverse of the probability of nonempty boxes $\lambda = 1/(1 - P_0)$ are different quantities, they turn out to coincide at the present level of approximation, simply because, in the adiabatic approximation, the probability of having an empty box relaxes as $P_0 \simeq 1 - 1/z^\star$ as a function of $z^\star$.

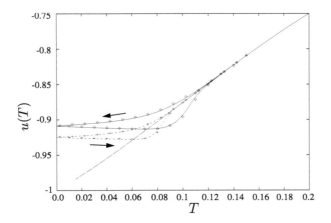

FIGURE 4.6

Hysteresis in heating and cooling. The cycles have been obtained by integrating numerically the adiabatic Eq. (4.38) and using the adiabatic approximation, Eq. (4.36). The cooling-heating rates are $3.3 \cdot 10^{-5}$ (MC step)$^{-1}$ (continuous lines) and $3.3 \cdot 10^{-6}$ (MC step)$^{-1}$ (dashed lines). The points (diamonds for fast cooling and crosses for slow cooling) are from Monte Carlo data for $N = 20,000$ particles. Reprinted figure with permission from [Franz & Ritort, 1995]. Copyright (1995) by the European Physical Society.

outcome of numerical simulations. The comparison of the time behavior of the internal energy obtained as an exact numerical solution, by numerical Monte Carlo simulations, and in the adiabatic approximation is shown in Fig. 4.5.

Arrhenius relaxation

Looking at the acceptance rate of the Monte Carlo dynamics, that is, the frequency by which a box becomes empty, this is equal to $P_1 = e^{-z}$. Its inverse is the relaxation time to equilibrium, following an Arrhenius law:

$$\tau_{\text{eq}} = e^z \simeq T \, e^{1/T} \qquad (4.44)$$

as one can also verify looking, e.g., at Eq. (4.39).

Hysteresis

Numerically integrating Eq. (4.17) one can study the dependence of the energy from the cooling rate and see what happens to the system when it is cooled down and reheated (at an equal rate, for instance). As one can see from Fig. 4.6, where the energy behavior is plotted in temperature at two different cooling (heating) rates, the effect of varying the rate is evident on the thermalization process. Moreover hysteresis effects are present: $u(T)$ does not follow the same path in cooling and heating because the system is out of equilibrium at a low temperature. As expected, the surface enclosed in the energy loop in cooling-heating decreases as the rate decreases (at equilibrium, i.e., for infinitely slow cooling, no hysteresis may be present).

4.1.4 Entropic barriers and a microcanonic derivation of the equation of motion

The same relaxation behavior, as found in Eqs. (4.34), (4.42), can be otherwise obtained constructing the equation of motion by means of purely configurational arguments [Crisanti & Ritort, 2003]. After quenching to zero temperature, the system starts to relax to its ground state. Since in this case there is no thermal activation, relaxation is purely driven by entropic barriers, i.e., flat directions in configurational space through which the system diffuses. Entropic relaxation is energy costless and its rate is, thus, only determined by the number of available configurations with energy smaller or equal to the actual energy. Let us denote by M_{occ} the number of occupied boxes (out of the total number M) and by $w(M_{occ})$ the number of configurations with M_{occ} occupied boxes. The typical time to increase, by one unit, the number of empty boxes is given by,

$$\tau \simeq \frac{w(M_{occ})}{w(M_{occ} - 1)} \tag{4.45}$$

For distinguishable particles, assuming there are many in every occupied urn, we have

$$w(M_{occ}) = \binom{N}{n_1 \dots n_{M_{occ}}} = \frac{N!}{\prod_{r=1}^{M_{occ}} n_r!} \simeq \frac{N^N}{\prod_{r=1}^{M_{occ}} n_r^{n_r}} \simeq M_{occ}^N$$

where we used the Stirling formula, the closure condition $\sum_r n_r = N$ and the further assumption that every occupied box has more or less the same number of particles $\sim N/M_{occ} = \lambda$. For a very long characteristic emptying time τ, longer than the observation time of an experiment, the system is stuck in one of the configurations with a given energy $U = -N + M_{occ}$.

In the large N, M_{occ} limit we, thus, obtain

$$\tau \simeq \left(\frac{M_{occ}}{M_{occ} - 1}\right)^N \simeq \exp\left(\frac{N}{M_{occ}}\right) \tag{4.46}$$

as was originally devised by Ritort [1995] from the observation of the results of numerical simulations. The logarithm of $w(M_{occ})$ is the configurational entropy of all configurations of equal energy that might be visited by the system (on longer time windows than the observation ones): $S_c = N \log(M_{occ}) = N \log(N + U)$. Using the relation $M_{occ} = N + U = N(1 - P_0)$ and Eq. (4.45), we find

$$\frac{dP_0}{dt} = -\frac{\Delta M_{occ}}{N \Delta t} = \frac{1}{\tau_{eq}} = \exp\left(-\frac{1}{1 - P_0}\right) \tag{4.47}$$

where $\Delta M_{\mathrm{occ}} = -1$, because at zero temperature the number of occupied boxes can only decrease by one unit.

This yields the result $P_0 \simeq 1 - 1/(\log t + \log(\log t))$ in agreement with Eqs. (4.34), (4.42).

Transition rate effective temperature in the backgammon model

Following the definition of effective temperature given in Eq. (3.161) we see that a $T_{\mathrm{e}}^{\mathrm{tr}}$ can be defined starting from the transition rate $1/\tau$ (cf. Eq. (4.45)), with $s_c = S_c/N = \log(N + U)$. Hence, one has both

$$\frac{1}{\tau(t)} \simeq e^{-N/M_{\mathrm{occ}}(t)} = e^{-\lambda(t)} \simeq e^{-z^\star(t)} \tag{4.48}$$

and

$$\frac{1}{\tau(t)} = \exp\left(-\frac{\partial s_c}{\partial u}\right) = e^{-\beta^\star} \tag{4.49}$$

implying

$$\beta_{\mathrm{e}}^{\mathrm{tr}} = z^\star \tag{4.50}$$

This has to be compared with Eq. (4.43) for the adiabatic effective temperature and they differ because of logarithmic corrections.

4.1.5   Backgammon random walker

An improved adiabatic approximation can be devised exploiting the mapping of urn problems to random walkers. The outcome is basically coinciding with the resolution presented in Sec. 4.1.2. We would like, however, to discuss this other approach, because a different point of view can help clarify the source of the entropic barriers and their role in the dynamics. For the case of the backgammon model, the random walker approach was implemented by Godrèche & Mézard [1995].

Let us start from the master equation

$$\frac{dP_n}{d\tau} = \mu_{n+1}P_{n+1} + \lambda_{n-1}P_{n-1} - (\mu_n + \lambda_n)P_n \tag{4.51}$$

describing a random walk with transition rates μ_n and λ_n. If we set $\mu_n = n$ and $\lambda_n = \lambda(\tau)$ and we put an absorbing site at $n = 0$, we are considering an asymmetric random walk. Rescaling, further, the time as $d\tau/dt = 1/\lambda(\tau(t))$, the equation of motion coincides with that of the backgammon problem at zero temperature, Eq. (4.14). The velocity v_n and diffusion coefficient D_n for the problem are

$$v_n = \lambda_n - \mu_n = \lambda(\tau) - n; \qquad D_n = (\lambda_n + \mu_n)/2 = (\lambda(\tau) + n)/2 \tag{4.52}$$

The bias is defined as $v_n/2D_n$. The parameter n is the size of a box (i.e., the number of particles contained in a box) and $\lambda(\tau)$ represents the mean size of the boxes (i.e., the total number of particles N divided by the number of filled-in boxes, $\lambda = N/M_{occ} = 1/(1 - P_0)$, cf. Eq. (4.15) at $T = 0$). If n is larger or smaller than the mean λ, the bias is negative or positive, at difference with a symmetric random walker, where both transition rates are equal (if n dependent or not is not relevant) and the bias is always zero. This property has fundamental consequences on the model behavior that we summarize in the following.

1. The random walker is located around $\lambda(t)$, because of a restoring potential $\sim 2n - 4\lambda \log(\lambda + n)$ (this will be evident in the continuous limit, see the Fokker-Planck approach, Eqs. (4.53), (4.55)). This is the source of the existence of entropic barriers (cf. Sec. 4.1.4): the probability for the random walker to be near the origin is much less than the probability of being in a position near λ, and if λ is far from the origin, the absorption phenomenon is strongly inhibited. Translated into the urn model language this is equivalent to saying that, provided a small number of filled-in boxes remains, the configurations with a completely empty box are far less numerous than those for which all boxes are occupied.

2. The random walker absorption at the origin occurs with low frequency and, therefore, its mean position, increases slowly. This also implies that the minimum of the restoring potential increases slowly. In other words, the relaxation of the energy $u = -1 + 1/\lambda$ is very slow.

3. The first two properties justify the resolution of Eq. (4.51) within the adiabatic approximation (cf. also Sec. 4.1.3), that consists in considering the system in a quasi-stationary state obtained by setting to zero the time derivative of P_n and neglecting the absorption at the origin. In the random walk language the two, separated, timescales at the basis of such an approximation are the average time for the random walker to be absorbed at the origin [long, of $\mathcal{O}\left(e^{\lambda}/\lambda\right)$], and the relaxation time to equilibrium at a given fixed value of λ [short, of $\mathcal{O}(1)$].

4. The time evolution is governed by rare events. Indeed, for long times (large λ), P_n possesses a Gaussian scaling form of width $\sqrt{\lambda}$, centered around λ, but the evolution of λ is driven by $P_1 = e^{-z^*} \sim e^{-\lambda}$ and lies in the nonuniversal tail of the scaling distribution.

Fokker-Planck backgammon description

The gradient expansion of Eq. (4.51) to second order yields a continuous Fokker-Planck equation:

$$\frac{\partial P(h,\tau)}{\partial \tau} = \frac{1}{2} \frac{\partial^2}{\partial h^2} \left[(h + \lambda) P(h,\tau) \right] - \frac{\partial}{\partial h} \left[(\lambda - h) P(h,\tau) \right] \quad (4.53)$$

$$= \frac{\partial^2}{\partial h^2} \left[D(h) P(h,\tau) \right] - \frac{\partial}{\partial h} \left[v(h) P(h,\tau) \right]$$

In the adiabatic approximation, the rate at which boxes become empty can be approximated by the equilibrium probability distribution in the origin, $P_{eq}(0)$. The equilibrium distribution for Eq. (4.53) turns out to be

$$P_{eq}(h) \simeq \frac{1}{D(h)} \exp \left\{ \int_0^h dh' \frac{v(h')}{D(h')} \right\} = \frac{\exp \{ \mathcal{V}(h) - \mathcal{V}(0) \}}{D(h)} \qquad (4.54)$$

with

$$\mathcal{V}(h) = 2h - 4\lambda \log(\lambda + h) \qquad (4.55)$$

$$\int_0^\infty dh \, P_{eq}(h) = \frac{1}{\lambda} = 1 - P_0 \qquad (4.56)$$

The function $\mathcal{V}(h)$ is the effective restoring potential, heuristically formalizing the concept of entropic barriers, and it is proportional to the integral of the bias $v(h)/(2D(h))$.

The above form of $P_{eq}(h)$, computed in the origin, yields an average period of time intercurrent between two events causing a box to become empty (i.e., an average absorption time) approximately proportional to $\tau \sim e^{a\lambda}$, with $a \simeq 0.77$. This must be compared with what we found in Sec. 4.1.4, cf. Eq. (4.46): $\tau \sim e^\lambda$ ($a = 1$). Although both the continuum and the discrete versions of the problem qualitatively lead to the same slow decay for the energy, $u(\tau) \simeq -1 + a/\log \tau$, there is a discrepancy that is a consequence of the nature of the rare events responsible for the relaxation dynamics of the backgammon model. We said above that, even though the P_n have Gaussian scaling forms, the dynamics of λ lies in the tails of the distribution, that are nonuniversal. Far away from the scaling region $n - \lambda \ll \sqrt{\lambda}$, then, no universal property can be expected.

To conclude, the random walker moves along the n-axis, in a confining potential centered around $\lambda(t)$, itself evolving with time and increasing each time that a box is emptied. Since the latter is a slow process at low temperature, the dynamics of λ is also very slow. On the contrary, the equilibration inside the potential $\mathcal{V}$ is fast. This defines two, well-separated, timescales, which are connected to the slow α and fast β processes, respectively (cf. Sec. 1.1).

4.2 Two-time dynamics and FDR effective temperature

The two-time observables studied in the backgammon model have been the energy-energy [Franz & Ritort, 1995, 1996; Godrèche & Luck, 1996, 1997] and the density-density [Franz & Ritort, 1997; Godrèche & Luck, 1999] correlation functions, as well as the relative response functions, energy-temperature and density-chemical potential. In both cases a typical aging regime has been identified and analytic expressions approximated both for this slow relaxation and the fast relaxation on short timescales have been computed.

To exemplify the computation we will use the density-chemical potential couple of the conjugated variable-field, allowing for computing fluctuation and dissipation at fixed temperature and, furthermore, providing a fast relaxation over short times in the correlation function, on timescales where the energy-energy correlation function is practically constant.

The density-density correlation function measures the correlation between the number of particles in a given box at time t_w (waiting time) and at time t (observation time):

$$C(t, t_w) = \langle \rho_i(t) \rho_i(t_w) \rangle - \langle \rho_i(t) \rangle \langle \rho_i(t_w) \rangle \tag{4.57}$$

where

$$\langle \rho_i(t) \rangle = \langle \rho(t) \rangle = \sum_{n=1}^{N} n \, P_n(t) = 1, \quad \forall t, \quad \forall i = 1, \dots, N \tag{4.58}$$

$$\langle \rho_i(t) \rho_i(t_w) \rangle = \sum_{n,m}^{1,N} n \, m \, w_n^{(m)}(t, t_w) P_m(t), \quad \forall i \tag{4.59}$$

where $w_n^{(m)}(t, t_w)$ is the probability to have n particles in the box at time t conditioned to have m particles in the same box at time t_w.

The probability $w_n^{(m)}(t, t_w)$ satisfies the same nonlinear differential equations of motion, Eq. (4.16), of P_n and its solution, in an implicit integral form, can, as well, be computed by the methods of generating functions (cf. Sec. 4.1.2). An approximated analytic solution can, then, be obtained, at long times.

The density response function measures the reaction of the density of a given box i at time t to a perturbation in chemical potential at time t_w. The chemical potential enters the Hamiltonian of the perturbed system as

$$\mathcal{H}_p = -\sum_{j=1}^{N} \delta_{n_j,0} - \mu \, n_i \tag{4.60}$$

This makes the occupation probability of box i depend on μ: $P_{n_i}^{(\mu)} = P_n^{(\mu)}$ (we drop the index i). The influence of the $-\mu n_i$ on $P_{j \neq i}$ vanishes in the thermodynamic limit as $1/N$.

One can, then, define the response function

$$G(t, t_w) = \left. \frac{\delta \langle \rho_i(t) \rangle}{\delta \mu(t_w)} \right|_{\mu \to 0} \tag{4.61}$$

The perturbed probability distributions $P_n^{(\mu)}$ satisfy a system of differential nonlinear equations that generalize Eq. (4.16). We will not report them here but their derivation is very clearly presented in [Godrèche & Luck, 1999] for the density response and in [Godrèche & Luck, 1997] for the energy response. We only stress that since the dynamics is based on the Metropolis acceptance rate $W(\beta \delta E) = \min[1, e^{-\beta \delta E}]$ (cf. Appendix 4.A) and because of the fact that W is not differentiable at $\delta E = 0$, we will have two different responses as the chemical potential tends to zero from the left or from the right. Indeed, in terms of $P_n^{(\mu)}$ the response function can be written as

$$G^{\pm}(t, t_{\mathrm{w}}) = \sum_{n=1}^{N} n \left. \frac{\delta P_n^{(\mu)}(t)}{\delta \mu(t_{\mathrm{w}})} \right|_{\mu \to 0^{\pm}} \tag{4.62}$$

where the observables $\delta P_n^{(\mu)}(t)/\delta \mu(t_{\mathrm{w}})\big|_{\mu \to 0^{\pm}}$ satisfy the dynamic Eq. (4.16). As usual, the equations of motion for $P_n^{(\mu)}$ can be solved (implicitly) by means of the method of generating functions and an analytic expression can be obtained for long times.

We report here the long time expressions for both correlation and response and we look at their dependence on the age of the system. In the next section, we will look at the FDR, that can be considered as a definition of effective temperature (cf. Sec. 2.8).

For long times both the correlation and the response show aging and have the following behavior:

$$C(t, t_{\mathrm{w}}) = C(t_{\mathrm{w}}, t_{\mathrm{w}}) \frac{h(t_{\mathrm{w}})}{h(t)} \tag{4.63}$$

$$G(t, t_{\mathrm{w}}) = G(t_{\mathrm{w}}, t_{\mathrm{w}}) \frac{h(t_{\mathrm{w}})}{h(t)} \tag{4.64}$$

The time sector function (cf. Secs. 2.8 and 3.3.4) is a function of time exclusively through λ

$$h(\lambda) = \exp \left[-\log \lambda + \int_1^{\lambda} d\lambda' \alpha(\lambda', \bar{\lambda}) \right] \tag{4.65}$$

$$\alpha(\lambda, \bar{\lambda}) \simeq \alpha(\lambda, \infty) \simeq \frac{1}{2} \left(1 + \frac{1}{\lambda} \right) \tag{4.66}$$

where $\bar{\lambda}$ is the asymptotic (equilibrium) value of $\lambda(t) = 1/(1 - P_0(t)) \to z/(1 - e^{-z})$, cf. Eq. (4.8). As temperature decreases the fugacity becomes very large (eventually infinite), cf. the inset of Fig. 4.2, and $\bar{\lambda} \simeq z$. In the aging regime that we are considering one can, therefore, take $\lambda(t)/\bar{\lambda} \ll 1$ if $T \gtrsim 0$.

Eventually one obtains $h(\lambda(t)) \simeq (e^{\lambda}/\lambda)^{1/2} \simeq (t)^{1/2}$ implying for the correlation function, and for the response as well,

$$C(t, t_{\mathrm{w}}) = C(t_{\mathrm{w}}, t_{\mathrm{w}}) \sqrt{\frac{t_{\mathrm{w}}}{t}} \tag{4.67}$$

The equal time values of C and G are given, respectively, by[4]

$$C(t_{\rm w}, t_{\rm w}) \simeq \lambda(t_{\rm w}) - 1 \tag{4.68}$$

$$G(t_{\rm w}, t_{\rm w}) \simeq \lambda(t_{\rm w})^2 e^{-\lambda(t_{\rm w})} \tag{4.69}$$

One can also consider the early times relaxation, for which the system looks like it is thermalizing at equilibrium (stationary regime, cf. Sec 2.7). Combining the two, separated, time regimes one obtains

$$C(t, t_{\rm w}) \simeq e^{(t-t_{\rm w})/\lambda(t_{\rm w})} + (\lambda(t_{\rm w}) - 1)\sqrt{\frac{t_{\rm w}}{t}} \tag{4.70}$$

representing, respectively, β and α relaxation in glasses.

4.2.1 Effective temperature(s) in the backgammon model

The FDR analysis is mainly carried out at zero temperature, where off-equilibrium properties are most evident and the analytical treatment easier [Franz & Parisi, 1997; Godrèche & Luck, 1997, 1999; Godrèche & Luck, 2002]. In this case the chemical potential scales as T so that $\tilde{\mu}\beta\mu$ is a finite quantity. The response function is, then, defined, as (cf. Eq. (4.62))

$$\tilde{G}^{\pm}(t, t_{\rm w}) = \frac{\delta \langle \rho_n(t) \rangle}{\delta \tilde{\mu}}\Big|_{\tilde{\mu}\to 0^{\pm}} \tag{4.71}$$

We notice that the two time functions all satisfy the scaling behavior in the aging regime:

$$\frac{C(t, t_{\rm w})}{C(t_{\rm w}, t_{\rm w})} \simeq \frac{\tilde{G}(t, t_{\rm w})}{\tilde{G}(t_{\rm w}, t_{\rm w})} \simeq \frac{\partial C(t, t_{\rm w})}{\partial t_{\rm w}} \left(\frac{\partial C(t, t_{\rm w})}{\partial t_{\rm w}}\Big|_{t=t_{\rm w}}\right)^{-1} \simeq \frac{h(t_{\rm w})}{h(t)} \tag{4.72}$$

This implies for the backgammon model the very often observed fact (cf. Sec. 2.8) that the FDR at any observation time t only depends on the waiting time. Using Eq. (4.69) we obtain the FDR (in the form T/T_e)

$$\lim_{T\to 0} \frac{T}{T_e} = \frac{\partial_{t_{\rm w}} C(t, t_{\rm w})|_{t=t_{\rm w}}}{\tilde{G}(t_{\rm w}, t_{\rm w})} \simeq \begin{cases} 1 - \frac{2}{\lambda(t_{\rm w})^2} - \frac{4}{\lambda(t_{\rm w})^3}, & \text{energy,} \\ 1 - \frac{2}{\lambda(t_{\rm w})^2} - \frac{2}{\lambda(t_{\rm w})^3}, & \text{density } \tilde{\mu} \to 0^+, \\ 1 - \frac{3}{\lambda(t_{\rm w})^2} - \frac{3}{\lambda(t_{\rm w})^3}, & \text{density } \tilde{\mu} \to 0^-, \end{cases} \tag{4.73}$$

As equilibrium is lost, that is, as soon as the early stationary regime is passed $(t \ll t_{\rm w})$ the FDR computed on different conjugated variable-field

---

[4]They are computed in the aging regime as well, by this meaning that all fast processes have already thermalized and that $t_{\rm w} \gg 1$.

couples differs (in the $\rho - \mu$ potential case they even depend on the sign of μ and on the microscopic rules of the dynamics[5]).

The existence of an effective temperature is closely related to the validity of some approximations used in the context of slowly relaxing systems such as, e.g., the adiabatic approximation, implementing the assumption of separation of timescales. Within the adiabatic approximation one obtains a Markovian description for the dynamics. This Markovian approach encodes within a single effective parameter ($z^\star$ or $T^\star$) all the complicated previous past of the system.

In the backgammon model, however, if one compares the "adiabatic" definition of effective temperature with the FDR definition, no unique long time behavior is found. A Markovian description for glassy dynamics is evidently too far from the realistic behavior.

To sum up, in the backgammon model the following definitions of effective temperature have been adopted:

- adiabatic, in Sec. 4.1.3,

- transition rate, in Sec. 4.1.4,

- FDR, in Sec. 4.2.1.

However, we stress that those three definitions do not yield a unique long time behavior (if not for extremely long times, for which the system is thermalized). In the harmonic oscillator spherical spin (HOSS) model for the case $\gamma \geq 1$, we found as well a similar situation in Chapter 3, Sec. 3.3. In the present formulation of the backgammon model there is no free γ-like parameter but this is present in its generalization allowing for different collective modes that we will present in the next section. We, hence, postpone the comparison between the long time relaxation behaviors of the HOSS and the backgammon models and the analysis of the source of the breaking down of the two temperature thermodynamics (see Sec. 4.3.2).

4.3 A model for collective modes: the backgammon model with quenched disorder

The purpose of this section is the study of an exactly solvable model where the dynamical relaxation of the different energy modes can be made explicitly clear. It is called the disordered backgammon model (DB model), since

---

[5]In the density-density case, one has to define two different response functions because of the choice of the Metropolis algorithm (dynamic dependence) and the two outcomes differ as soon as they are off equilibrium. The details of the dynamic rules persist in the long time, α relaxation.

it consists in a generalization of the backgammon model to allow different energies for different states [Leuzzi & Ritort, 2002]. Again, just like its predecessor, the slow relaxation of this model is due to entropic barriers. For the DB model we show the existence of an energy threshold $\epsilon^*(t)$ which separates equilibrated from nonequilibrated modes. With its associated threshold $\epsilon^*(t)$, the model provides a microscopic realization that is reminiscent of some phenomenological models proposed in the past such as the trap model on a tree considered by Bouchaud & Dean [1995]. The advantage in the DB model is that now one can exhaustively investigate the distinct relaxation of each one of the different energy modes.

Let us take N particles which can occupy N boxes, each one labeled by an index r which runs from 1 to N. Suppose now that all particles are distributed among the boxes. A given box r contributes to the Hamiltonian with an energy $-\epsilon_r$ only when it is empty. In this case the total Hamiltonian of the system reads

$$\mathcal{H} = -\sum_{r=1}^{N} \epsilon_r \delta_{n_r,0} \tag{4.74}$$

where $\delta_{i,j}$ is the Kronecker delta and n_r denotes the occupancy or number of particles in box r. The ϵ_r are quenched random variables extracted from a distribution $g(\epsilon)$. We consider the Monte Carlo mean-field dynamics of Sec. 4.1.2 where a particle is randomly chosen in a departure box d and a move to an arrival box a is proposed.[6] The proposed change is accepted according to the Metropolis rule with probability $W(x) = \text{Min}(1, \exp(-\beta x))$ where the energy variation is

$$x = \epsilon_a \delta_{n_a,0} - \epsilon_d \delta_{n_d,1} \tag{4.75}$$

In the dynamics, the total number of particles is conserved so that the occupancies satisfy the condition $\sum_r n_r = N$.

The interesting case corresponds to the situation where $g(\epsilon)$ is only defined for $\epsilon \geq 0$. In this case, the dynamics turns out to be extremely slow at low temperatures, similar to what happens for the original backgammon model. The difference lies in the type of ground state. The ground state of Eq. (4.74) corresponds to the case where all particles occupy a single box, the one with the smallest value of ϵ (ϵ_m), labeled by r_m. The ground state energy is given by

$$U_{GS} = -\sum_{r=1}^{N} \epsilon_r + \epsilon_\text{m} = -\sum_{r \neq r_\text{m}} \epsilon_r \tag{4.76}$$

In contrast to the backgammon model, this ground state is nondegenerate. Since all the ϵ are positive, no other configuration can have a lower energy. If $g(\epsilon)$ is a continuous distribution, instead, the ground state is also unique. It is

---

[6] *Ball-to-box* statistics, i.e., the model belongs to the Ehrenfest calls according to the classification reported in the chapter introduction.

easy to understand that, during the dynamical evolution at zero temperature, all boxes with high values of ϵ become empty quite soon and the dynamics involves boxes with progressively lower values of ϵ. The asymptotic dynamics is then determined by the behavior of the distribution $g(\epsilon)$ in the limit $\epsilon \to 0$. If $g(\epsilon) \sim \epsilon^{\nu}$, for $\epsilon \to 0$, the asymptotic long time properties only depend on ν, as we will show in Sec. 4.3.2. Note that the normalization of the $g(\epsilon)$ imposes $\nu > -1$. This classification includes also the original backgammon model where there is no disorder at all. In that case $g(\epsilon) = \delta(\epsilon - 1)$ and the distribution has a finite gap at $\epsilon = 0$. The behavior corresponding to this singular energy distribution can be eventually obtained from the one with a regular $g(\epsilon)$ in the limit $\nu \to \infty$.

One important aspect of the model, cf. Eq. (4.74) is that, in the presence of disorder, it is not invariant under an arbitrary constant shift of the energy levels. Actually, by changing $\epsilon_r \to \epsilon'_r = \epsilon_r + c$ with $c \geq 0$, the model turns out to be a combination of the model characterized by the distribution $g(\epsilon)$ plus the original backgammon model. After shifting, the new distribution $g(\epsilon' - c)$ has a finite gap (equal to c plus the gap of the original distribution). The new model corresponds again to the $\nu \to \infty$ case and the asymptotic dynamical behavior coincides with that of the backgammon model without quenched disorder. As we will see in Sec. 4.3.2, the present model is characterized by an energy threshold $\epsilon^{\star}$ which drives relaxation to the stationary state. Only when the energy threshold can go to zero are we able to see a different asymptotic behavior from the one already probed. For all models with a finite gap, $\epsilon^{\star}$ cannot be smaller than the gap, hence it asymptotically sticks to the gap and the relaxation behavior of the DB model with a finite gap corresponds to that presented in Sec. 4.1.

For what concerns the dynamics, the most relevant feature of this model is that a description in the framework of an adiabatic approximation turns out to be independent from the type of distribution $g(\epsilon)$ (and hence on ν) despite the fact that the asymptotic long time behavior of the effective temperature and of the internal energy depend on the value of ν.

4.3.1   Observables and equilibrium

Now we focus on the static solution of the model described by the Hamiltonian Eq. (4.74). Like in the original backgammon model, we define the occupation probabilities, P_n, that a box contains n particles,

$$P_n = \frac{1}{N} \sum_{r=1}^{N} \delta_{n_r, n} \qquad (4.77)$$

and the corresponding densities that a box of energy ϵ contains k particles

$$g_n(\epsilon) = \frac{1}{N} \sum_{r=1}^{N} \delta(\epsilon_r - \epsilon)\delta_{n_r,n} \quad k \geq 0 \tag{4.78}$$

$$g(\epsilon) = \frac{1}{N} \sum_{r=1}^{N} \delta(\epsilon_r - \epsilon) \tag{4.79}$$

The P_n and the g_n are related by

$$P_n = \int_0^\infty g_n(\epsilon) \, d\epsilon \quad k \geq 0 \tag{4.80}$$

and the conservation of particles reads

$$\sum_{k=0}^{\infty} P_n = 1, \quad \sum_{k=0}^{\infty} g_n(\epsilon) = g(\epsilon) \tag{4.81}$$

The energy density can be expressed in terms of the density $g_0(\epsilon)$ as

$$u = - \int_0^\infty d\epsilon \, \epsilon \, g_0(\epsilon) \tag{4.82}$$

This set of observables depends on time through the time evolution of the occupancies n_r of all boxes. We now analyze the main equilibrium properties of the model.

The solution of the thermodynamics proceeds similarly as for the case of the original backgammon model. The partition function can be computed in the grand partition ensemble. It reads

$$\mathcal{Z}_{GC} = \sum_{N=0}^{\infty} \mathcal{Z}_C(N) z^N \tag{4.83}$$

where $z = \exp(\beta\mu)$ is the fugacity, μ is the chemical potential and $\mathcal{Z}_C(N)$ stands for the canonical partition function of a system with N particles. The canonical partition function can be written as

$$\mathcal{Z}_C = \sum_{n_r=0}^{N} \frac{N!}{\prod_{r=1}^{N} n_r!} \exp(\beta \sum_{r=1}^{N} \epsilon_r \delta_{n_r,0}) \delta \left(N, \sum_{r=1}^{N} n_r \right) \tag{4.84}$$

where $\delta_{i,j}$ and $\delta(i,j)$ are both the Kronecker delta. Introducing this expression in Eq. (4.83) we can write down $\mathcal{Z}_{GC}$ as an unrestricted sum over all the occupancies n_r

$$\mathcal{Z}_{GC} = N! \sum_{n_r=0}^{\infty} \prod_{r=1}^{N} \frac{z^{n_r}}{n_r!} \exp(\beta\epsilon_r \delta_{n_r,0}) \tag{4.85}$$

The overcounting term $N!$ in the numerator is eliminated as in Eq. (4.4) and the final result is

$$\mathcal{Z}_{GC} = \exp\left\{\sum_{r=1}^{N} \log\left[\sum_{n=0}^{\infty} \frac{z^n}{n!} \exp\left(\beta\epsilon_r \delta_{n,0}\right)\right]\right\} = \exp\sum_{r=1}^{N} \log\left(e^{\beta\epsilon_r} + e^z - 1\right) \tag{4.86}$$

yielding the grand-canonical potential energy per box

$$\frac{\Omega}{N} = -\frac{T}{N}\log(\mathcal{Z}_{GC}) = -T\sum_{r=1}^{N} \log\left(e^{\beta\epsilon_r} + e^z - 1\right) \tag{4.87}$$

$$= -T\int_0^\infty g(\epsilon)\log\left(e^{\beta\epsilon} + e^z - 1\right)\, d\epsilon$$

The Helmholtz free energy is then $F = \Omega + N\mu$. The fugacity z is determined by the conservation condition (4.13) yielding, in the presence of disorder, the closure condition

$$\int_0^\infty \frac{g(\epsilon)}{e^{\beta\epsilon} + e^z - 1}\,d\epsilon = \frac{1}{ze^z} \tag{4.88}$$

generalization of Eq. (4.5). This equation gives the fugacity z as a function of β and, from Eq. (4.87) and its derivatives allow us to build the whole thermodynamics. In particular, the equilibrium expressions for $g_n(\epsilon)$ are

$$\bar{g}_n(\epsilon) = \frac{z^n g(\epsilon)\exp(\beta\epsilon\delta_{n,0})}{n!(e^{\beta\epsilon} + e^z - 1)} \tag{4.89}$$

The corresponding P_n are obtained integrating Eq. (4.89) in ϵ [see Eq.(4.80)]. Using the closure relation Eq. (4.88), the integration leads to the expression

$$P_n(z) = \delta_{n,0}(1 - \frac{e^z - 1}{ze^z}) + (1 - \delta_{n,0})\frac{z^{n-1}}{n!e^z} \tag{4.90}$$

i.e., Eq. (4.8). Starting from Eq. (4.82) the equilibrium energy density is obtained as

$$\bar{u} = -\int_0^\infty d\epsilon \frac{\epsilon\, g(\epsilon)e^{\beta\epsilon}}{e^{\beta\epsilon} + e^z - 1} \tag{4.91}$$

All these expressions can be evaluated at finite temperature. Note that, although the value of $P_n(z)$ in Eq. (4.90) is independent of the disorder distribution $g(\epsilon)$, it directly depends on that distribution through the equilibrium value of z [which obviously depends on the $g(\epsilon)$].

Thermodynamics at low temperature

Of particular interest for the dynamical behavior of the model are the low-temperature properties. A perturbative expansion can be carried out close to $T \to 0$ to find the leading behavior of different thermodynamic quantities. Let us start analyzing the closure condition (4.88). Imposing the transformation $s = \beta\epsilon$, Eq. (4.88) can be rewritten as

$$T \int_0^\infty \frac{g(Ts)}{e^s + e^z - 1} ds = \frac{1}{ze^z} \tag{4.92}$$

In the limit $T \to 0$ the fugacity z depends on the behavior of $g(Ts)$, i.e., on the behavior of $g(\epsilon)$ for $\epsilon \to 0$. Assuming $g(\epsilon) \sim \epsilon^\nu$ for $\epsilon \ll 1$ we define the function $c(\epsilon)$ through the relation $g(\epsilon) = \epsilon^\nu c(\epsilon)$, where $c(\epsilon)$ is a smooth function of ϵ with a finite $c(0)$. The integral can be expanded around $T = 0$ by taking successive derivatives of the function c:

$$ze^z T^{\nu+1} \int_0^\infty \frac{s^\nu}{e^s + e^z - 1} \sum_{n=0}^\infty \frac{c^{(n)}(0)(Ts)^n}{n!} ds = 1 \tag{4.93}$$

Using the asymptotic result $z \to \infty$, as $T \to 0$, everything reduces to estimate the following integral in the large z limit:

$$\int_0^\infty ds \, \frac{s^{\nu+n}}{e^s + e^z - 1} \sim z^{\nu+n+1} e^{-z} \tag{4.94}$$

The term $n = 0$ in the series yields the leading behavior for z, which turns out to be

$$z \sim \beta^{\frac{\nu+1}{\nu+2}} \tag{4.95}$$

In a similar way the energy can be computed to leading order in T,

$$u = u_{GS} + aT + \mathcal{O}(T^2) \tag{4.96}$$

providing a finite specific heat at low temperatures. Notice that the above expression does not depend on ν and, hence, on $g(\epsilon)$. In Sec. 4.3.4 we will show explicitly such a behavior for two specific DB models, one with $\nu = 1$ [defined in Eq. (4.115)] and the other with $\nu = 0$ [Eq. (4.116)]. Solving Eq. (4.88) numerically for $z(T)$ in each specific model and inserting $z(T)$ in the expression (4.91) for the equilibrium energy density, we get the energy dependence on the temperature. In all cases, at equilibrium, the energy is linear for very low T as predicted in Eq. (4.96). It yields, therefore, a finite specific heat as in any classical model with Maxwell-Boltzmann statistics (see Figs. 4.7-4.8).

4.3.2   Dynamics of the disordered backgammon model

Here we consider the dynamical equations for the occupation probabilities P_n and their associated densities $g_n(\epsilon)$. The dynamical equations in this model are derived in a similar way as for the standard backgammon model (Sec. 4.1.2) [Franz & Ritort, 1996]. The main difference is that in the DB model the equations for the occupancy probabilities P_n do not generate a

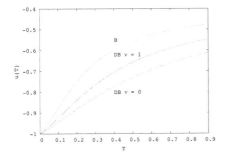 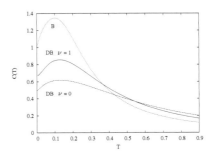

FIGURE 4.7

Energy vs. T at equilibrium for two DB models ($\nu = 1$ and $\nu = 0$), and for the standard backgammon model, "B" ($\nu \to \infty$). Reprinted figure with permission from [Leuzzi & Ritort, 2002]. Copyright (2002) by the American Physical Society.

FIGURE 4.8

Specific heat for DB models with $\nu = 0, 1$ and for the backgammon model ("B"). For all cases (disordered or not) the specific heat turns out to be finite at $T = 0$. Reprinted figure with permission from [Leuzzi & Ritort, 2002].

closed hierarchy of equations. Only for $T = 0$ is such a closed hierarchy obtained. As we will see later, this has important consequences to probe the zero temperature relaxation.

A hierarchy of equations can only be obtained at the level of the occupation probability densities $g_n(\epsilon)$. A detailed derivation of these equations is reported in Appendix 4.A. Here we show the final outcome,

$$\frac{\partial g_0(\epsilon)}{\partial t} = g_1(\epsilon) \left[1 + \int_\epsilon^\infty d\epsilon' g_0(\epsilon') \left(e^{-\beta(\epsilon'-\epsilon)} - 1 \right) \right] \tag{4.97}$$
$$- g_0(\epsilon) \left[e^{-\beta\epsilon} + P_1 \left(1 - e^{-\beta\epsilon} \right) + \int_0^\epsilon d\epsilon' g_1(\epsilon') \left(e^{-\beta(\epsilon-\epsilon')} - 1 \right) \right] \quad n = 0$$

$$\frac{\partial g_1(\epsilon)}{\partial t} = 2g_2(\epsilon) \left(1 + \int_0^\infty d\epsilon g_0(\epsilon) e^{-\beta\epsilon} - P_0 \right) \tag{4.98}$$
$$- g_1(\epsilon) \left[2 + \int_\epsilon^\infty d\epsilon' g_0(\epsilon') \left(e^{-\beta(\epsilon'-\epsilon)} - 1 \right) \right]$$
$$+ g_0(\epsilon) \left[e^{-\beta\epsilon} + P_1 \left(1 - e^{-\beta\epsilon} \right) + \int_0^\epsilon d\epsilon' g_1(\epsilon') \left(e^{-\beta(\epsilon-\epsilon')} - 1 \right) \right] \quad n = 1$$

$$\frac{\partial g_n(\epsilon)}{\partial t} = (n+1)g_{n+1}(\epsilon) \left(1 + \int_0^\infty d\epsilon g_0(\epsilon) e^{-\beta\epsilon} - P_0 \right) \tag{4.99}$$
$$- g_n(\epsilon) \left[1 + n + n \left(\int_0^\infty d\epsilon g_0(\epsilon) e^{-\beta\epsilon} - P_0 \right) \right] + g_{n-1}(\epsilon) \quad n > 1$$

The equations for the P_n are directly obtained by integrating the $g_n(\epsilon)$ according to (4.80). They are

$$\frac{dP_n(t)}{dt} = (n+1)\left[P_{n+1}(t) - P_n(t)\right] + P_{n-1}$$
(4.100)

$$+ \left(\int_0^\infty d\epsilon g_0(\epsilon)e^{-\beta\epsilon} - P_0\right)\left[\delta_{n,1} - \delta_{n,0} - nP_n(t) + (n+1)P_{n+1}(t)\right] \quad \forall n$$

with $P_{-1} = 0$. It is easy to check that the equilibrium solutions Eq. (4.89) are indeed stationary solutions. As previously said, for $T > 0$, the equations for the P_n do not generate a hierarchy by themselves but depend on the $g_n(\epsilon)$ through the distribution $g_0(\epsilon)$ in Eq. (4.100).[7] Nevertheless, a remarkable aspect is that they generate a well-defined hierarchy at $T = 0$ which coincides with the equations of the original backgammon model Eq. (4.14) with $T = 0$.

It is easy to understand why at $T = 0$ the dynamical equations are independent of the density of states $g(\epsilon)$. For $T = 0$, all moves of particles between departure and arrival boxes with different energies ϵ_d and ϵ_a depend on the precise values of these energies only when the departure box contains a single particle and the arrival box is empty, but such a move does not lead to any change in any of the P_n. The dynamical equations for the P_n remain, therefore, independent of $g(\epsilon)$. Obviously, this does not hold for other observables such as the energy u [cf. Eq. (4.91)] and higher moments of $g_n(\epsilon)$.

The analysis of the dynamical equations at $T = 0$ decomposes into two parts. On the one hand, the equations for the P_n coincide with those of the original non-disordered backgammon model [Eq. (4.14) in the limit $\beta \to \infty$]. Consequently, the same adiabatic approximation used for the P_n in the original backgammon model is still valid for the DB model. On the other hand, in order to analyze the behavior of the energy one must analyze the behavior of the hierarchy of equations for the $g_n(\epsilon)$ which is quite complicated. We will then approach the analysis of these equations within the framework of a *generalized* adiabatic approximation.

The key idea behind the adiabatic approximation is that, while P_0 constitutes a slow mode, the other P_n with $n > 0$ are fast modes. Hence, relying once again on the separation of timescales assumption, they can be considered as if they were in equilibrium at the hypersurface in phase space $P_0 = \text{constant}$, this constant being given by the actual value of P_0 at time t. In the original backgammon model P_0 is related to the internal energy, $P_0 = -u$, hence thermalization of the fast modes P_n ($n > 0$) occurs on the hypersurface of constant energy. For the DB model this is not true: the hypersurface where equilibration of fast modes occurs does not coincide with the constant energy hypersurface simply because the energy and P_0 are different quantities (we will see that even their asymptotic time behavior is different).

---

[7]Unless $g(\epsilon) = \delta(\epsilon - 1)$. In that case $\int_0^\infty d\epsilon g_0(\epsilon)e^{-\beta\epsilon} = P_0 e^{-\beta}$ and the backgammon equations of motion are constructed.

At $T = 0$ the equation for P_0 [Eq. (4.100) for $n = 0$] is still Eq. (4.18) $\partial P_0/\partial t = P_1(1 - P_0)$. Formally, the same manipulation can be carried out in terms of the time-dependent fugacity $z^\star$ as in Eqs. (4.35, 4.38,4.41). The occupation probabilities are, then, given by

$$P_n \simeq \delta_{n,0}\left(1 - \frac{1}{z^\star}\right) + (1 - \delta_{n,0})\frac{(z^\star)^{n-1}}{n!\exp(z^\star)} \qquad (4.101)$$

as long as $z^\star \gg 1$. That is, for the energy density, $u - u_{GS} = 1/z^\star$.

If local equilibrium is reached on the hypersurface of constant P_0 we can relate P_1 to P_0 using Eq. (4.101). The simplest way of dealing with Eq. (4.18) is to write both P_1 and P_0 in terms of the time-dependent fugacity $z^\star$ writing down a dynamical equation for $z^\star$, Eq. (4.38), whose solution is Eq. (4.41)

What is different now is the time behavior of the effective inverse temperature that [cf. Eq. (4.95)], displays the following time dependency for very long times:

$$\beta^\star \sim (z^\star)^{\frac{\nu+2}{\nu+1}} \sim (\log t)^{\frac{\nu+2}{\nu+1}} \qquad (4.102)$$

The effective temperature depends on the properties of the disorder distribution $g(\epsilon)$ in the limit $\epsilon \to 0$ through the value of the exponent ν. Clearly, when the density of levels decreases as we approach $\epsilon = 0$, the relaxation turns out to be slower; the limiting case being the original backgammon model for which $\nu \to \infty$, cf. Eq. (4.43). In the other limit $\nu \to -1$, when disorder becomes unnormalized, the inverse effective temperature diverges very fast. Already from Eq. (4.96) one can anticipate that the same asymptotic behavior holds for the energy [see Eq. (4.111)]. Hence, ν interpolates between fast relaxation ($\nu = -1$) and very slow relaxation ($\nu = \infty$). A relaxation slower than logarithmic is not possible in the present model.

A comparison can be made with the qualitatively similar effective temperature time dependence in the HOSS model of Chapter 3. There, at $T_0 = 0$, [8] the time behavior of the effective temperature obtained in a number of different approaches, cf. Eqs. (3.130, 3.141, 3.146, 3.158), was

$$T_e \simeq \tilde{K}\mu_2(t) \simeq \frac{KD}{D + J^2}\frac{1}{(\log t)^{1/\gamma}}$$

where we used Eq. (3.74) for the static value of the generalized spring constant and Eq. (3.119) for the dynamic evolution of μ_2 in the aging regime. The two models have the same long time behavior provided the identification

$$\gamma = \frac{\nu + 1}{\nu + 2} \qquad (4.103)$$

---

[8] That is when the configurational constraint is absent, $m_0 = 0$ and the HOSS model represent a strong glass. The same argument holds, however, qualitatively also in the fragile glass case, at $T = T_0$.

is put forward. We notice that the admissible interval of ν values goes from -1 (fast relaxation) to ∞ (i.e., a model equivalent to the original backgammon model). The relative values of the HOSS γ exponent are then $\gamma \in [0 : 1]$. We stress that this is the interval of values of the exponent γ for which no unique effective temperature can encode the off-equilibrium dynamics in the aging regime of the HOSS model. In Sec. 4.2, we actually saw that in the backgammon model ($\gamma = 1$) the differently defined effective temperatures did not coincide, as well.

Generalized adiabatic solution for the $g_n(\epsilon)$

Equations (4.97)-(4.99) for the $g_n(\epsilon)$ at $T = 0$ become

$$\frac{\partial g_0(\epsilon)}{\partial t} = g_1(\epsilon) \left[1 - \int_\epsilon^\infty d\epsilon' g_0(\epsilon') \right] - g_0(\epsilon) \int_\epsilon^\infty d\epsilon' g_1(\epsilon') \qquad (4.104)$$

$$\frac{\partial g_1(\epsilon)}{\partial t} = 2g_2(\epsilon)(1 - P_0) - g_1(\epsilon) \left[2 - \int_\epsilon^\infty d\epsilon' g_0(\epsilon') \right] + g_0(\epsilon) \int_\epsilon^\infty d\epsilon' g_1(\epsilon')$$

$$(4.105)$$

$$\frac{\partial g_n(\epsilon)}{\partial t} = (n+1)g_{n+1}(\epsilon)(1 - P_0) - g_n(\epsilon)\left[1 + n(1 - P_0)\right] + g_{n-1}(\epsilon) \quad n > 1$$

$$(4.106)$$

To solve the dynamical equations for the $g_n(\epsilon)$ in the adiabatic approximation we note that, contrary to the global quantities P_n, they cannot be equilibrated among all different modes. The reason is that, due to the entropic character of the relaxation, very low energy modes are rarely involved, because the time needed to empty one further box increases progressively with time and therefore they cannot be considered effectively thermalized.

Note that in the original backgammon model all boxes have the same energy, hence there is a unique class of modes. For the general disordered model we, instead, expect the existence of a time-dependent energy scale ϵ^* separating equilibrated from nonequilibrated modes. The mechanism of relaxation is the one proceeding through collective modes (see the introduction of the chapter and Fig. 4.1) of different energy: the lower the energy the longer the lifetime of that mode and the slower the relaxation. At zero temperature there is no thermal activation and the equilibrated modes are in the sector $\epsilon \gg \epsilon^*$ while the nonequilibrated modes are in the other sector $\epsilon \ll \epsilon^*$. [9]

The value of ϵ^* can be easily guessed making use of a simple microcanonical argument for relaxation rate, following exactly the steps of Sec. 4.1.3, with the difference that now, in the presence of modes of different energy, the structural relaxation is led by the energy scale ϵ^*. Indeed, if the threshold ϵ^* plays the role of an energy barrier and T^* accounts for the effective

---

[9]The same separation of modes, in thermalized and nonthermalized modes, has been implemented in the disordered version of the HO model that we presented in Sec. 3.7, see the work of Garriga & Ritort [2005].

thermal activation due to entropic effects, we obtain, for the typical relaxation time, $\tau_{eq} \simeq \exp(\beta^\star \epsilon^\star)$. This expression is only valid to the leading order. As we can see in Appendix 4.C [Eq. (4.C.9)], or later here, from Eq. (4.112), there are sub-leading corrections to this expression arising for the fact that the relaxation time is more properly described by the expression $\tau_{eq} \simeq \exp(\beta^\star \epsilon^\star)/(\beta^\star \epsilon^\star)$ [see Eq. (4.112)]. Hence, at a given timescale t (i.e., the time elapsed since the system was quenched) all modes where $\tau_{eq} \ll t$ are equilibrated at the temperature of the thermal bath, that in this case is zero, and therefore are frozen. Modes with $\tau_{eq} \gg t$, although dynamically evolving, are also blocked because the barriers (in this case entropic barriers) are too high to allow for relaxation within the timescale t.

Only those modes whose characteristic time is $\tau_{eq} \sim t$ are relaxing at a given timescale t. We get for the time-dependent energy scale $\epsilon^\star$ and the effective temperature the relation $\epsilon^\star \sim T^\star \log t$ and, using Eq. (4.102), this yields the leading behavior

$$\epsilon^\star \sim (\log t)^{-\frac{1}{\nu+1}} \tag{4.107}$$

According to this, one can impose the following Ansatz solution for the $g_n(\epsilon)$. If $\bar{g}_n$ stands for the equilibrium density at $T = 0$ [i.e., according to Eq. (4.89), $\bar{g}_n = g(\epsilon)\, \delta_{n,0}$] we have,

$$\Delta g_n(\epsilon) \equiv g_n(\epsilon) - \bar{g}_n(\epsilon) = \frac{\Delta P_n}{\epsilon^\star} \hat{g}_n\left(\frac{\epsilon}{\epsilon^\star}\right) \tag{4.108}$$

where $\Delta P_n \equiv P_n - P_n^{eq} = P_n - \delta_{n,0}$ and $\hat{g}_n(x)$ decays pretty fast to zero for $x > 1$ and the condition $\int_0^\infty dx\, \hat{g}_n(x) = 1$ is imposed on the scaling function $\hat{g}_n$. The prefactor $\Delta P_n/\epsilon^\star$ is introduced to fulfill condition (4.80).

This expression tells us the following: above $\epsilon^\star$ the $g_n(\epsilon)$ have relaxed to their corresponding equilibrium distributions at the temperature of the bath. On the other hand, in the sector of the energy spectrum where $\epsilon < \epsilon^\star$, the densities g_n are still relaxing [especially in the region $\epsilon/\epsilon^\star \sim \mathcal{O}(1)$]. Since the relaxation is driven by the shift in time of the threshold energy $\epsilon^\star$, the proposed Ansatz scaling solution seems quite reasonable.

In Appendix 4.B we show how this Ansatz closes the set of equations (4.104) reproducing also the leading asymptotic behavior for $\epsilon^\star$ and $z^\star$:

$$\epsilon^\star \simeq \frac{1}{(\log t)^{\frac{1}{\nu+1}}}$$

$$z^\star \simeq \frac{1}{(\epsilon^\star)^{1+\nu}} \simeq \log t \tag{4.109}$$

For later use we define the following function

$$G_n(\epsilon) \equiv \frac{\Delta g_n(\epsilon)\, \epsilon^\star}{\Delta P_n} = \hat{g}_n\left(\frac{\epsilon}{\epsilon^\star}\right) \tag{4.110}$$

which scales as a function of $\epsilon/\epsilon^\star$. The scaling relation (4.108) yields the leading asymptotic behavior of all observables different from the occupation probabilities P_n. For instance, the energy is given by $u = -\int_0^\infty d\epsilon\, \epsilon\, g_0(\epsilon)$; using the scaling relation (4.108) and the asymptotic expression (4.108) we get for the leading term

$$u - u_{GS} \sim -\int_0^\infty d\epsilon\, [g_0(\epsilon) - g(\epsilon)]\, \epsilon \sim (\epsilon^\star)^{\nu+2} \sim T^\star \simeq \frac{1}{\log t^{\frac{\nu+2}{\nu+1}}} \qquad (4.111)$$

Note that the asymptotic scaling behavior of the energy is the same as for the effective temperature $T^\star$ [cf. Eq. (4.96)], in agreement with the quasi-equilibrium hypothesis: $u - u_{GS} = aT^\star$.

An important result is that the threshold $\epsilon^\star$ decays slower to zero than the effective temperature. A case where this difference can be clearly appreciated corresponds to the case where the density of states vanishes exponentially fast $g(\epsilon) \sim \exp(-A/\epsilon)$. In this case $\epsilon^\star$ decays slower than logarithmically, namely like $1/\log(\log t)$ (see Appendix 4.B for details).

4.3.3   Relaxational spectrum in equilibrium

One of the crucial characteristics behind the applicability of the adiabatic approximation is that the long time behavior at zero temperature has to display a correspondence with the low temperature relaxational properties of the equilibrium state.

To analyze the spectrum of relaxation times $\tau_{eq}(\epsilon)$ to equilibrium we expand up to the first order in the perturbation theory the dynamical equations for the $g_n(\epsilon)$ around their equilibrium solutions $\bar{g}_n(\epsilon)$. Using the expansion $g_n(\epsilon) = \bar{g}_n(\epsilon) + \delta g_n(\epsilon)$ we get a set of equations for the variations $\delta g_n(\epsilon)$. These are shown in Appendix 4.C.

A complete derivation of the relaxation time $\tau_{eq}(\epsilon)$ in equilibrium is complicated. But it is easy to see that, as $T \to 0$, the relaxation time is asymptotically strongly peaked around the threshold energy $\epsilon^\star$. For $\epsilon \gg \epsilon^\star$ the relaxation time is small because the population of high energy boxes in equilibrium is rather small. On the other hand, for $\epsilon/\epsilon^\star \ll 1$ the relaxation is estimated to be finite and independent of T.[10] Starting from Eqs. (4.C.1)-(4.C.3) for $\delta g_0(\epsilon)$ and $\delta g_1(\epsilon)$ and making use of the adiabatic Ansatz Eq. (4.108), for $\epsilon \simeq \bar{\epsilon}^\star$ one has

$$\tau_{eq}(\bar{\epsilon}^\star) \sim \frac{e^{\beta \bar{\epsilon}^\star}}{\beta \bar{\epsilon}^\star} \qquad (4.112)$$

where $\bar{\epsilon}^\star(T) \sim T^{1/(2+\nu)}$ is the asymptotic temperature dependence of the threshold energy at low temperature. For the temperature dependence on the relaxation time this yields

$$\tau_{eq}(T) \sim T^\gamma \exp\left[\frac{1}{T^\gamma}\right] \qquad (4.113)$$

---

[10]This result is derived in the aforementioned Appendix 4.C where we show that the maximum relaxation time occurs for ϵ around $\epsilon^\star$.

where $\gamma(\nu)$ is given by Eq. (4.103). The above formula shows that there is activated behavior as a function of the temperature but with a relaxation time increasing slower than Arrhenius as $T \to 0$. Note that for the standard backgammon model corresponding to $\gamma = 1$ we obtained an Arrhenius behavior [cf. Eq. (4.44)] and in the opposite limit, $\gamma \to 0$, the relaxation time is finite at any temperature. This generalized Arrhenius law is the one that we were finding in Chapter 3 for the HOSS model in the strong glass case.[11] Given the interval of possible ν values we are in the presence of sub-Arrhenius relaxation, that is, even though exponential, the increase of the relaxation time, as temperature is increased, is slower than the typical increase for strong glasses ($\gamma = 1$). We will analyze in the next section two models whose distribution behavior at small energies is led, respectively, by $\nu = 0$ ($\gamma = 1/2$) and $\nu = 1$ ($\gamma = 2/3$).

4.3.4 Specific examples of continuous energy distribution

Here we report a numerical test of the main results obtained in the previous sections. In particular, we show the existence of the threshold energy $\epsilon^\star$ separating equilibrated from nonequilibrated energy modes. We show the comparison among two models each characterized by a different distribution of the disorder [Fig. 4.9] and the original backgammon model. The temperature is zero in all cases. All three distributions were chosen to satisfy the conditions

$$\int_0^\infty d\epsilon \, g(\epsilon) = \int_0^\infty d\epsilon \, \epsilon \, g(\epsilon) = 1 \tag{4.114}$$

so that that the ground state has energy $u_{GS} = -1$ in the limit $N \to \infty$ for all three cases. The models are the following ones:

- **Case A:** *Non-disordered model with a gap* [Fig. 4.9 (left)]. This is the original backgammon model (cf. Sec. 4.1) where $g(\epsilon) = \delta(\epsilon - 1)$. Therefore $\epsilon^\star = 1$ and the threshold energy is time independent; this case corresponds to $\nu \to \infty$. For very large times the energy is expected to decay like $u + 1 \sim T^\star \sim 1/\log t$. The same behavior is expected for any disorder distribution $g(\epsilon)$ with a finite gap.

- **Case B:** *Disordered model without gap and $g(0) = 0$* [Fig. 4.9 (center)]. We consider the distribution

$$g(\epsilon) = \frac{\pi}{2} \epsilon \exp\left(-\frac{\pi}{4}\epsilon^2\right) \tag{4.115}$$

This case corresponds to $\nu = 1$. The energy threshold $\epsilon^\star$ scales like $1/\sqrt{\log t}$ and the effective temperature and the energy scale like $u + 1 \sim T^\star \sim 1/(\log t)^{\frac{3}{2}}$.

---

[11] That is, assuming that the configurational constraint was absent: $m_0 = T_0 = 0$, cf. Eq. (3.124)

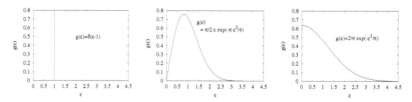

FIGURE 4.9

Probability distribution of the energy weights of the model. (left) The standard backgammon model has no disordered distribution, all boxes have the same weight. (center) The probability distribution function of a DB with $\nu = 1$: at low energy the density goes to zero. (right) DB model with $\nu = 0$: the probability of having boxes with energies arbitrarily close to zero is finite. Reprinted figure with permission from [Leuzzi & Ritort, 2002].

- **Case C:** *Disordered model without gap and finite $g(0)$* [Fig. 4.9 (right)]. We consider the distribution

$$g(\epsilon) = \frac{2}{\pi} \exp\left(-\frac{\epsilon^2}{\pi}\right) \qquad (4.116)$$

This case corresponds to $\nu = 0$. The energy threshold $\epsilon^\star$ scales like $1/\log t$ and the effective temperature and the energy scale like $u + 1 \sim T^\star \sim 1/(\log t)^2$.

In Fig. 4.10 the decay of the energy for all three models is plotted. The reproduced data come from simulations performed for $N = 10^4, 10^5, 10^6$ particles [Leuzzi & Ritort, 2002] (the number of particles is identical to the number of states) showing that finite-size effects are not big in the asymptotic regime. We show data for one sample and $N = 10^6$. We plot the energy as function of time starting from a random initial condition [particles randomly distributed among states: $u(t = 0) = -1/e$]. Relaxation is faster for Case C and slower for the standard backgammon model (Case A).

The different asymptotic behaviors are shown in Fig. 4.11. There we plot $[u(t) - u_{GS}](\log t)^{1/\gamma}$ with γ defined in Eq. (4.103). To avoid finite-size corrections when the energy is close to its ground state the quantity computed is, cf. Eq. (4.76), $u_{GS} = \frac{1}{N}(-\sum_{r=1}^{N} \epsilon_r + \epsilon_m)$ where ϵ_m is the minimum value among the randomly extracted ϵs. The different curves saturate at finite values, corresponding to their asymptotic leading constant. Note that convergence is slow, showing the presence of sub-leading logarithmic corrections to the leading behavior.[12]

---

[12]The distribution probabilities were numerically computed by binning the ϵ axis from $\epsilon = 0$ up to $\epsilon = \epsilon_{max}$ where ϵ_{max} is the maximum value of ϵ_r among all the N boxes. One hundred bins are enough to see the behavior of the time evolution of the different distributions [Leuzzi & Ritort, 2002].

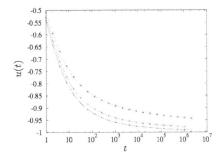

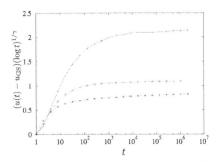

FIGURE 4.10

Energy as a function of time for cases A, B and C (see text). The lower curve represents the $\nu = 0$ DB model (C), the middle curve the $\nu = 1$ DB model (B) and the upper curve the standard backgammon model (A). Reprinted figure with permission from [Leuzzi & Ritort, 2002].

FIGURE 4.11

Rescaled energy $(u - u_{GS})(\log t)^{1/\gamma}$ vs. time, with $\gamma \equiv \frac{\nu+1}{\nu+2}$, for the three different model cases of Fig. 4.9. The upper curve refers to Case C, the middle one to Case B and the lower one to the standard backgammon model (Case A). The slowest relaxation is the one of the model with $\nu = 0$. Reprinted figure with permission from [Leuzzi & Ritort, 2002].

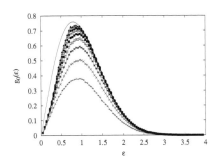

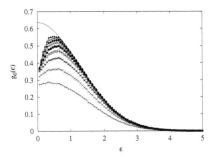

FIGURE 4.12

Distribution $g_0(\epsilon)$ for Case B at times 2^k with $k = 4, 6, 8, 10, 12, 14, 16, 18, 20$ (from bottom to top). The continuous line represents $g(\epsilon)$ given by Eq. (4.115). Reprinted figure with permission from [Leuzzi & Ritort, 2002].

FIGURE 4.13

Distribution $g_0(\epsilon)$ for Case C for times 2^k with $k = 4, 6, 8, 10, 12, 14, 16, 18, 20$ (bottom to top). The continuous line represents $g(\epsilon)$, given in Eq. (4.116). Reprinted figure with permission from [Leuzzi & Ritort, 2002].

In Figs. 4.12 and 4.13 we show the $g_0(\epsilon)$ for cases B and C, respectively. Note that the $g_0(\epsilon)$ converge to the asymptotic result $g(\epsilon)$ for $\epsilon > \epsilon^\star$, in agreement with the adiabatic solution (4.108) while they are clearly different for $\epsilon < \epsilon^\star$. The value of $\epsilon^\star$ where $g_0(\epsilon)$ deviates from the asymptotic curve $g(\epsilon)$

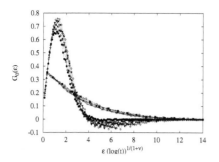

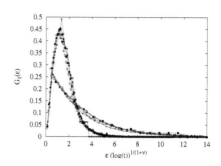

FIGURE 4.14

Distribution $G_0(\epsilon)$ vs. $\epsilon\sqrt{\log t}$ for Case B (peaked datasets) and vs. $\epsilon \log t$ for Case C (smooth datasets). Times are $t = 2^k$ with $k = 6, 8, 10, 12, 14, 16$. The different curves superimpose, verifying Eq. (4.118). Reprinted figure with permission from [Leuzzi & Ritort, 2002].

FIGURE 4.15

Distribution $G_1(\epsilon)$ as a function of $\epsilon\sqrt{\log t}$ for Case B (peaked datasets) and as a function of $\epsilon \log t$ for Case C (smooth datasets). Times are $t = 2^k$ with $k = 6, 8, 10, 12, 14$. Reprinted figure with permission from [Leuzzi & Ritort, 2002].

shifts slowly to zero [like $1/(\log t)^{\frac{1}{2}}$ or $1/\log t$ for Cases B and C, respectively], as can be seen in Figs. 4.12, 4.13. Other probability densities (for instance g_1) decay very fast to zero (already for $t = 2^{17}$ there are no occupied boxes with more than one particle).

In Figs. 4.14 and 4.15 one can observe how the adiabatic Ansatz, Eqs. (4.108), (4.110), for the densities g_0 and g_1 in the two model cases B and C, is verified. Fig. 4.14 plots $G_0(\epsilon)$ for both models. Fig. 4.15 plots $G_1(\epsilon)$ for both models. Using Eq. (4.107) together with $z^\star = \log t + \log(\log t)$ yields

$$G_0(\epsilon) = \Delta g_0(\epsilon) \frac{\log t + \log(\log t)}{\log t^{\frac{1}{\nu+1}}} = \hat{g}_0\left(\frac{\epsilon}{\epsilon^\star}\right) \qquad (4.117)$$

$$G_1(\epsilon) = \Delta g_1(\epsilon) \frac{t}{(\log t)^{\frac{\nu}{\nu+1}}} = \hat{g}_1\left(\frac{\epsilon}{\epsilon^\star}\right) \qquad (4.118)$$

where the scaling functions $\hat{g}_n(x)$ are defined in Eq. (4.108). Looking at Figs. 4.14, 4.15 one can observe that the scaling is pretty well satisfied and that the $\hat{g}_n(x)$ indeed vanishes for $x \simeq 1$ yielding an estimate for $\epsilon^\star$ in both cases. The threshold is $\epsilon^\star \simeq 6/\sqrt{\log t}$ for Case B and $\epsilon^\star \simeq 12/\log t$ for Case C. Note also that the quality of the collapse of the G_0 is slightly worse for Case B than for Case C (see Fig. 4.14). This is due to the stronger sub-leading corrections to the shift of $\epsilon^\star$ which decays slower to zero for case B. Hence, the asymptotic regime is reached only for later times. Indeed, as Fig. 4.12 shows, the value of $\epsilon^\star$ obtained within the timescales considered has not yet reached the maximum of the distribution $g(\epsilon)$, so that the asymptotic behavior $g(\epsilon^\star) \sim \epsilon^\star$ is still far

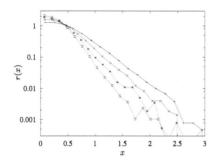

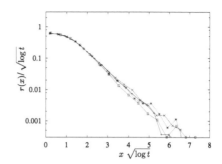

FIGURE 4.16

The rate of accepted changes $r(x)$ versus the energy variation x for different times $t = 10^2, 10^3, 10^4, 10^5$ (from top to bottom) computed as explained in the text. Reprinted figure with permission from [Leuzzi & Ritort, 2002].

FIGURE 4.17

Scaling plot for $r(x)/\sqrt{\log t}$ versus $x\sqrt{\log t}$ for different times $t = 10^2, 10^3, 10^4, 10^5$. Reprinted figure with permission from [Leuzzi & Ritort, 2002].

away. Yet, it is remarkable how well the scaling *Ansatz* of Eqs. (4.108, 4.110) makes the numerical data of Figs. 4.14-4.15 collapse.

4.3.5 A method to determine the threshold energy scale

Is there a general method to determine the energy scale $\epsilon^\star$ without having any precise information about the adiabatic modes present in the system?

In the previous sections we addressed this question by proposing an adiabatic scaling Ansatz to the dynamical equations. Here we propose a general method to determine the energy scale $\epsilon^\star$ from first principles without the necessity of knowing the nature of the slow modes present in the system. Obviously for models such as the standard backgammon model this energy scale plays no role since we know from the beginning that relaxation takes place on a single energy scale.

Consider the following quantity $r(x)$ defined as the rate with which a first accepted energy change x occurs at time t. Let us consider the case of zero temperature where this probability density is defined only for $x \leq 0$. For the $T > 0$ case, the interested reader can look at Appendix 4.D. The distribution $p(x)$ denotes the probability of proposing an energy change at time t (the move is not necessarily accepted), and it is proportional to r:

$$r(x) = \frac{p(x)}{A_0} \quad \text{if } x \leq 0 \tag{4.119}$$

where $A_0 = \int_{-\infty}^{0} p(x)\,dx$ is the acceptance rate, i.e., the inverse of the characteristic relaxation time to equilibrium. The expression for $p(x)$ [and therefore

$r(x)]$ can be exactly computed. Note that computing $p(x)$ yields all information about the statistics of energy changes, in particular the evolution equation for the energy.[13] On the contrary, the time evolution for the energy does not necessarily yield the distribution $p(x)$. For the DB it can be exactly derived (cf. Appendix 4.D). Here we quote the result,

$$p(x) = (1 - P_0)\theta(-x)g_1(-x) \qquad (4.120)$$
$$A_0 = 1 - P_0 \qquad (4.121)$$

Using the scaling Ansatz of Eq. (4.108) for g_1 we obtain the simple scaling relation,

$$r(x) = \frac{P_1}{\epsilon^*}\hat{g}_1\left(\frac{-x}{\epsilon^*}\right) = \frac{1}{\epsilon^*}\hat{P}\left(\frac{x}{\epsilon^*}\right), \quad \text{with } x < 0 \qquad (4.122)$$

A collapse of different $r(x)$ for different times can be used to determine the time evolution of ϵ^*. In Fig. 4.16 we show the scaling of $r(x)$ for the model B for $N = 10^4$ and different times $t = 10^2, 10^3, 10^4, 10^5$. Starting from a random initial configuration, statistics has been collected over approximately $30,000$ jumps for every time. In Fig. 4.17 we check the scaling relation (4.122) plotting $r(x)\epsilon^*$ as a function of x/ϵ^* where we have taken $\epsilon^* \sim 1/\sqrt{\log t}$. Note also that the range where $r(x)$ is finite corresponds to the region where $\epsilon \sim \epsilon^*$. In Fig. 4.17 this corresponds to $\epsilon^* \simeq 6/\sqrt{\log t}$ in agreement with what was observed in Figs. 4.14, 4.15.

The scaling works pretty well, showing how this method could be used to guess the time evolution of the energy threshold ϵ^* in general glass models in those cases where different normal modes take place.

---

[13] Actually, in equilibrium at finite temperature $p(x)$ satisfies detailed balance $p(x) = p(-x)\exp(-\beta x)$.

4.A Occupation probability density equations

In this appendix we derive the equations of motion for the occupation probability densities for box-energy between ϵ and $\epsilon + d\epsilon$. First we start from the densities of having zero particles in a box of energy ϵ.

In Table 4.1 we list the processes contributing to the evolution of the occupation probability density of boxes containing zero particles. In the left column we show the processes involved in terms of occupation numbers of the departure box and of the arrival box. In the right column we write the corresponding contribution of a given process to the variation of the occupation density, $\Delta g_0(\epsilon)$.

The particle for which a jump is proposed is chosen in box d with probability n_d/N. The arrival box is chosen with uniform probability $1/N$. The total difference per particle in the probability density of empty boxes of energy ϵ is then

$$\Delta g_0(\epsilon) = \frac{1}{N} \sum_{p=0}^{N} \sum_{a=0}^{N} \frac{n_p}{N} \frac{1}{N} \{ \delta_{n_d,1} \delta_{n_a,0} \left[\delta(\epsilon - \epsilon_d) - \delta(\epsilon - \epsilon_a) \right]$$
$$\times \left[1 + \left(e^{-\beta(\epsilon_a - \epsilon_d)} - 1 \right) \theta(\epsilon_a - \epsilon_d) \right]$$
$$+ \delta_{n_d,1}(1 - \delta_{n_a,0})\delta(\epsilon - \epsilon_d) - (1 - \delta_{n_d,1})\delta_{n_a,0} e^{-\beta\epsilon_a} \delta(\epsilon - \epsilon_a) \}$$

Using Eqs. (4.77-4.80) and the following identities,

$$\frac{1}{N} \sum_{a=0}^{N} \delta_{n_a,0} \theta(\epsilon_a - \epsilon) \left[e^{-\beta(\epsilon_a - \epsilon)} - 1 \right] = \int_{\epsilon}^{\infty} d\epsilon' \, g_0(\epsilon') \left[e^{-\beta(\epsilon' - \epsilon)} - 1 \right]$$

(4.A.2)

$$\frac{1}{N} \sum_{d=0}^{N} n_d \delta_{n_d,1} \theta(\epsilon - \epsilon_d) \left[e^{-\beta(\epsilon - \epsilon_d)} - 1 \right] = \int_{0}^{\epsilon} d\epsilon' \, g_1(\epsilon') \left[e^{-\beta(\epsilon - \epsilon')} - 1 \right]$$

(4.A.3)

$$\frac{1}{N} \sum_{a=0}^{N} \delta_{n_a,0} \delta(\epsilon - \epsilon_a) e^{-\beta\epsilon_a} = g_0(\epsilon) e^{-\beta\epsilon}$$

(4.A.4)

we get the equation of motion for $g_0(\epsilon)$ [namely Eq. (4.97)]:

$$\frac{\partial g_0(\epsilon)}{\partial t} = \lim_{N \to \infty} \frac{\Delta g_0(\epsilon)}{1/N} = g_1(\epsilon) \left[1 + \int_{\epsilon}^{\infty} d\epsilon' g_0(\epsilon') \left(e^{-\beta(\epsilon' - \epsilon)} - 1 \right) \right]$$
$$- g_0(\epsilon) \left[e^{-\beta\epsilon} + P_1 \left(1 - e^{-\beta\epsilon} \right) + \int_{0}^{\epsilon} d\epsilon' g_1(\epsilon') \left(e^{-\beta(\epsilon - \epsilon')} - 1 \right) \right]$$

We then consider the evolution of the probability density for boxes containing one particle. In Table 4.2 we list the processes contributing to the evolution of such occupation probability densities.

TABLE 4.1

Processes involved in the dynamics of probability density $g_0(\epsilon)$ of empty boxes at energy ϵ.

occupation		contribution to $\Delta g_0(\epsilon)$
$n_d = 1$	$n_a = 0$	$\delta_{n_d,1}\delta_{n_a,0}\left[\delta(\epsilon - \epsilon_d) - \delta(\epsilon - \epsilon_a)\right]\left[1 + \left(e^{-\beta(\epsilon_a - \epsilon_d)} - 1\right)\theta(\epsilon_a - \epsilon_d)\right]$
	$n_a > 0$	$\delta_{n_d,1}(1 - \delta_{n_a,0})\delta(\epsilon - \epsilon_d)$
$n_d > 1$	$n_a = 0$	$-(1 - \delta_{n_d,1})\delta_{n_a,0}e^{-\beta\epsilon_a}\delta(\epsilon - \epsilon_a)$

TABLE 4.2

Processes contributing to the dynamics of $g_1(\epsilon)$.

occupation		contribution to $\Delta g_1(\epsilon)$
$n_d = 1$	$n_a = 0$	$\delta_{n_d,1}\delta_{n_a,0}\left[\delta(\epsilon - \epsilon_d) - \delta(\epsilon - \epsilon_a)\right]\left[1 + \left(e^{-\beta(\epsilon_a - \epsilon_d)} - 1\right)\theta(\epsilon_a - \epsilon_d)\right]$
	$n_a = 1$	$-\delta_{n_d,1}\delta_{n_a,1}\left[\delta(\epsilon - \epsilon_d) + \delta(\epsilon - \epsilon_a)\right]$
	$n_a > 1$	$-\delta_{n_d,1}(1 - \delta_{n_a,1} - \delta_{n_a,0})\delta(\epsilon - \epsilon_d)$
$n_d = 2$	$n_a = 0$	$\delta_{n_d,2}\delta_{n_a,0}\left[\delta(\epsilon - \epsilon_d) + \delta(\epsilon - \epsilon_a)\right]e^{-\beta\epsilon_a}$
	$n_a = 1$	$\delta_{n_d,2}\delta_{n_a,1}\left[\delta(\epsilon - \epsilon_d) - \delta(\epsilon - \epsilon_a)\right]$
	$n_a > 1$	$\delta_{n_d,2}(1 - \delta_{n_a,1} - \delta_{n_a,0})\delta(\epsilon - \epsilon_d)$
$n_d > 2$	$n_a = 0$	$(1 - \delta_{n_d,2} - \delta_{n_d,1})\delta_{n_a,0}\delta(\epsilon - \epsilon_a)e^{-\beta\epsilon_a}$
	$n_a = 1$	$-(1 - \delta_{n_d,2} - \delta_{n_d,1})\delta_{n_a,1}\delta(\epsilon - \epsilon_a)$

Departure boxes are chosen with probability n_d/N. Arrival boxes are chosen with uniform probability $1/N$.

Using again Eqs. (4.77-4.80) and Eqs. (4.A.2-4.A.4) we are able to derive the equation of motion for the probability density of boxes with one particle and energy equal to ϵ:

$$\frac{\partial g_1(\epsilon)}{\partial t} \tag{4.A.5}$$

$$= 2g_2(\epsilon)\left(1 + \int_0^\infty d\epsilon\, g_0(\epsilon)e^{-\beta\epsilon} - P_0\right)$$

$$- g_1(\epsilon)\left[2 + \int_\epsilon^\infty d\epsilon'\, g_0(\epsilon')\left(e^{-\beta(\epsilon'-\epsilon)} - 1\right)\right]$$

$$+ g_0(\epsilon)\left[e^{-\beta\epsilon} + P_1\left(1 - e^{-\beta\epsilon}\right) + \int_0^\epsilon d\epsilon'\, g_1(\epsilon')\left(e^{-\beta(\epsilon-\epsilon')} - 1\right)\right]$$

For densities of boxes with $k > 1$ particles the scheme of the contributions is presented in Table 4.3:

Combining all the contributions we obtain for $g_k(\epsilon)$ Eq. (4.99)

$$\frac{\partial g_k(\epsilon)}{\partial t} = (k+1)g_{k+1}(\epsilon)\left(1 + \int_0^\infty d\epsilon\, g_0(\epsilon)\, e^{-\beta\epsilon} - P_0\right)$$

$$- g_k(\epsilon)\left[1 + k + k\left(\int_0^\infty d\epsilon\, g_0(\epsilon)\, e^{-\beta\epsilon} - P_0\right)\right] + g_{k-1}(\epsilon)$$

TABLE 4.3

List of the processes involved in the dynamics of the probability density $g_k(\epsilon)$, for $k > 1$.

occupation		contribution to $\Delta g_k(\epsilon)$
$n_d = h < k$	$n_a = k-1$	$\delta_{n_d,h}\delta_{n_a,k-1}\delta(\epsilon - \epsilon_a)$
	$n_a = k$	$-\delta_{n_d,h}\delta_{n_a,k}\delta(\epsilon - \epsilon_a)$
$n_d = k$	$n_a = 0$	$-\delta_{n_d,k}\delta_{n_a,0}\delta(\epsilon - \epsilon_a)e^{-\beta\epsilon_a}$
	$0 < n_a = h < k-1$	$-\delta_{n_d,k}\delta_{n_a,h}\delta(\epsilon - \epsilon_a)$
	$n_a = k-1$	$-\delta_{n_d,k}\delta_{n_a,k-1}[\delta(\epsilon - \epsilon_d) - \delta(\epsilon - \epsilon_a)]$
	$n_a = k$	$-\delta_{n_d,k}\delta_{n_a,k}[\delta(\epsilon - \epsilon_d) + \delta(\epsilon - \epsilon_a)]$
	$n_a > k$	$-\delta_{n_d,k}\left(1 - \sum_{h=0}^{k}\delta_{n_a,h}\right)\delta(\epsilon - \epsilon_d)$
$n_d = k+1$	$n_a = 0$	$-\delta_{n_d,k+1}\delta_{n_a,0}\delta(\epsilon - \epsilon_d)e^{-\beta\epsilon_a}$
	$0 < n_a < k-1$	$\delta_{n_d,k+1}\delta_{n_a,h}\delta(\epsilon - \epsilon_d)$
	$n_a = k-1$	$\delta_{n_d,k+1}\delta_{n_a,k-1}[\delta(\epsilon - \epsilon_d) + \delta(\epsilon - \epsilon_a)]$
	$n_a = k$	$\delta_{n_d,k+1}\delta_{n_a,k}[\delta(\epsilon - \epsilon_d) - \delta(\epsilon - \epsilon_a)]$
	$n_a > k$	$\delta_{n_d,k+1}\left(1 - \sum_{h=0}^{k}\delta_{n_a,h}\right)\delta(\epsilon - \epsilon_d)$
$n_d > k$	$n_a = k-1$	$\left(1 - \sum_{h=1}^{k+1}\delta_{n_d,h}\right)\delta_{n_a,k-1}\delta(\epsilon - \epsilon_a)$
	$n_a = k$	$-\left(1 - \sum_{h=1}^{k+1}\delta_{n_d,h}\right)\delta_{n_a,k}\delta(\epsilon - \epsilon_a)$

4.B Ansatz for the adiabatic approximation

In this appendix we show that the Ansatz solution (4.108) is asymptotically a solution of the Eqs. (4.97)-(4.99) at $T = 0$ yielding the leading behavior of $\epsilon^{\star}$. We start by rewriting Eq. (4.108) in the following way

$$\Delta g_k(\epsilon) = \frac{\Delta P_k}{\epsilon} \, r_k\left(\frac{\epsilon}{\epsilon^{\star}}\right) \qquad (4.B.1)$$

where $\Delta P_k \equiv P_k - \delta_{k,0}$, $\Delta g_k(\epsilon) \equiv g_k(\epsilon) - \delta_{k,0}\, g(\epsilon)$, $r_k(x) = x\,\hat{g}_k(x)$ and $\int_0^{\infty} dx\; \hat{g}_k(x) = \int_0^{\infty} dx\; r_k(x)/x = 1$. Here we will perform the analysis for the case $k = 0$. The equations for $k > 0$ can be done in a similar fashion. Substituting this expression into Eq. (4.104) we get

$$\frac{\partial g_0(\epsilon)}{\partial t} = \frac{\partial \Delta P_0}{\partial t}\frac{1}{\epsilon} r_0\left(\frac{\epsilon}{\epsilon^{\star}}\right) - \frac{\Delta P_0}{(\epsilon^{\star})^2}r'_0\left(\frac{\epsilon}{\epsilon^{\star}}\right)\frac{d\epsilon^{\star}(t)}{dt} = \qquad (4.B.2)$$

$$-\frac{\Delta P_1}{\epsilon}r_1\left(\frac{\epsilon}{\epsilon^{\star}}\right)\left[\int_0^{\epsilon} d\epsilon'g(\epsilon') - \Delta P_0 \int_{\epsilon}^{\infty} d\epsilon'\frac{1}{\epsilon'}r_0\left(\frac{\epsilon}{\epsilon^{\star}}\right)\right]$$

$$+\Delta P_1\left[g(\epsilon) + \frac{\Delta P_0}{\epsilon}r_0\left(\frac{\epsilon}{\epsilon^{\star}}\right)\int_{\epsilon}^{\infty} d\epsilon'\frac{1}{\epsilon'}r_1\left(\frac{\epsilon}{\epsilon^{\star}}\right)\right] \quad .$$

where $r'_0(x)$ stands for the first derivative of $r_0(x)$. Note that the scaling function r_0 does not depend on time, hence there is no time derivative of it in that expression. Introducing Eq. (4.18) in the first term of the left-hand side of (4.B.2), we multiply the whole equation by $\epsilon/\Delta P_0$ to obtain,

$$\Delta P_1 \, r_0(x) + x \, r_0'(x) \frac{\partial \log(\epsilon^\star)}{\partial t} \tag{4.B.3}$$

$$= \Delta P_1 \left\{ \frac{r_1(x)}{\Delta P_0} \int_0^\epsilon d\epsilon' g(\epsilon') - r_1(x) \int_x^\infty dx' \hat{g}_0(x) - \left[\frac{\epsilon g(\epsilon)}{\Delta P_0} + r_0(x) \right] \int_x^\infty dx' \hat{g}_1(x) \right\}$$

where $\hat{g}_k(x) = (r_k(x)/x)$. From this equation we can guess the scaling behavior of all quantities in the asymptotic long time limit $\epsilon^\star \to 0$. In the sector $\epsilon \leq \epsilon^\star$, we use $g(\epsilon) \sim \epsilon^\nu$ obtaining $\int_0^\epsilon d\epsilon' g(\epsilon') \sim \epsilon^{\nu+1}$. Assuming all terms of the same order, we get for $\epsilon \sim \epsilon^\star$

$$\Delta P_0 \sim (\epsilon^\star)^{\nu+1} \tag{4.B.4}$$

$$\Delta P_1 \sim -\frac{\partial \log(\epsilon^\star)}{\partial t} \tag{4.B.5}$$

Using the standard adiabatic approximation $P_0 = 1 - 1/z^\star, P_1 = 1/e^{z^\star}$, cf. Eq. (4.101), we obtain Eq. (4.109). Note that the set of equations for h_k is still impossible to solve. Only in certain regimes such as $\epsilon \ll \epsilon^\star$ it may be possible to obtain results. There is a set of equations which couples the different h_k. But this set of equations is time-independent and should yield all the scaling functions $\hat{g}_k(x)$ once appropriate treatment is taken of the amplitude constant which fixes the leading behavior of $\epsilon^\star$.

We also consider, as an example, the case in which the probability distribution of the quenched disorder becomes exponentially high at high values of ϵ and zero for low values, namely we choose

$$g(\epsilon) = \exp\left(-\frac{A}{\epsilon}\right) \tag{4.B.6}$$

For this choice $\int_0^\epsilon d\epsilon' g(\epsilon') \sim -\epsilon \exp\left(-\frac{A}{\epsilon}\right) - A \, \Gamma\left(0, \frac{A}{\epsilon}\right)$, where the generalized Euler function $\Gamma(0, x)$ goes to zero as $x \to \infty$. In order to estimate $\epsilon^\star$ from Eq. (4.B.3) we notice now that for P_1 Eq. (4.B.5) is still valid, while for ΔP_0 we obtain

$$\Delta P_0 \sim -\epsilon^\star \exp\left(-\frac{A}{\epsilon^\star}\right) \tag{4.B.7}$$

eventually yielding

$$\epsilon^\star(t) \sim \frac{A}{\log\left(\log t\right)} \tag{4.B.8}$$

4.C Approach to equilibrium of occupation densities

We present the equations of motion for the occupation densities in the asymptotic regime. The values of the densities are expanded to the first order around their equilibrium values: $g_n = g_n^{\mathrm{eq}} + \delta g_n$:

$$
\frac{\partial \delta g_0(\epsilon)}{\partial t} = \delta g_1(\epsilon) \left\{ 1 + \int_\epsilon^\infty d\epsilon' \bar{g}_0(\epsilon') \left[e^{-\beta(\epsilon'-\epsilon)} - 1 \right] \right\} \tag{4.C.1}
$$

$$
-\delta g_0(\epsilon) \left[e^{-\beta\epsilon} + P_1^{\mathrm{eq}} \left(1 - e^{-\beta\epsilon} \right) + z \int_0^\epsilon d\epsilon' \, \bar{g}_0(\epsilon') \left(e^{-\beta\epsilon} - e^{-\beta\epsilon'} \right) \right]
$$

$$
+ \bar{g}_0(\epsilon) \left\{ z \int_\epsilon^\infty d\epsilon' \, \delta g_0(\epsilon') \left(e^{-\beta\epsilon'} - e^{-\beta\epsilon} \right) \right.
$$

$$
\left. - \left(1 - e^{-\beta\epsilon} \right) \int_0^\infty d\epsilon' \delta g_1(\epsilon') - \int_0^\epsilon d\epsilon' \, \delta g_1(\epsilon') \left[e^{-\beta(\epsilon-\epsilon')} - 1 \right] \right\}
$$

$$
\frac{\partial \delta g_1(\epsilon)}{\partial t} = \frac{2}{z} \delta g_2(\epsilon) - \delta g_1(\epsilon) \left[2 + \int_\epsilon^\infty d\epsilon' g_0(\epsilon') \left[e^{-\beta(\epsilon'-\epsilon)} - 1 \right] \right] \tag{4.C.2}
$$

$$
+ \delta g_0(\epsilon) \left[e^{-\beta\epsilon} + P_1^{\mathrm{eq}} \left(1 - e^{-\beta\epsilon} \right) + z \int_0^\epsilon d\epsilon' g_0(\epsilon') \left(e^{-\beta\epsilon} - e^{-\beta\epsilon} \right) \right]
$$

$$
- \bar{g}_0(\epsilon) \left\{ z^2 e^{-\beta\epsilon} \int_0^\infty d\epsilon' \, \delta g_0(\epsilon') \left(1 - e^{-\beta\epsilon'} \right) + z \int_\epsilon^\infty d\epsilon' \, \delta g_0(\epsilon') \left(e^{-\beta\epsilon'} - e^{-\beta\epsilon} \right) \right.
$$

$$
\left. - \left(1 - e^{-\beta\epsilon} \right) \int_0^\infty d\epsilon' \, \delta g_1(\epsilon') - \int_0^\epsilon d\epsilon' \, \delta g_1(\epsilon') \left[e^{-\beta(\epsilon-\epsilon')} - 1 \right] \right\}
$$

$$
\frac{\partial \delta g_n(\epsilon)}{\partial t} = \delta g_{n+1}(\epsilon) \frac{k+1}{n} - \delta g_n(\epsilon) \left(1 + \frac{n}{z} \right) + \delta g_{n-1}(\epsilon) \tag{4.C.3}
$$

$$
- g_0^{\mathrm{eq}}(\epsilon) \frac{z^{n+1}}{n!} e^{-\beta\epsilon} \left(1 - \frac{n}{z} \right) \int_0^\infty d\epsilon' \, \delta g_0(\epsilon') \left(1 - e^{-\beta\epsilon} \right), \quad n > 1
$$

In the above equations, β is the inverse thermal bath temperature and z is the equilibrium fugacity at that temperature.

As T goes to zero ($\beta \to \infty$, $z(\beta) \to \infty$) the equations for the first order perturbation to equilibrium can be closed:

$$
\frac{\partial \delta g_0(\epsilon)}{\partial t} = \delta g_1(\epsilon) - \int_\epsilon^\infty d\epsilon' \left[\delta g_1(\epsilon) \, g(\epsilon') + g(\epsilon) \, \delta g_1(\epsilon') \right] \tag{4.C.4}
$$

$$
\frac{\partial \delta g_1(\epsilon)}{\partial t} = -2 \, \delta g_1(\epsilon) + \int_\epsilon^\infty d\epsilon' \left[\delta g_1(\epsilon) \, g(\epsilon') + g(\epsilon) \, \delta g_1(\epsilon') \right] \tag{4.C.5}
$$

$$
\frac{\partial \delta g_n(\epsilon)}{\partial t} = -\delta g_n(\epsilon) + \delta g_{n-1}(\epsilon) \qquad n > 1 \tag{4.C.6}
$$

In order to estimate the relaxation characteristic time to equilibrium at low temperature, we can expand Eqs. (4.C.1)-(4.C.3). First we introduce the asymptotic threshold energy $\epsilon^*(T)$ as the energy discriminating between thermalized and nonthermalized collective modes at temperature T. If we define it through the relation $\epsilon^\star(T) = Tz(T)$ and we use the relation (4.95) obtained by doing a low T expansion then we get,

$$\epsilon^\star(T) = z_0 T^{\frac{1}{2+\nu}} \tag{4.C.7}$$

where z_0 is the coefficient of the leading term of $z(T)$ at low T [see Eq. (4.95)]: $z(T) = z_0 T^{\frac{1+\nu}{2+\nu}}$.

Then we expand Eqs. (4.C.1), take $\epsilon \simeq \epsilon^\star$ and introduce the following adiabatic Ansatz,

$$\delta g_n(\epsilon) \equiv g_n(\epsilon) - g_n^{\mathrm{eq}}(\epsilon) = \frac{\Delta P_n(T,t)}{\epsilon^\star(T)} \hat{g}_n\left(\frac{\epsilon}{\epsilon^\star(T)}\right) \tag{4.C.8}$$

Note that this solution is equivalent to the Ansatz Eq. (4.108) introduced for the asymptotic dynamics at zero temperature but with a static $\epsilon^\star(T)$ now replacing the dynamical threshold. Let us now consider Eq. (4.C.1) for $\delta g_0(\epsilon)$. Because $\delta P_k = \int d\epsilon \, \delta g_k(\epsilon)$, it can be shown that the slowest mode corresponds to $k = 0$, i.e., $\delta g_0(\epsilon) \gg \delta g_k(\epsilon)$ for $k > 0$. Therefore, the second term in the right-hand side of Eq. (4.C.1) dominates the first and the second terms. Introducing Eq. (4.C.8) into Eq. (4.C.1) we find that the relaxation time behaves like,

$$\tau_{\mathrm{eq}}(\epsilon^\star) \propto \frac{e^{\beta\epsilon^\star}}{\beta\epsilon^\star} \tag{4.C.9}$$

For $\epsilon \gg \epsilon^\star$ the relaxation time is much smaller, since those are the modes with lower energy barriers.

4.D   Probability distribution of proposed energy updates

In this appendix the probability distribution of proposed energy updates is built. In Table 4.4 we summarize all the processes contribuiting to it, together with their probabilities.

The probability distribution $p(x)$ of proposed energy updates is the average of all possible changes, each computed with its probability:

$$p(x) \equiv \overline{\delta\left(E' - E - x\right)} \tag{4.D.1}$$

where x is the proposed update, E is the energy of the system before the

TABLE 4.4
Contributions to the probability distribution $p(x)$ of proposed energy updates.

occupation		contrib. to $E' - E$	probability
$n_d = 1$	$n_a = 0$	$-\epsilon_d + \epsilon_a$	$g_1(\epsilon_d)\, g_0(\epsilon_a)$
	$n_a > 0$	$-\epsilon_d$	$g_1(\epsilon_d)\, [g(\epsilon_a) - g_0(\epsilon_a)]$
$n_d > 1$	$n_a = 0$	ϵ_a	$g_0(\epsilon_a)\frac{1}{N}\sum_p n_p[g(\epsilon_d) - g_1(\epsilon_d)]$
	$n_a > 0$	0	$[g(\epsilon_a) - g_0(\epsilon_a)]\frac{1}{N}\sum_p n_p[g(\epsilon_d) - g_1(\epsilon_d)]$

updating and E' the energy afterwards. This means

$$
\begin{aligned}
p(x) = {} & \int_0^\infty d\epsilon \int_0^\infty d\epsilon'\, g_1(\epsilon)\, g_0(\epsilon')\, \delta(x + \epsilon - \epsilon') \\
& + \int_0^\infty d\epsilon \int_0^\infty d\epsilon'\, g_1(\epsilon)\, [g(\epsilon') - g_0(\epsilon')]\, \delta(x + \epsilon) \\
& + \int_0^\infty d\epsilon \int_0^\infty d\epsilon'\, g_0(\epsilon')\frac{1}{N}\sum_p n_p\, [g(\epsilon) - g_1(\epsilon)]\, \delta(x - \epsilon') \\
& + \int_0^\infty d\epsilon \int_0^\infty d\epsilon'\, [g(\epsilon') - g_0(\epsilon')]\frac{1}{N}\sum_p n_p\, [g(\epsilon) - g_1(\epsilon)]\delta(x) \\
= {} & \int_x^\infty d\epsilon\, g_1(\epsilon - x)\, g_0(\epsilon) + (1 - P_0)\, g_1(-x)\, \theta(-x) \qquad \text{(4.D.2)} \\
& + (1 - P_1)\, g_0(x)\, \theta(x) + (1 - P_0)(1 - P_1)\, \delta(x)
\end{aligned}
$$

where the $\theta(x)$ function includes $x = 0$. The term with $\delta(x)$ is the term responsible for diffusive motion of particles. Such a contribution does not actually give any contribution to the relaxation of the system and therefore we will not consider it from now on.

The normalized distribution of accepted changes of energy difference x is given by

$$
r(x) = \frac{p(x)W(\beta x)}{A_0} \qquad (4.D.3)
$$

where $W(\beta x)$ is the Metropolis function

$$
W(\beta x) = \begin{cases} e^{-\beta x} & \text{if } x > 0 \\ 1 & \text{if } x \le 0 \end{cases} \qquad (4.D.4)
$$

The normalization factor A_0 is the acceptance rate of the Monte Carlo dynamics:

$$
A_0 = \int_{-\infty}^\infty dx\, W(\beta x)\, p(x) \qquad (4.D.5)
$$

as it was defined in Sec. 3.1.1, Eq. (3.34), now yielding

$$
A_0 = \int_0^\infty d\epsilon\, g_0(\epsilon) \int_0^\epsilon d\epsilon'\, g_1(\epsilon') + (1 - P_0)P_1 + \frac{1 - P_1}{z e^z} \qquad (4.D.6)
$$

where we used the identity

$$g_0(\epsilon') \, g_1(\epsilon) e^{-\beta(\epsilon'-\epsilon)} = g_0(\epsilon) g_1(\epsilon') \tag{4.D.7}$$

and the closure Eq. (4.88), otherwise written as

$$\int_0^\infty d\epsilon \, g_0(\epsilon) \, e^{-\beta\epsilon} = \frac{1}{ze^z} \tag{4.D.8}$$

The first term in Eqs. (4.D.6, 4.D.2) is nonzero only if $x = \epsilon_a - \epsilon_d = \epsilon - \epsilon' > 0$, and does not contribute at zero temperature where only negative or null changes in energy are accepted.

As $T \sim 0$, indeed, the distribution $r(x)$ of accepted changes, contributing to the relaxation, tends to

$$r(x) \simeq \frac{\theta(-x)(1 - P_0) \, g_1(-x)}{A_0} = \theta(-x) \, g_1(-x) \tag{4.D.9}$$

where $A_0 = 1 - P_0 \simeq 1/z$.

Using the same notation we can write the energy evolution as

$$\frac{\partial E}{\partial t} = \int_{-\infty}^\infty dx \, x \, W(\beta x) \, p(x) \tag{4.D.10}$$

$$= -P_1 \, E - \int_0^\infty d\epsilon \, g_1(\epsilon) \, \epsilon + (1 - P_1) \int_0^\infty d\epsilon \, g_0(\epsilon) e^{-\beta\epsilon} \epsilon$$

$$+ \int_0^\infty d\epsilon' \int_{\epsilon'}^\infty d\epsilon g_1(\epsilon') g_0(\epsilon) \left[e^{-\beta(\epsilon-\epsilon')} - 1 \right] (\epsilon - \epsilon')$$

The right-hand side of this equation can be equivalently obtained following the procedure presented in Appendix 4.A. Indeed, by definition of energy density, it is

$$\frac{\partial E}{\partial t} = -\int_0^\infty d\epsilon \, \epsilon \, \frac{\partial g_0(\epsilon)}{\partial t} \tag{4.D.11}$$

Inserting Eq. (4.97) in Eq. (4.D.11) we get Eq. (4.D.10) back.

5

Glassiness in a directed polymer model

A content person owns half the world

Dutch proverb

Polymer physics is a mature branch of chemistry and industry (e.g., plastics, nylon, rubber), and of biochemistry (DNA, RNA, microtubules). The theoretical description of polymers is a major branch in theoretical physics, to which many excellent books and reviews have been devoted, see, e.g., de Gennes [1979]; Doi & Edwards [1986]. There are many forms of polymers: linear polymers, cross-linking polymers, and, of particular interest in biophysics, heterogeneous polymers.

The first aging experiments were performed on polymers [Struik, 1978], but mostly the phenomenon of aging has been investigated in different systems. Solvable models for aging in heteropolymer systems exist, see, e.g., Montanari *et al.* [2004]; Müller *et al.* [2004] and references therein.

The glass transition is caused by the appearance of a multitude of long-lived states, which prevents exploration of the whole phase space. These effects are so strong that, in practice, one can only observe precursor effects. Experimentally, one observes a dynamical freezing around the tunable glass temperature T_g, see Sec. 1.1, set by the cooling rate.

The ergodic theorem says that time-averages may be replaced by ensemble averages. It is widely believed that the inherent dynamical nature of the glass transition implies that there is neither need nor chance for a thermodynamic explanation. However, since so many decades in time are involved, this is an unsatisfactory point of view. In previous chapters we have discussed exactly solvable models with glassy behavior. Here we discuss this picture of the glassy transition in a polymer model, introduced by Nieuwenhuizen [1997b]. We consider a linear monomeric polymer, in an idealized geometry and in the presence of a disordered substrate. This setup induces important simplifications in the statics and dynamics. As in the two previous chapters, the model is designed with a simple statics and a glassy dynamics and can be solved analytically.

5.1   The directed polymer model

Consider a directed polymer (or an interface without overhangs) in $d = 1 + 1$ dimensions, described by a height function $z(x)$. The x-coordinates are discrete and lie in the region $1 \leq x \leq L$, and the discrete z-coordinates lie in $1 \leq z \leq W$.

The directed polymer can locally be flat ($z(x + 1) = z(x)$; no energy cost) or make a single step ($z(x+1) - z(x) = \pm 1$) at an energy cost J. We consider the restricted solid-on-solid approximation, where larger steps are assumed to cost infinite energy. For an introduction, see Forgacs *et al.* [1991]. The partition sum of this system, subject to periodic boundary conditions, can be expressed in the $W \times W$ matrix $\mathcal{T}$ that transfers the system from x to $x + 1$

$$Z = \operatorname{Tr} e^{-\beta \mathcal{H}} = \operatorname{Tr} \mathcal{T}^L \tag{5.1}$$

with transfer matrix

$$(\mathcal{T})_{z',z} = \delta_{z',z} + (\delta_{z',z+1} + \delta_{z',z-1})e^{-\beta J} \tag{5.2}$$

We may write the partition sum as

$$Z = \sum_{w=1}^{W} \Lambda_w^L \tag{5.3}$$

where Λ_w are the eigenvalues of $\mathcal{T}$.

For this pure system, at temperature $T = 1/\beta$, Fourier analysis allows us to use the Fourier index k as the label w, and it gives the eigenvalues

$$\Lambda(k) = 1 + 2e^{-\beta J} \cos k \tag{5.4}$$

Imposing, for simplicity, periodic boundary conditions, gives the allowed k-values

$$k_n = \frac{2\pi}{W} n, \qquad n = 1, 2, \cdots W. \tag{5.5}$$

The partition sum thus is

$$Z = \sum_{n=1}^{W} \Lambda^L(k_n) \approx \frac{W}{2\pi} \int_{-\pi}^{\pi} dk \Lambda^L(k) \tag{5.6}$$

where we asssumed that W is large, and we shifted the integration interval. We clearly need the largest eigenvalues, which occur at small momentum,

$$\Lambda(k) \approx \exp\left[-\beta f_B - \frac{\Gamma k^2}{2\pi^2}\right] \tag{5.7}$$

The term of order k^0 is

$$f_B(T) = -T \ln(1 + 2e^{-\beta J}) \tag{5.8}$$

while Γ is the stiffness coefficient,

$$\Gamma(T) = \frac{2\pi^2 e^{-\beta J}}{1 + 2e^{-\beta J}} \tag{5.9}$$

This brings us to

$$Z = \frac{W}{\sqrt{2\pi\Gamma L}} e^{-L\beta f_B} \tag{5.10}$$

The free energy, thus, reads

$$F = -T \ln Z = L f_B + T \ln \frac{\sqrt{2\pi\Gamma L}}{W} \tag{5.11}$$

The limit $L \to \infty$, at fixed W, allows us to interpret f_B of Eq. (5.8) as the bulk free energy density.

The internal energy density is

$$u_B = \frac{\partial \beta f_B}{\partial \beta} = \frac{2Je^{-\beta J}}{1 + 2e^{-\beta J}} \tag{5.12}$$

the entropy density is

$$s_B = -\frac{\partial f_B}{\partial T} = \frac{2\beta Je^{-\beta J}}{1 + 2e^{-\beta J}} + \ln(1 + 2e^{-\beta J}) \tag{5.13}$$

and the specific heat per monomer reads

$$c_B = \frac{\partial u_B}{\partial T} = \frac{2\beta^2 J^2 e^{-3\beta J}}{(1 + 2e^{-\beta J})^2} \tag{5.14}$$

5.1.1   Disordered situation and Lifshitz-Griffiths singularities

For the rest of the chapter we shall consider the situation of randomly located potential barriers parallel to the x-axis, so that the random potential is correlated, viz., $V(x, z) = V(z)$. Hereto we assume binary disorder: $V(z) = 0$ with probability p or $V(z) = V_1 > 0$ with probability $1 - p$. We shall denote

$$p = \exp(-\mu) \tag{5.15}$$

The transfer matrix now reads

$$(T)_{z',z} = \delta_{z',z} e^{-\beta V(z)} + (\delta_{z',z+1} + \delta_{z',z-1})e^{-\beta J} \tag{5.16}$$

As before, Eq. (5.1) is dominated by the largest eigenvalues. In this disordered setup, they can be identified explicitly, since they occur due to Lifshitz-Griffiths singularities [Lifshitz, 1964; Griffiths, 1964; Nieuwenhuizen & Luck, 1989; Nieuwenhuizen, 1989a,b, 1998a]. These singularities are lanes of width $\ell \gg 1$ in which all $V(z) = 0$, bordered by regions with $V(z) \neq 0$. These

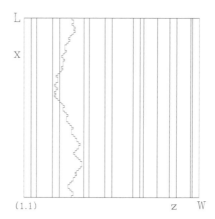

FIGURE 5.1

A directed polymer can move on a substrate with parallel potential barriers. For entropic reasons, it prefers to lie in wide lanes between the barriers.

dominant configurations are the "states" or "components" of our system. In spin-glass theory such states are called "TAP-states" [Thouless *et al.*, 1977], while in the next chapter they will be identified with the "inherent structures." The interest of the model lies, among others, in the fact that these dominant states can be identified explicitly. The situation is depicted in Fig. 5.1.

In disordered one dimensional media the eigenfunctions are exponentially localized. The ones with a large eigenvalue are related to a large pure region with $V(z) = 0$. These are the so-called Lifshitz, or Griffiths, singularities. Let such a disorder-free lane have width $\ell_a \gg 1$ and be located at $z_a \leq z \leq z_a + \ell_a$. These states can, thus, be labeled by $a = (z_a, \ell_a)$. Outside this region, the eigenfunction will decay exponentially over a fixed length, the localization length; so it will be small both at $z = z_a$ and $z = z_a + \ell_a$. This eigenfunction will, therefore, have the approximate form $\sin[\pi(z - z_a)/\ell_a]$ inside the lane.

Since $k \to \pi/\ell_a$, the free energy of a lane follows as $F_a = F_{\ell_a}$, where

$$\beta F_\ell \equiv -L \ln \Lambda \left(\frac{\pi}{\ell} \right) \approx \beta f_B L + \frac{\Gamma L}{2\ell^2} \tag{5.17}$$

The number of regions with ℓ successive sites with $V(z) = 0$, surrounded by sites with $V(z) = V_1$, can be estimated by its ensemble average,

$$\mathcal{N}_\ell = W(1 - p)^2 p^\ell = W(1 - p)^2 e^{-\mu \ell} \tag{5.18}$$

where we used the relation between p and μ, Eq. (5.15). For large L, we may restrict the partition sum to these dominant states. We, thus, evaluate, instead of Eq. (5.1), the "TAP" partition sum

$$Z = \sum_\ell \mathcal{N}_\ell e^{-\beta F_\ell} = (1-p)^2 W \sum_\ell e^{-\beta F_\ell - \mu \ell} \qquad (5.19)$$

Since the optimal ℓ is large, the total free energy can be read off,

$$\beta F = -\ln Z = L\beta f_B + \frac{\Gamma L}{2\ell^2} + \mu\ell - \ln[(1-p)^2 W] \qquad (5.20)$$

which has to be optimized in ℓ (as usual, we may neglect algebraic prefactors of Z that arise from the saddle point integrations). This brings us to

$$\ell_*(T) = \left(\frac{\Gamma(T)L}{\mu} \right)^{1/3}. \qquad (5.21)$$

Since the relevant ℓ is of order $L^{1/3}$, an interesting scenario occurs when we consider the width W of the system to scale as a stretched exponential in the height of the system, viz.,

$$(1-p)^2 W = \exp(\lambda L^{1/3}) \qquad (5.22)$$

When L is known, λ sets the width W. For large L, and $\lambda = \mathcal{O}(1)$, the geometry is then very asymmetric, much, much wider than long, thus quasi one dimensional. For moderate L, however, this need not be the case, e.g., for $L = 10^6$ and $\lambda = 0.13815$, or for $L = 10^9$ and $\lambda = 0.02072$, one would even have $W = L$.

The states with width ℓ thus have a configurational entropy

$$S_c(\ell) \equiv \ln \mathcal{N}_\ell = \lambda L^{1/3} - \mu\ell \qquad (5.23)$$

The largest ℓ which occurs in the system can be estimated by setting $S_c(\ell) = 0$ or $\mathcal{N}_\ell \approx 1$, yielding

$$\ell_{max} = \frac{\lambda L^{1/3}}{\mu} \qquad (5.24)$$

It is a geometrical length, defined by the aspect of the model, and independent of T. Its effect is that Eq. (5.21) holds if $\ell_* \leq \ell_{max}$, or else it is replaced by ℓ_{max}. We may now write Eq. (5.23) as

$$S_c(\ell) = \mu(\ell_{max} - \ell) \qquad (5.25)$$

Let us introduce a scaled stiffness coefficient γ by setting $\Gamma = \lambda^3 \gamma^3 / \mu^2$, viz.,

$$\gamma(T) = \frac{1}{\lambda} \left[\Gamma(T)\mu^2 \right]^{1/3} = \frac{(2\pi^2\mu^2)^{1/3}}{\lambda} \frac{e^{-\beta J/3}}{(1 + 2e^{-\beta J})^{1/3}} \qquad (5.26)$$

The free energy of the widest state then reads

$$\beta F = L\beta f_B + \frac{1}{2}\lambda L^{1/3}\gamma^3(T) \qquad (5.27)$$

It follows from Eq. (5.21) that at low enough T the optimal width is smaller than ℓ_{max}, with a ratio set by γ,

$$\ell_* = \gamma(T)\ell_{max} \tag{5.28}$$

The free energy of this phase is

$$\beta F = L\beta f_B + \lambda L^{1/3}[\frac{3}{2}\gamma(T) - 1] \tag{5.29}$$

In general, one has the optimal length

$$\ell_* = \min(\gamma(T), 1)\ell_{\max} \tag{5.30}$$

In contrast to realistic glasses, our model has no equivalent of a crystal state. Let us recall that this neither occurs for some binary model glasses in a certain parameter regime, e.g., [Kob & Andersen, 1994; Parisi, 1997b], see also Secs. 1.1, 7.3 and Appendix 6.A.

5.1.2   Static phase diagram

In the temperature interval where $\gamma(T) > 1$, the polymer lies in the non-degenerate widest lane present in the large but finite system. When $\gamma(T) < 1$, it lies in one of the $\mathcal{N}_{\ell_*} \gg 1$ optimal states, each of which has a higher free energy than the widest lane; this free energy loss is more than compensated for by their configurational entropy. The system, thus, undergoes a static glass transition at the "Kauzmann" temperature where $\gamma = 1$,

$$T_K = \frac{J}{\ln(2\pi^2\mu^2\lambda^{-3} - 2)} \tag{5.31}$$

i.e., where the thermally optimal lane width is the largest lane width available in the system. Clearly, for this to happen, we have to demand that $\gamma(T = \infty) > 1$, so that the argument of the logarithm in Eq. (5.31) exceeds unity. For not-too-small and not-too-large temperatures (see later), our model has unfamiliar properties. The condition $\gamma(T = \infty) > 1$ is the key condition which ensures that there is a high temperature ideal glass phase where the widest state of the system dominates. Otherwise, there would only be the regime of many relevant states with a finite configurational entropy. With our setup of having a high temperature phase with a dominant single state, we are in an "inverse temperature world" as compared to usual glasses.

For a reason that will become clear below, we shall demand that

$$\pi^2\mu^2 > 2\lambda^3 \tag{5.32}$$

which is equivalent to $T_K < J/\ln 2$.

For any finite L, there is also a very low temperature regime $T < 1/\ln L$, where the interface is essentialy straight and can lie anywhere in the system, and a high temperature regime $T > \ln L$, where the potential barriers are ineffective, and the interface shape is truly random. These two regimes are of no interest for us and will be disregarded.

From Eq. (5.17), the internal energy of a state of width ℓ follows as

$$U_\ell = \left.\frac{\partial \beta F_\ell}{\partial \beta}\right|_\ell = u_B L + \frac{L}{2\ell^2}\frac{\partial \Gamma}{\partial \beta} = u_B L - \frac{3\lambda^3\gamma^2 L T^2}{2\mu^2\ell^2}\frac{\partial \gamma}{\partial T} \qquad (5.33)$$

Its entropy is

$$S_\ell = -\left.\frac{\partial F_\ell}{\partial T}\right|_\ell = s_B L - \frac{L}{2\ell^2}\frac{\partial T\Gamma}{\partial T} = s_B L - \frac{\lambda^3\gamma^2 L}{2\mu^2\ell^2}(\gamma + 3T\frac{\partial \gamma}{\partial T}) \qquad (5.34)$$

To obtain the thermodynamic values, for $T < T_K$ these results are to be taken at $\ell_* = \gamma(T)\ell_{max}$, while for $T > T_K$ they apply to the largest lane in the system, $\ell_* = \ell_{max}$. The internal energy of the whole system is, thus,

$$U = \begin{cases} u_B L - \frac{3}{2}\lambda L^{1/3}T^2\frac{\partial \gamma}{\partial T} & T < T_K \\ u_B L - \frac{3}{2}\lambda L^{1/3}\gamma^2 T^2\frac{\partial \gamma}{\partial T} & T > T_K \end{cases} \qquad (5.35)$$

and the total thermodynamic entropy is

$$S = S_{\ell_*} + S_c(\ell_*) = \begin{cases} s_B L + \lambda L^{1/3}(1 - \frac{3}{2}\gamma - \frac{3}{2}T\frac{\partial \gamma}{\partial T}) & T < T_K \\ s_B L - \lambda L^{1/3}(\frac{1}{2}\gamma^3 + \frac{3}{2}\gamma^2 T\frac{\partial \gamma}{\partial T}) & T > T_K \end{cases} \qquad (5.36)$$

At the transition point ($\gamma = 1$), the function $\gamma(T)$ has a finite derivative, and both the energy and the entropy are, thus, continuous, a general property of glassy systems. The free energy, Eq. (5.27) above T_K, or Eq. (5.29) below, is a thermodynamic potential, and reproduces these results by direct differentiation.

On the side $\gamma < 1$ the free energy (5.29) deviates quadratically from Eq. (5.27), leading to a higher specific heat. It reads

$$C = \frac{dU}{dT} = \begin{cases} c_B L - 3\lambda L^{1/3}(T\frac{\partial \gamma}{\partial T} + \frac{1}{2}T^2\frac{\partial^2 \gamma}{\partial T^2}) & T < T_K \\ c_B L - 3\lambda L^{1/3}\gamma^2(T\frac{\partial \gamma}{\partial T} + \frac{1}{2}T^2\frac{\partial^2 \gamma}{\partial T^2} + \frac{1}{\gamma}\left(T\frac{\partial \gamma}{\partial T}\right)^2 & T > T_K \end{cases}$$
$$(5.37)$$

The result for $T > T_K$ is the "field cooled" or thermodynamic value. There is a discontinuity

$$\Delta C = C(T_K - 0) - C(T_K + 0) = 3\lambda L^{1/3}\left(T\frac{\partial \gamma}{\partial T}\right)^2 . \qquad (5.38)$$

The "zero field cooled" or component averaged specific heat of this phase is a short time value, taken at fixed ℓ. It reads

$$\overline{C} = \sum_a p_a C_{\ell_a} = c_B L - 3\lambda L^{1/3}\frac{\gamma^2\ell_{max}^2}{\ell^2}\left[T\frac{\partial \gamma}{\partial T} + \frac{1}{2}T^2\frac{\partial^2 \gamma}{\partial T^2} + \frac{1}{\gamma}(T\frac{\partial \gamma}{\partial T})^2\right]$$
$$(5.39)$$

For $T > T_K$ one has $\ell = \ell_{max}$, so the component average indeed coincides with the thermodynamic average, there being only one component. But for $T < T_K$ this value continues smoothly, since $\ell_* = \gamma\ell_{max}$ there. Thus, there is no jump in this quantity, $\Delta\overline{C} = 0$. The general relation $|\Delta\overline{C}| \geq |\Delta C|$, based on the smaller timescale involved in $\overline{C}$, is obeyed.

When considered as a function $\beta = 1/T$, the specific heat makes a downward jump on cooling through $\beta_K = 1/T_K$, as it occurs in realistic glasses.

5.1.3   Dual view in temperature

In our directed polymer model, the moderately high temperature interval $T_K < T \ll \ln L$ is dominated by the widest lane present in the system, and the moderately low temperature interval $1/\ln L \ll T < T_K$ by a large set of narrower, thermodynamically optimal states. This situation, due to the directedness of the polymers and the geometry of the model, is the reverse of what happens usually in glasses: finite configurational entropy above the Kauzmann transition, and vanishing below.

Such an "inverse temperature world" is not completely unexpected, as it is known to occur in real polymer systems, where a true inverse transition occurs, and is called *inverse freezing* [Rastogi *et al.*, 1999; Greer, 2000; van Ruth & Rastogi, 2004].[1]

When considered as a function of $\beta = 1/T$ the situation is reminiscent of the p-spin interaction spin-glass and of the random energy model [Derrida, 1980, 1981] (also refer to Sec. 7.2). The very-high β regime extends up to $\beta \sim \ln L$, in which a gradual freezing takes place in "TAP" states of width ℓ^* much larger than unity. This smeared transition is related to the sharp dynamical transition at some temperature $T_d > T_K$ of mean-field spin-glass models, as we mentioned in Sec. 1.1 and we will analyze again in Secs. 7.2-7.3. In the regime $\beta_K < \beta \lesssim \ln L$ the system is frozen in TAP states of appropriate degeneracy; a similar static phenomenon was found [Kirkpatrick & Wolynes, 1987b; Thirumalai & Kirkpatrick, 1988] for spin-glass models in the regime $T_K < T < T_d$. Like in these models, below the "Kauzmann" temperature β_K there is only an essentially nondegenerate state. This is a manifestation of the "entropy crisis" of glasses and glassy systems, cf. Sec. 1.4.

---

[1]Inverse transitions were already hypothesized by [Tammann, 1903], even though almost a century has passed before experimental evidence for such an intuition was obtained. Glassy models displaying inverse transitions have been studied recently by Schupper & Shnerb [2004]; Crisanti & Leuzzi [2005]; Sellitto [2006]; Leuzzi [2007].

5.2 Directed polymer dynamics

The internal energy of one polymer can, in principle, be monitored as a function of time in a numerical experiment. One then obtains essentially a noisy telegraph signal. Each plateau describes trapping of the polymer in one lane for some definite time. The variance of the noise in the internal energy on this plateau is equal to $T^2\overline{C}$, implying fluctuations of the order $L^{1/2}$. From time to time the polymer moves to another lane, causing additional noise. The variance of the total noise equals T^2C, and it indeed exceeds $\overline{C}$ by an amount of order $L^{1/3}$.

We now consider moderately long time dynamics. On appropriate timescales, our system can be viewed as a one dimensional set of deep states (traps) labeled by a, located in lanes of width ℓ_a around the location z_a of the center of gravity, with associated free energies $F_a \equiv F_{\ell_a}$ given by Eq. (5.17).

These minima are separated by very wide regions (separation $\sim \exp L^{1/3}$) with a fully random potential that builds a barrier. Between traps a and $a+1$ there is a free energy barrier determined by the intermediate state of highest free energy. Let us call its free energy B_a: it will typically lie at a distance $L^{1/3}$ below the maximal free energy[2]

$$Lf_{\text{max}} = -LT \ln \Lambda_{min} = -LT \ln(1 - 2e^{-\beta J}) \qquad (5.40)$$

The free energy barrier for the polymer to move from state a to state $a+1$ is, thus, $B_a - F_{\ell_a}$, while the barrier for moving to the left is $B_{a-1} - F_{\ell_a}$.

For a statistical description of dynamics, we consider a statistical ensemble of many independent polymers, of which the units make random thermally activated moves. On appropriate timescales one then gets a master equation for the probability $p_a(t)$ that the center of a polymer is inside the ath state:

$$t_0 \frac{dp_a(t)}{dt} = e^{\beta(F_{\ell_{a-1}} - B_{a-1})} p_{a-1} + e^{\beta(F_{\ell_{a+1}} - B_a)} p_{a+1}$$
$$- e^{\beta(F_{\ell_a} - B_{a-1})} p_a - e^{\beta(F_{\ell_a} - B_a)} p_a \qquad (5.41)$$

Here t_0 is the attempt time for a move of one polymer to another deep state. This model is a combination of the random jump-rate model and the random bond models studied by Haus *et al.* [1982] and Nieuwenhuizen & Ernst [1985]. For a review see Haus & Kehr [1987]. The stationary state is independent of the barriers B_a:

$$p_a^{\text{eq}} = \frac{e^{-\beta F_{\ell_a}}}{Z_{\text{eq}}} \qquad (5.42)$$

---

[2]For $T > T^* \equiv J/\ln 2$ things get more complicated since $\Lambda_{min} < 0$. Under the condition (5.32) it holds that $T^* > T_K$, implying that this issue only shows up in the phase where the interface is located in the widest lane.

The denominator Z_{eq} is equal to the thermal TAP partition sum, Eq. (5.19). So the master equation (5.41) exactly reproduces the Gibbs distribution discussed before.

Let us now consider the motion of polymers in an equilibrium ensemble, after making some nonessential simplifications of the system. First we slightly modify the actual height of the barriers by setting $B_a = Lf_{max}$, thus neglecting their $L^{1/3}$ deviations. Next we assume, for some fixed ℓ_0 ($1 \ll \ell_0 \ll L^{1/3}$), that all relevant deep traps (and, further, a lot of shallow traps) are located at positions that are multiples of $W_{\ell_0} = \exp\mu\ell_0$, and we only consider those states. Their number is $\mathcal{N}_{\ell_0} = \exp(\lambda L^{1/3} - \mu\ell_0)$. After these simplifications we have arrived at the random jump-rate model,

$$\frac{dp_a}{dt} = \Gamma_{a-1}p_{a-1} + \Gamma_{a+1}p_{a+1} - 2\Gamma_a p_a \qquad (5.43)$$

in one dimension, with a lattice distance W_{ℓ_0}. The jump rates are set by an attempt time t_0 and a suppression rate due to the barrier,

$$\Gamma_a = \frac{1}{t_0}e^{-L\beta f_{max}+\beta F_a} \qquad (5.44)$$

In such models a result of Haus *et al.* [1982] says that in the stationary ensemble the linear diffusion law holds exactly at all times,

$$\frac{1}{2}\langle(\delta z(t)^2)\rangle \equiv \langle(z(t) - z(0))^2\rangle = Dt \qquad (5.45)$$

Moreover, the diffusion coefficient is explicitly known,

$$\frac{1}{D} = \frac{1}{W_{\ell_0}^2}\left\langle\frac{1}{\Gamma}\right\rangle = \frac{1}{W_{\ell_0}^2}\sum_a p_a^{eq}\frac{1}{\Gamma_a} \qquad (5.46)$$

In our situation we have

$$\left\langle\frac{1}{\Gamma}\right\rangle = t_0 e^{L\beta f_{max}}\frac{\sum_a e^{-2\beta F_{\ell_a}}}{\sum_a e^{-\beta F_{\ell_a}}} = t_0 e^{L\beta f_{max}}\frac{\int_0^{\ell_{max}} d\ell \mathcal{N}_\ell e^{-2\beta F_\ell}}{\int_0^{\ell_{max}} d\ell \mathcal{N}_\ell e^{-\beta F_\ell}} \qquad (5.47)$$

where F_{ℓ_a} is expressed by Eq. (5.17) and Eq. (5.26).

Both integrals are dominated by a similar optimum, the one in the denominator by ℓ_* from (5.30) and the one in the numerator by

$$\ell'_* = \min(2^{1/3}\gamma, 1)\,\ell_{max}. \qquad (5.48)$$

The diffusion coefficient can, thus, be expressed as

$$D = \frac{W_{\ell_0}^2}{t_0}e^{L\beta[f_B(T)-f_{max}(T)]} \times e^{\lambda L^{1/3}\sigma(T)} \qquad (5.49)$$

(we again neglect algebraic prefactors that arise from the integrations). There occur three regimes

$$\sigma(T) = \begin{cases} \frac{3}{2}(2^{1/3} - 1)\gamma(T), & 0 < \gamma(T) < 2^{-1/3} \\ 1 - \frac{3}{2}\gamma(T) + \gamma^3(T), & 2^{-1/3} < \gamma(T) < 1 \\ \frac{1}{2}\gamma^3(T), & \gamma(T) > 1 \end{cases} \qquad (5.50)$$

Since the barriers have a height that deviates by the order $L^{1/3}$ from a fixed value $L(f_{max} - f_B)$, we introduce the logarithmic time variable θ by

$$t(\theta) = t_0 e^{\beta L(f_{max} - f_B)} \times e^{\lambda L^{1/3}\theta} \qquad (5.51)$$

where θ ranges from $\sim -L^{2/3}$, where $t \sim t_0$, up to $\mathcal{O}(1)$, where $t \sim t_0 \exp(L + L^{1/3})$ and the interesting physics occurs.

To get an impression of the dynamics, we now make an intuitive step: we assume that the dynamics of individual polymers can be related to the ℓ-dependence of the numerator of Eq. (5.47), which shows up as a denominator in D, Eq. (5.49), due to the relation (5.46) (in doing so, we consider the denominator of (5.47), and, thus, the numerator of D, merely as a normalization). This view on the dynamics differs somewhat from the one by Nieuwenhuizen [1997b].

Let us consider the regime $T < T_K$. At given θ, the polymers have a time $t(\theta)$ to make moves. The typical deviation follows from Eqs. (5.45) and (5.49) as

$$\ln \frac{|\delta z|}{W_0} \sim \frac{1}{2}\lambda L^{1/3}(\theta + \sigma - \frac{\ell'_*}{\ell_{max}} - \frac{\gamma^3 \ell^2_{max}}{\ell'^2_*} + \frac{\ell}{\ell_{max}} + \frac{\gamma^3 \ell^2_{max}}{\ell^2}) \qquad (5.52)$$

Given the fact that, viewed statistically, the polymer will start from a rather narrow lane, it will, in the course of time, cover wider and wider lanes. Thus we can relate Eq. (5.52) to the distance between optimal states $W_\ell = \exp \mu\ell$ already reached at time $t(\theta)$, and their number, $\sim \exp S_c^{inter}(\ell, \theta)$, which follows by setting $|\delta z| \sim W_\ell \exp S_c^{inter}$. It thus defines the *intercluster configurational entropy*, i.e., the configurational entropy of the states of width ℓ already reached at time $t(\theta)$:

$$S_c^{inter}(\ell, \theta) = \frac{1}{2}\lambda L^{1/3}(\theta + \sigma - \frac{\ell'_*}{\ell_{max}} - \frac{\gamma^3 \ell^2_{max}}{\ell'^2_*} - \frac{\ell}{\ell_{max}} + \frac{\gamma^3 \ell^2_{max}}{\ell^2}) \qquad (5.53)$$

For short times, that is, for large negative θ, indeed, only small $\ell \leq \ell_{max}\sqrt{\gamma^3/|\theta|}$ will bring a nonnegative value, as only these states can be reached. The largest ℓ reached at time θ, $\ell_{dyn}(\theta)$, is found from $S_c^{inter} = 0$, i.e., from the cubic equation

$$\frac{\ell_{dyn}}{\ell_{max}} - \frac{\gamma^3 \ell^2_{max}}{\ell^2_{dyn}} = \theta + \sigma - \frac{\ell'_*}{\ell_{max}} - \frac{\gamma^3 \ell^2_{max}}{\ell'^2_*} \qquad (5.54)$$

The equilibrium relation $\ell_{dyn} = \ell'_*$ tells us that, for a time set by

$$\theta_* = -\sigma + 2\frac{\ell'_*}{\ell_{max}} \qquad (5.55)$$

that is, for

$$\theta_*(T) = \begin{cases} (\frac{3}{2} + 2^{-2/3})\gamma, & 0 < \gamma(T) < 2^{-1/3} \\ 1 + \frac{3}{2}\gamma - \gamma^3, & 2^{-1/3} < \gamma(T) < 1 \\ 2 - \frac{1}{2}\gamma^3(T), & \gamma(T) > 1 \end{cases} \qquad (5.56)$$

the next thermodynamically optimal state will typically have been reached, so that, then, each of the polymers has swept a region of phase space in accord with the equilibrium prediction, and has achieved thermodynamic equilibrium by itself.

We consider the states with ℓ up to $\ell_{dyn}(\theta)$ as one cluster. The configurational entropy of these clusters is $S_c^{dyn}(\theta) = \ln\mathcal{N}_{dyn} = \mu[\ell_{max} - \ell_{dyn}(\theta)]$. Since ℓ_{dyn} grows in time, the configurational entropy decreases. At $\theta = \theta_*$, where $\ell_{dyn}(\theta_*) = \ell_*(T)$, it equals the equilibrium value $S_c(\ell_*(T))$, defined by Eq. (5.23).

This supports the view presented in Chapter 1, Sec. 1.4 that, in the course of time, states with lower free energy and lower configurational entropy become relevant. Let us recall that our system has a mean-field nature because we take L large. Moreover, we look at time scales that depend exponentially on L, see Eq. (5.51). Because of this, the system visits deep states separated by very large energy barriers, that allow a description in terms of configurational entropy.[3]

This dynamical behavior of individual polymers may be expressed in terms of their dynamical partition sum at timescale $t(\theta)$, where the system is split up in independent clusters $c = 1, \cdots, \mathcal{N}_{\ell_{dyn}(\theta)}$ of width $W_{\ell_{dyn}(\theta)}$ and fixed common length L:

$$Z(\theta) = \sum_{c=1}^{\mathcal{N}_{\ell_{dyn}(\theta)}} Z_c[W_{\ell_{dyn}(\theta)}] \qquad (5.57)$$

where each of the $Z_c(W)$s is as in Eq. (5.1), with a running value for W but fixed L. This approach results in the dynamical free energy

$$\beta F_{dyn}(\theta) = L\beta f_B + \lambda L^{1/3} \left[\frac{\gamma^3 \ell_{max}^2}{2\ell_{dyn}^2(\theta)} - 1 + \frac{\ell_{dyn}(\theta)}{\ell_{max}} \right] \qquad (5.58)$$

Let us consider $T < T_K$. The dynamical free energy has a minimum, which is approached in the limit $\theta \to \theta_*(T)$. The value of the minimum coincides with the static value, Eq. (5.29). At that timescale a polymer will typically have found a state of thermodynamically optimal width $\ell_{dyn}(\theta_*) = \gamma(T)\ell_{max}$. For $T > T_K$, at time $t(\theta_*)$, it will have found the widest lane and, thus, be in equilibrium.

---

[3]For the p-spin spin-glass model, such a scenario was anticipated in [Nieuwenhuizen, 1998a].

5.3 Cooling and heating setups

Contrary to most glassy systems, our model has a regime with many thermo-dynamic states below T_K, and a single one above T_K. In order to compare with cooling experiments in realistic glasses, we must consider a heating exper-iment (which is a cooling experiment in the variable β). Let the temperature change slowly with time, $T = T(\theta)$. It defines the inverse function $\theta(T)$, that characterizes the heating trajectory. Due to the L-dependence in Eq. (5.51), θ will start at $\sim -L^{2/3}$ for small t, but when $\theta = \mathcal{O}(1)$, it need not be a monotonically increasing function of T. Approaching T_K from below under appropriate conditions, a thermal glass transition (cf. Sec. 1.1.1) will occur at some temperature $T_g < T_K$ (it is a freezing transition in terms of β). This temperature is set by $\ell_{\mathrm{dyn}}(\theta(T_g)) = \gamma(T_g)\ell_{\max}$, where, starting from small widths, the dynamically achieved width equals the thermodynamically opti-mal width. At T_g the internal energy is continuous, as it is at T_K. For $T < T_g$ the specific heat takes the equilibrium value from Eq. (5.37),

$$C = c_B L - 3\lambda L^{1/3} \left(T\frac{\partial\gamma}{\partial T} + \frac{1}{2}T^2 \frac{\partial^2\gamma}{\partial T^2} \right) \tag{5.59}$$

Having fallen out of equilibrium, for $T > T_g$ it will take the component average value from Eq. (5.39),

$$\overline{C} = c_B L - 3\lambda L^{1/3}\frac{\gamma^2\ell_{\max}^2}{\ell_{\mathrm{dyn}}^2(\theta)} \left[T\frac{\partial\gamma}{\partial T} + \frac{1}{2}T^2 \frac{\partial^2\gamma}{\partial T^2} + \frac{1}{\gamma}\left(T\frac{\partial\gamma}{\partial T}\right)^2 \right] \tag{5.60}$$

Around the thermal glass transition one has $\ell_{\mathrm{dyn}}(\theta) = \gamma\ell_{\max}$. There is a jump in the specific heat

$$\Delta C = 3\lambda L^{1/3}\frac{1}{\gamma}\left(T\frac{\partial\gamma}{\partial T}\right)^2 \tag{5.61}$$

which differs from Eq. (5.38) by the fact that $1/\gamma(T_g) > 1$ whenever $T_g < T_K$.

There is no symmetry between cooling and heating experiments. In order to have a decreasing $T(t)$, Eq. (5.51) tells us that θ must be of the order $-L^{2/3}$, whereas it is typically of the order of unity for heating. Equating Eq. (5.53) to zero, then leads to widths $\ell \sim \mathcal{O}(L^0)$, thus completely reinitializing the relaxation. Such a phenomenon was observed upon heating in spin glasses and explained in terms of hierarchy of phase space [Lefloch *et al.*, 1992].

5.3.1 Poincaré recurrence time

One may still consider the Poincaré time, where, statistically, an individual polymer has had enough time to visit the whole system. It basically suffices that it finds the widest lane, so we have to consider Eq. (5.54) for $\ell_{dyn} = \ell_{max}$, yielding

$$\theta_{\text{Poincare}} = 1 - \gamma^3(T) - \sigma(T) + \frac{\ell_*'(T)}{\ell_{max}} + \gamma^3(T)\frac{\ell_{max}^2}{\ell_*'(T)^2} \qquad (5.62)$$

that becomes

$$\theta_{\text{Poincare}}(T) = \begin{cases} 1 + \frac{3}{2}\gamma - \gamma^3, & 0 < \gamma(T) < 2^{-1/3} \\ 2 - \frac{1}{2}\gamma^3 & \gamma(T) > 2^{-1/3} \end{cases} \qquad (5.63)$$

Comparing with the parameter for achieving thermodynamic equilibrium, θ_*, from Eq. (5.56), we see that the time a given polymer needs to visit the whole system, after having found its thermodynamically optimal state, is characterized by

$$\theta_{\text{Poincare}} - \theta_* = 1 - \frac{\ell_*'(T)}{\ell_{max}} - \gamma^3\left(1 - \frac{\ell_{max}^2}{\ell_*'(T)^2}\right) \qquad (5.64)$$

Obviously, this vanishes when $\ell_*' = \ell_{max}$, that is, for $\gamma > 2^{-1/3}$, while otherwise it equals

$$\theta_{\text{Poincare}} - \theta_* = 1 - 2^{-2/3}\gamma - \gamma^3 = (2^{-1/3} - \gamma)(\gamma^2 + 2^{-1/3}\gamma + 2^{1/3}) \qquad (5.65)$$

Conclusion

Our dynamical analysis puts forward the picture that the hierarchical structure of phase space, here having the structure of a one level tree and reminiscent of one step replica symmetry breaking, is a dynamical effect. At a given timescale $t(\theta)$, only nearby states can be reached, having widths up to ℓ_{dyn} determined by Eq. (5.54). The degeneracy of these regions leads to a configurational entropy $S_c(\theta) = \mu[\ell_{max} - \ell_{dyn}(\theta)]$. At the value $\theta_*(T) = -\sigma(T) + 2\ell_*'(T)/\ell_{max}$, where $\sigma(T)$ is defined in Eq. (5.50) and $\ell_*'(T)$ in Eq. (5.48), the thermodynamically optimal width $\ell_*(T)$ has been reached: many states for $T < T_K$ and the widest state for $T \geq T_K$.

6

Potential energy landscape approach

> In just the same way the thousands of successive positions of a runner are contracted into one sole symbolic attitude, which our eye perceives, which art reproduces, and which becomes for everyone the image of a man who runs.
>
> Henri-Luis Bergson

In the previous chapters we have been analyzing how phenomena occurring in glassy materials can be reproduced by means of very simple models starting from a couple of basic ingredients, such as the separation of timescales between fast and slow processes and some kind of collective process for the relaxation of the slow modes. The previous description, however, is limited to the search for the fundamental mechanisms behind the slowing down of the relaxation and the fall out of equilibrium of the slow degrees of freedom inducing the glass transition. Those models are very helpful because, since they are simple, a lot of computation can be carried out and a rather straightforward connection between basic mechanisms and glass behavior can be obtained. They cannot, however, explain *how* these mechanisms arise in real systems. To get this information one should try to devise models that are direct representations of the intermolecular forces and chemical properties of the components of glass formers in nature. Unfortunately, moving to the level of a more faithful microscopic description implies a substantial loss in the power of theoretical predictability, unless further assumptions are introduced and numerical simulations are carried out to guide our intuition of physical phenomena.

In this chapter we will, indeed, show and discuss a very broadly diffused method to approach the study of the glassy behavior of models that are more realistic with respect to those met in previous chapters. These are systems whose space of states is complicated, both diversified and highly degenerate, and whose dynamics becomes slower and slower as temperature decreases, eventually leading to an arrest, right because of the complexity in the organization of the states. The price to pay to implement this "rugged landscape" description, as we will see in detail, will be to assume the existence of a fictitious space of the states, somehow related to the original one, and study the dynamics and the related glassy properties in this substitutive ensemble. We will, thus, deal with a symbolic dynamics, whose equivalence to the original one is the fundamental assumption of the whole approach.

The characteristics of a glassy system arise from the complex topography of the multidimensional function representing the collective potential energy yielding a nontrivial partition function and thermodynamic potential. The spatial atomic patterns in crystals and in amorphous systems share the common basic attribute that both represent minima in the *free* energy. At low enough temperature, where vibrations are minimal, one can try to assume that they are approximately represented by the minima of the *potential* energy function describing the interactions. The presence of distinct processes acting on two different timescales would mean that the deep and wide (i.e., wider than the crystal ones) local minima are geometrically organized to create a two length-scale potential energy pattern. Lowering the temperature of the liquid glass former, the bifurcation takes place as soon as it becomes "viscous." By definition the temperature at which it occurs is the dynamic glass transition temperature T_d (Sec. 1.1). The viscosity above which the decoupling of timescales occurs is usually estimated of the order of 10^{-2} Poise. This has to be compared with the order of magnitude, $\eta \sim 10^{13}$ Poise, at which the glass transition temperature, T_g, is operatively defined, allowing for a probe range for theories for the glass formation of about fifteen orders of magnitude (see Chapter 1, in particular Secs. 1.1, 1.2).

In the general case, decreasing the temperature, the free energy local minima can, in principle, be split into smaller local minima or disappear. However, if we can assume that the possible birth/death of minima is not so dramatic that they lose their identity almost everywhere in the configurational space, we can set a one-to-one correspondence between metastable states and *inherent structures* (IS) [Stillinger & Weber, 1982, 1984; Stillinger, 1995; Sastry *et al.*, 1998], i.e., between the minima of the free energy and the minima of the potential energy (see Fig. 6.1). Upon such an assumption one can, therefore, study the dynamical evolution of a glass former in its equilibrium and aging regime by means of a symbolic dynamics through ISs, rather than the true dynamics through metastable states at finite temperature.

In this point of view an approximate approach to the problem is to divide the complicated multidimensional landscape in structures formed by large deep basins and to describe the dynamics of the processes taking places as *intra-basin* and *inter-basin*. The potential energy landscape derived this way is, indeed, a description viewpoint. It helps to classify some static and kinetic phenomena associated with the glass transition according to a topographic analysis of the potential energy function. It was initially devised by Goldstein [1969] as an alternative approach to the study of the glassy state, that was potentially able to overcome the problems and inconsistencies of the *free volume* theory [Williams *et al.*, 1955; Ferry, 1961],[1] and the descriptive limitations of

---

[1]We do not consider the free volume theory in the present book. The interested reader can consult the papers of Turnbull & Cohen [1961, 1970] or the recent books of Wales [2003]

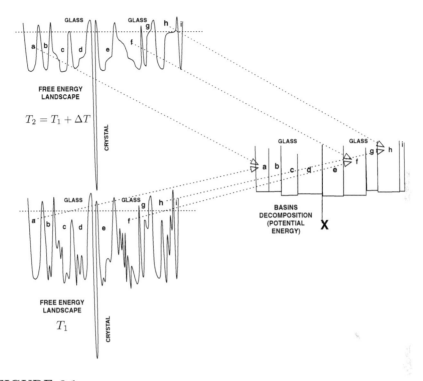

FIGURE 6.1

The figure illustrates, in a rather simplified one dimensional projection in the generalized coordinates, the free energy landscape of a glassy system (left) and its PEL basin decomposition (right). In going from the FEL to the potential energy basins (labeled by the values of their minima) the assumption is made that the basins of the true free energy landscape do not split (or disappear) as the temperature is lowered, e.g., from T_2 to T_1 in the plot, so that the same basin decomposition holds for all temperatures. What changes with the temperature is only the probability of visiting a basin with a given minimal energy (cf. Eq. (6.12)). Even though in reality upon changing T, minima can split, merge, appear, or disappear, the assumption of robustness of the basin decomposition turns out to be quite reliable, at least for the classes of models reported in the present chapter.

Adam-Gibbs-Di Marzio "entropic" theory (see Chapter 1, Sec. 1.5 and also Chapter 7) employed in the 1960s to study the amorphous materials and the viscous liquids.

In order for the dynamics in the potential energy landscape to significantly represent the actual dynamics of the system at finite temperature, not only should there be a one-to-one correspondence between ISs and real minima

---

and Binder & Kob [2005].

of the free energy landscape (FEL) at finite temperature, but also these ISs should be visited with the same frequency with which the corresponding FEL minima are visited.

In this chapter we will present the potential energy landscape theory, mainly constructed by Stillinger and Weber after the above-reported intuition. We will look at the quality of such a scheme in representing finite temperature systems and we will consider some applications to model systems and the connections with the out-of-equilibrium thermodynamics introduced in Chapter 2 and extensively investigated in Chapter 3. For a widening of the subject we might suggest that the reader start from the recent reviews of Debendetti & Stillinger [2001] and Sciortino [2005], as well as from the dedicated monograph by Wales [2003].

6.1   Potential energy landscape

In a D-dimensional system of N particles, each displaying n inner degrees of freedom the potential energy landscape (PEL) (or surface or hypersurface) is a function of $M = D \times N \times n$ variables embedded in a $DNn + 1$ dimensional space. The state of the system is represented as a state-point $\mathbf{r} \in \Re^M$ moving on the surface with a M-dimensional velocity whose average value is temperature dependent. However, the potential energy $\Phi(\mathbf{r})$ is not a function of temperature. Minima correspond to mechanically stable arrangements of N particles in the D-dimensional real space (no force or torque). Any small displacement from such an arrangement gives rise to restoring forces. Lowest lying minima are those whose neighborhoods would be selected for occupation by the system if it were cooled slowly enough to maintain the thermal equilibrium at any temperature (adiabatic cooling). These are not the minima representing a glass, that are, by definition, stuck out of equilibrium. The only exception might be the case of an ideal glassy phase, the transition to which is (would be) a true thermodynamic one (see Secs. 1.4 and 6.1.4)

One can make a configurational mapping of sets of molecular positions to minima of the PEL and separate the statistical mechanics description of the many-body problem in two distinct parts: one taking care of mechanically stable packings (in the absence of thermal excitation) and the other one dealing with the vibrational thermal excitations around those packings. In Fig. 6.2 we show a schematic representation of the PEL decomposition in IS energies and shapes.

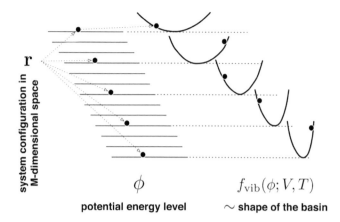

FIGURE 6.2

One dimensional reduction of the PEL description of the system. The configuration of a system of N particles (and their n inner degrees of freedom) moving in a D-dimensional space, is represented by the $M = DNn$ dimensional vector r. Different r can be grouped together according to the basin of the PEL to which they belong. Basins whose minimal energy (the inherent structure energy ϕ) is the same, contribute in the same way to the thermodynamic description, they display the same vibrational free energy. In other words, the shape of the basin, for given values of, e.g., temperature and volume, is assumed to depend only on ϕ. The potential energy level corresponding to the crystal state is not drawn.

6.1.1 Steepest descent

The potential energy function is supposed to be bounded and differentiable whenever there is no overlap of particles. It can include contributions from electrostatic multipoles and polarization effects, covalent and hydrogen bonding, intermolecular force fields, short-range electron-cloud-overlap repulsions and longer range dispersion attractions. It is, indeed, a very complicated object to devise and draw.

Not knowing the PEL, Φ, nor the FEL, *a priori*, in order to establish the mapping between configurational space of the real system and inherent structures one has, then, to define an operative procedure: the *steepest descent* method. Calling **r** the M-dimensional vector of the coordinates of the N molecules' configuration of the glass former, the steepest descent paths are solutions of the dissipative equation (in unitary mass and damping coefficient)

$$\ddot{\mathbf{r}} + \dot{\mathbf{r}} = -\nabla\Phi\left(\mathbf{r}\right) \tag{6.1}$$

in the limit case where $\ddot{\mathbf{r}} \to 0$. This highly dissipative motion represents a very effective and rapid quench, such that the system is never able to overcome any barrier in the process and it can only go to the underlying minimum of the PEL. In practice, this operative procedure establishes a mapping between

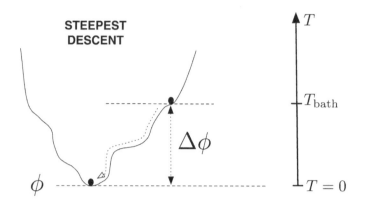

FIGURE 6.3

Steepest descent procedure: any point in the basin is connected to the minimum ϕ if the dynamics expressed by Eq. (6.2) is applied. The basin is composed by the set of points in configurational space connected to the configuration of minimal potential energy without overcoming any energy barrier. On the right a temperature axis is displayed. The system point representing the system at the temperature of the heat bath is connected by steepest descent to the minimum of its basin, as if performing an instantaneous quench to zero temperature: no energy barrier can be overcome and the system description is, thus, decomposed by sets of points connected to the same minimum in potential energy.

the continuous coordinates space to the discrete set of minima of the PEL. See Fig. 6.3.

One can define an inherent structure as that minimum below an actual configuration of the system evolving in time at some temperature T, that is, the minimum of the potential energy reached by steepest descent. In this respect we must stress that by "instantaneous quenching", i.e., cooling with an infinite rate, we mean to decrease the kinetic energy in a continuous way as the evolution time is stopped. The purely mechanical dynamics that defines operatively an IS, is performed, even conceptually, "out of the real time", on an auxiliary time variable s keeping track of the iteration step. The configurations are mapped onto IS by solving the multidimensional equation

$$\frac{\partial \mathbf{r}}{\partial s} = -\nabla\Phi(\mathbf{r}) \qquad (6.2)$$

If we take a given $\mathbf{r}$, describing the real system at time t, as the initial condition ($s = 0$) for the above dynamics, the solution $\mathbf{r}(s)$ identifies the corresponding IS in the asymptotic limit in s.

The introduction of ISs allows, at low enough temperature, a decomposition of the partition function into an IS part, connected to the potential energy levels corresponding to the configurations of the system at temperature T, and a part connected to the thermal excitation of the configurations in a single minimum. This decomposition holds both for undercooled liquids and for solid glasses. Actually, we might turn this around and say that an undercooled, or viscous, liquid is defined as a liquid whose thermodynamic description can be decomposed as above.

6.1.2 Features of the PEL description borrowed from vitreous properties

The first basic assumption behind the FEL-PEL correspondence is that the existence of large energy barriers between minima, much larger than the thermal excitation, is intrinsic in glasses. Moreover, as we will see in detail later on, the mechanical stability of a glass at low temperature is also taken for granted: a glass state point is always considered as near to a potential energy minimum. This property is translated also to undercooled liquids, even though no mechanically stable structure can be identified in real space. The thermodynamic properties can be approximately described in terms of a spectrum of *harmonic* vibrational frequencies (see Sec. 6.1.5). In this approximation, the flow description is consistent with the PEL description only at low temperature and only when the (undercooled) liquid is very viscous and can its flow be represented by means of movements through PEL minima. At higher temperatures the whole description becomes inconsistent but can be cured by introducing anharmonic corrections (cf., e.g., [Büchner & Heuer, 1999; La Nave *et al.*, 2003a; Keyes & Chowdhary, 2004]).

The PEL region representing a glass former displays a large number of minima, unlike the region representing crystals. The decrease of entropy recorded in undercooled liquids is associated with the progressive ordering of the system in configuration space, i.e., in the progressive population of basins with lower energy and lower degeneracy. When the characteristic time of this population evolution towards the bottom of the PEL is long, the system properties turn out to depend on its history.

The function Φ depends on all degrees of freedom of all molecules. However, in a transition over a barrier only a certain number of molecules cooperatively rearrange their positions to move the system point to a nearby minimum. Molecules in a small region of the real space rearrange their positions and this is described as a transition between two "connected" minima in the M-dimensional PEL. Unless the temperature is very low, rearrangements can occur in different real space regions at the same moment. In this case the system point in the PEL is no more near a minimum most of the time. Anyway, a correspondence can be set with the PEL minimum "lying below" the state point. This means that the system would reach a minimum in a steepest descent procedure taking place with relatively small changes of most of the coordinates. The points connected in this way to a minimum, belong to its basin of attraction, i.e., to its inherent structure. A transition between two configurations pertaining to the same IS - an *inter-basin* transition - is due to a local change.

6.1.3   Inter- and intra-basins transitions: scales separation

In a real glass, the presence of distinct processes, acting on different timescales, can be obtained from a careful analysis of the relaxation response function above T_{g}. We limit ourselves to a two timescale approach. This means that the deep and wide local minima below T_{d} are geometrically organized into a two length-scales potential energy pattern. As a consequence, the system shows α and β processes. The α processes represent the escape from one deep minimum within a large scale valley to another valley. This escape requires a lengthy directed sequence of elementary transitions producing a very large activation energy, much higher than for β processes. Moreover, the high-lying minima between any two valleys, among which the system is making a transition, are many and degenerate. This implies a large activation entropy for the *inter-basin* transition between two deep (amorphous state-related) minima. The β processes are instead related to elementary relaxations between neighboring minima (*intra-basin* dynamics). Usually one puts together all kinds of β processes in the short timescale, since they are in any case much shorter than the observation time considered.[2]

Processes are characterized by their vibrational frequencies, or characteristic times, and by their length-scales. Vibrations can take place also in materials with no regular lattice, but when the relaxation time for the structure is such that vibrations are damped in a time equal to a cycle or less, then the idea of vibration itself loses its meaning. In viscous liquids, a decoupling takes place in vibrations of low frequencies, damped, and vibrations of high frequencies, active. A threshold between the two regimes is given by frequencies of the order of magnitude of the inverse relaxation time. The modification in a cooperative rearranging region (CRR) takes some time to occur. This time must be less than the relaxation time, for the description in terms of the PEL to hold, and this is exactly the case for highly viscous undercooled liquids and glasses.

A separation of length-scales also takes place. Processes with short-length scales ("rattling in the cage", i.e., high frequency vibrations) are decoupled from processes with long length scales (diffusive motion, with low frequency, highly damped in the amorphous phase). These different length-scales go hand-in-hand with different timescales.

From old neutron scattering experiments [Sjölander, 1965] it was possible already for Goldstein [1969] to estimate that the separation of timescales can

---

[2]In many amorphous materials, besides the relatively fast "rattling-in-the-cage" β processes (sometimes called β_{fast}), also the so-called Johari-Goldstein β processes [Johari & Goldstein, 1971] are detected. Even if considered fast with respect to the almost frozen α processes, their relaxation times to equilibrium usually display an Arrhenius behavior in temperature. They are, indeed, experimentally clearly identified only at low temperature and can be related to activated processes, though of a local nature and, thus, not truly contributing to the structural relaxation. See also Fig. 1.1.

appear for viscosity values above the order 10^{-2} Poise in normal liquids, this way defining T_d (cf. Sec. 1.1).

6.1.4 Inherent structures distribution: formal treatment

We now derive the expression of the thermodynamic potential for viscous liquid based on the PEL approach. We stress that decoupling of timescales is assumed even though the equilibrium for both scales is still supposed. The out-of-equilibrium case, valid (possibly) for the glass state description will be discussed in Sec. 6.3.

Let us call $R(I)$ the ensemble of $\mathbf{r}$ configurations related, through steepest descent, to the inherent structure I. $R(I)$ is connected but not necessarily convex. Indeed, a saddle point can occur within it and, in this case, the quench paths bifurcate.

The partition function of N particles in a D dimensional space (we set the number n of degrees of freedom per particle equal to one) is

$$Z_N = \int d\mathbf{r}\, \exp[-\beta\Phi(\mathbf{r})] \frac{1}{\lambda^{DN} N!} = \sum_I \int_{R(I)} d\mathbf{r}\, \exp[-\beta\Phi(\mathbf{r})] \frac{1}{\lambda^{DN} N!} \quad (6.3)$$

The parameter λ is the thermal wavelength coming from the integration over the momenta of the molecules. In a monoatomic system of just one type of particles with a mass m it is, e.g., $\lambda = \sqrt{2\pi\beta\hbar^2/m}$. We notice, however, that it plays no role in the PEL formulation. Instead of single inherent structures, it is more meaningful to consider equivalence classes, i.e., to group all the basins with the same minima together. Generically, there will be $N!/\sigma$ minima in each class, where σ is a model parameter depending on the kind of potential.[3]

Moreover, one can write $\Phi(\mathbf{r}) = \Phi_I + \Delta_I(\mathbf{r})$ and

$$Z_N = \sum_I {}' e^{-\beta\Phi_I} \int_{R(I)} \frac{d\mathbf{r}}{\lambda^{DN}} \exp[-\beta\Delta_I(\mathbf{r})] \quad (6.4)$$

where the prime on the sum means that possible global (crystal) minima are not counted (we are not describing the crystallization process). The minimum of the potential energy is Φ_I, whereas $\Delta_I(\mathbf{r})$ is the difference between the potential energy of a generic configuration $\mathbf{r}$ belonging to a basin $R(I)$, and the relative IS energy Φ_I. We now introduce the potential energy per particle, $\phi = \Phi/N$ and the density of distinct packing mechanically stable states

$$\Omega(\phi) \equiv \sum_I {}' \delta(\phi - \Phi_I/N)/\sigma(I) \quad (6.5)$$

---

[3]Limiting case values are $\sigma = 1$ for wall forces and $\sigma = N$ for periodic boundary conditions (otherwise $1 < \sigma < N$). The factorial $N!$ is due to permutation of identical particles [Stillinger & Weber, 1982].

with $\phi_m \leq \phi \leq \phi_{th}$, where ϕ_m is the crystal minimum and ϕ_{th} is the value of the "threshold" states above which no minima occur anymore.

The density Ω grows exponentially with N. If we take a system of N particles with short-range interactions it is likely that the move from one state to another one in the PEL can be caused by sequences of local rearrangements in the real space. If we take a large system we can divide it in cells large enough that any local set of rearrangements in one cell will not interfere with others. The total number of configurations explored this way can then be considered as multiplicative over the cells (at leading order) and this observation leads to an exponential behavior in N for the total number of distinguishable potential energy minima:

$$\Omega(\phi) \sim e^{s_c(\phi)N/k_B} \tag{6.6}$$

where $s_c(\phi)$ takes into account all the minima of the PEL at a given ϕ level.

The vibrational part, at a given, narrow, interval of values of ϕ, can be averaged over all possible packings displaying the same energy. This leads to the definition of the vibrational contribution of the free energy:

$$f_{\text{vib}}(\phi, \beta) \equiv -\lim_{N \to \infty} \frac{1}{N\beta} \log \left\langle \int_{R(I)} d\mathbf{r} \exp[-\beta \Delta_I(\mathbf{r})] \right\rangle \tag{6.7}$$

This assumption is equivalent to stating that the value of potential energy uniquely characterizes the properties (i.e., the shape) of the basin in the PEL (cf. Fig. 6.2).

Eventually, using the definitions of Eqs. (6.5) and (6.7) we obtain

$$Z_N \sim \lambda^{-DN} \int_{\phi_m}^{\phi_{th}} d\phi \exp \{N[-\beta\phi + s_c(\phi)/k_B - \beta f_{\text{vib}}(\phi, \beta)]\} \tag{6.8}$$

and, approximating further the integrand by its maximum as $N \to \infty$ (saddle point approximation), the free energy per particle of the undercooled liquid is

$$f(\beta, \bar{\phi}) = \lim_{N \to \infty} -\frac{1}{\beta N} \log Z_N = \bar{\phi} + f_{\text{vib}}(\beta, \bar{\phi}) - Ts_c(\bar{\phi}) \tag{6.9}$$

where $\bar{\phi}$ is the solution of the saddle point equation

$$T \frac{\partial s_c(\phi)}{\partial \phi} = 1 + \frac{\partial f_{\text{vib}}(\beta, \phi)}{\partial \phi} \tag{6.10}$$

with

$$\partial_\phi^2 s_c(\phi) < \beta \partial_\phi^2 f_{\text{vib}}(\beta, \phi) \tag{6.11}$$

The value $\bar{\phi}$ is the average potential energy per particle obtained by quenching to zero temperature a collection of system configurations randomly selected from the equilibrium state at temperature $T = 1/\beta$. The free energy f, Eq. (6.9), can, thus, be derived if the depth ($\bar{\phi}$), the shape ($\sim f_{\text{vib}}$) and the number (e^{Ns_c/k_B}) of the basins of the PEL are known. We should stress that, even though Eq. (6.9) is already a guess, based on the assumption that the PEL minima are in a one-to-one correspondence with the real - unknown - FEL minima (cf. Fig. 6.1), further approximation will be necessary to actually compute it, especially for what concerns the shape, that is, the expression for f_{vib}, Eq. (6.7) (cf. Sec. 6.1.5). We also notice that $s_c(\phi)$ is independent of the temperature as a function of the potential energy and that it acquires a temperature dependence exclusively through the insertion of the saddle point value $\bar{\phi}(T)$ in Eq. (6.9).

For monoatomic substances in the fluid phase, e.g., liquified noble gases and molten alkali metals, the distribution of distinguishable minima (equivalence classes) in ϕ shows a high peak at the amorphous packing value of potential energy. The dominating amorphous minima are narrowly distributed in ϕ. The liquid phase displays the free energy Eq. (6.9) plus some additional T-dependent terms [Stillinger & Weber, 1982].

In glasses, however, this cannot hold, since no thermalization takes place. It becomes infeasible even among the distribution of ϕ values limited to the amorphous interval. On the contrary, the glass transition (unlike the melting transition) turns out to be "nonuniversal," the sharpness of the changes in thermodynamic observables (such as the specific heat, cf. Figs. 1.4, 1.5) is smeared out and, depending on the different histories of the material, the degree of sharpness can be different. As we underlined from the very beginning (Chapter 1, Sec. 1.1) the glass transition, indeed, is a purely kinetic, off-equilibrium transition.

The configurational entropy of the inherent structures might, in principle, be obtained from the molecular structure and from the interactions. In practice, one needs to numerically simulate the dynamics of glass former models on a computer to get such information.

In cooling, the (temperature-dependent) inherent structure energy $\bar{\phi}(T)$ tends to the minimum value of the potential energy, ϕ_{m}. Furthermore, we observe that Eq. (6.10) is consistent at any temperature only if the slope of the configurational entropy at the minimum ϕ_{m} tends to infinity right as $T \rightarrow 0$ (the right-hand side does not diverge). Instead, if $\partial s_c(\phi)/\partial \phi|_{\phi=\phi_{\text{m}}}$ is finite, this implies that $\bar{\phi} \rightarrow \phi_{\text{m}}$ at some finite temperature "T_{is}," below which Eq. (6.10) becomes meaningless. In the latter case, below that temperature, the system would be stuck in an amorphous packing belonging to one of the sub-extensively many low-lying minima of the PEL. Consistent with previous definitions, and provided that the excess entropy is identified with the configurational entropy (consult again the discussion in Sec. 1.4), the temperature in question coincides with the Kauzmann temperature T_K and the lack of analyticity occurring in Eq. (6.10) signals a thermodynamic transition to an "ideal glass" phase. We stress, however, that this is true under the assumptions that (i) in the metastable undercooled liquid below the melting temperature (but above the glass transition) the vibrational modes yield the same entropic contribution as the vibrational modes in the underlying stable crystal, and (ii) the liquid-crystal excess entropy can be correctly extrapolated even far below the glass temperature T_g, down to temperatures at which it vanishes. These assumptions cannot be experimentally verified and, furthermore, they are not always satisfied in numerical simulations. As a counterexample the reader can refer to a soft spheres model with a mean-field attraction studied by Shell *et al.* [2003] where it is explicitly shown that the ideal glass transition point and the Kauzmann locus (at which the excess entropy vanishes) are distinct items.

From Eqs. (6.9)-(6.10) one can define the IS probability distribution

$$p(\phi, T) = \frac{1}{Z_N(T)} \Omega(\phi) \exp\{-N\beta [\phi + f_{\text{vib}}(\phi, T)]\} \qquad (6.12)$$

where $Z_N(T)$ is the partition function. Eq. (6.12) represents the probability that an equilibrium configuration at $T = 1/\beta$ belongs to a basin associated with an IS structure with an energy density in the interval $[\phi, \phi+d\phi]$ [Stillinger & Weber, 1982; Sciortino *et al.*, 1999]. The quantity $k_B \log \Omega(\phi) = N s_c(\phi)$ is the configurational entropy of the states at energy ϕ. Integrating over all energy levels, the temperature-dependent configurational entropy counting all basins involved in the zero temperature symbolic representation of the FEL minima at finite temperature, is defined as

$$s_c^{\text{tot}}(T) = \int d\phi \, p(\phi, T) \, s_c(\phi) \qquad (6.13)$$

We recall that, in order to write down $p(\phi, T)$, the basin information must depend only on the potential energy level: minima of equal ϕ all contribute with the same f_{vib} to the system. At very low temperatures, even below T_g, a further approximation is sometimes made, namely that $f_{\text{vib}}(\phi, T) \sim f_{\text{vib}}(T)$, because fluctuations inside one IS are small [Sciortino *et al.*, 1999; Kob *et al.*, 2000]. In that case the shape of the basin depends exclusively on the temperature: the internal (vibrational) states of every IS produce the same (vibrational) free energy f_v at given T. We will consider this very special case in the following, when the out-of-equilibrium framework will be considered, but we remark for now that this hypothesis is already violated in very standard models for glass formers such as, e.g., the Lennard-Jones binary mixture [Sciortino & Tartaglia, 2001]. See also [Starr *et al.*, 2001; Mossa *et al.*, 2002; Giovambattista *et al.*, 2003; Sciortino *et al.*, 2003].

6.1.5   Harmonic approximation

The vibrational contribution to the free energy can be approximated by considering the quadratic approximation of the potential energy around an inherent structure configuration of energy ϕ. In the base of normal mode eigenvectors, the dynamics of the system can be described as a set of M harmonic oscillators and the partition function of a basin, averaged over all degenerate basins, can be written as

$$Z_{\text{basin}}(\phi; T) = e^{-\beta\Phi} \left\langle \prod_{i=1}^{M} \frac{1}{\beta\hbar\omega_i(\phi)} \right\rangle \tag{6.14}$$

Its logarithm yields the free energy contribution of the single Φ-basin: $\Phi + F_{\text{vib}}(\phi, T)$. The vibrational part, Eq. (6.7) turns out, thus, to be

$$F_{\text{vib}} = N f_{\text{vib}}(\phi; T) = -k_B T \log Z_{\text{basin}}(\phi; T) - \Phi \tag{6.15}$$

$$= -k_B T \log \left\langle \exp\left[-\sum_{i=1}^{M} \log \beta\hbar\omega_i(\phi) \right] \right\rangle \simeq k_B T \left\langle \sum_{i=1}^{M} \log \beta\hbar\omega_i(\phi) \right\rangle$$

Apart from constants and temperature factors, F_{vib} contains the information of the shape of the mean basin at ϕ:

$$\mathcal{S}(\phi) \equiv \left\langle \sum_{i=1}^{M} \log \frac{\omega_i(\phi)}{\omega_0} \right\rangle = +\beta F_{\text{vib}} - M \log \beta\hbar\omega_0 \tag{6.16}$$

where ω_0 is the frequency unit (making the argument of the logarithm dimensionless). This quantity is usually found to be linear or almost linear in ϕ in numerical simulations of viscous liquid models, e.g., in Lennard-Jones (LJ) binary mixtures [Sciortino *et al.*, 1999; Sastry, 2001], in the Lewis-Wahnström (LW) orthoterphenyl model [Mossa *et al.*, 2002], and in the simple point charge extended (SPC/E) water model [Starr *et al.*, 2001; Giovambattista *et al.*, 2003; Sciortino *et al.*, 2003]:

$$\mathcal{S}(\phi) = a + b\,\phi \tag{6.17}$$

where the coefficients are volume dependent. Following this observation, then, in the harmonic approximation f_{vib} turns out to be linear in ϕ, to a very good approximation, in a large number of significant cases.

6.2 Thermodynamics in supercooled liquids

We consider in this section how the IS-vibrational separation of the PEL theory is implemented in thermodynamics quantities and, in particular, how an equation of state can be constructed, displaying some universal features common to the vast majority of undercooled liquids. We, thus, start looking at the pressure and its relationship with temperature and volume (or density).

6.2.1 Inherent structure pressure

As any other observable, in the PEL formalism, the pressure can be considered as the sum of contributions due both to the vibrational displacements around some minimal energy packing and to the configurational contribution due to the separation of the space of metastable states into inherent structures. The inherent structure pressure contribution P_{is} is, therefore, an interesting parameter that can be directly measured in numerical simulations [Roberts et al., 1999; Utz et al., 2001; La Nave et al., 2002; Shell et al., 2003]. Indeed, a typical behavior has emerged, common to several glass former models, such as the extended simple point charge (SPC/E) effective pair water model [Roberts et al., 1999], models for ethan, n-pentene and cyclopentene [Utz et al., 2001], a monoatomic Lennard-Jones (LJ) model with a cutoff [Sastry et al., 1997] and the Lewis-Wahnström (LW) model for orthoterphenyl [La Nave et al., 2002], to mention a few. To calculate P_{is}, at fixed values of temperature and density/volume, after having sampled a large number of inherent structures by steepest descent in molecular dynamics, the standard virial expression can be used. Even though an inherent structure is defined as a mechanically stable configuration of molecules and, thus, the latter do not experience any force, a virial function and a corresponding pressure can still be computed [Sastry et al., 1997]. In all cases, the density dependence of the IS pressure displays a qualitatively similar behavior (see Fig. 6.4):

1. It is positive for large density.

2. As the density decreases it becomes negative: the system moves from a state of positive pressure to a state of tension.

3. Decreasing ρ further, the IS pressure reaches a minimum at a given value (usually referred to as the "Sastry density," ρ_S) that describes the situation of maximum feasible tension. Below this point the system fractures and the PEL is no more homogeneous, presenting empty regions. Below ρ_s the system is mechanically unstable.

4. As the density goes to zero, $P_{is} \rightarrow 0^-$.

In terms of the PEL formalism the IS pressure can be reasonably defined as minus the volume derivative of the average IS energy $\bar{\phi}$ [Shell et al., 2003]. We can verify this definition identifying the IS pressure contribution to the total pressure, that is, differentiating Eq. (6.9) with respect to the volume:

$$P = -\frac{\partial f}{\partial v} = \rho^2 \frac{\partial f}{\partial \rho} = \rho^2 \left[\frac{\partial f_{vib}}{\partial \rho} - T \frac{\partial s_c}{\partial \rho} \right] \tag{6.18}$$

where v is the specific volume V/N and $\rho = 1/v$. We have used Eq. (6.10), where now the vibrational free energy and the configurational entropy also depend on the specific volume (or the density). The above expression can be

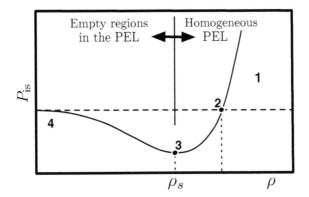

FIGURE 6.4
Inherent structure pressure as a function of density: typical behavior for glass form-ing undercooled liquids. The horizontal dashed line indicates the zero pressure level. The numbers refer to the description in the text.

rewritten as $P = P_{\text{vib}} + P_{\text{is}}$, with

$$P_{\text{vib}} = \rho^2 \left[\frac{\partial f_{\text{vib}}}{\partial \rho} - T \frac{\partial s_c}{\partial \rho} \frac{\partial f_{\text{vib}}}{\partial \phi} \bigg|_\rho \left(1 + \frac{\partial f_{\text{vib}}}{\partial \phi} \bigg|_\rho \right)^{-1} \right] \qquad (6.19)$$

$$= - \frac{\partial f_{\text{vib}}}{\partial v} - \frac{\partial f_{\text{vib}}}{\partial \phi} \bigg|_v \frac{\partial \bar{\phi}}{\partial v} \bigg|_{s_c}$$

$$P_{\text{is}} = -\rho^2 \frac{\partial s_c}{\partial \rho} \left(\frac{\partial s_c}{\partial \phi} \bigg|_\rho \right)^{-1} = - \frac{\partial \bar{\phi}}{\partial v} \bigg|_{s_c} \qquad (6.20)$$

From the last line of Eq. (6.19) we see that, indeed, P_{is} coincides with the definition of the (negative) derivative of the mean potential energy value of the wells visited at temperature T with respect to the volume.

6.2.2 Random energy model and Gaussian approximation

A very popular and pioneering model in the study of the slowly relaxing amorphous systems, that comes from the spin-glass theory, is the random energy model (REM) of Derrida [1980].[4] We report some of its features in Sec. 7.2 but, for present purposes, we recall the three basic ingredients of the model:

---

[4]This model is a simplification of the p-spin spin-glass model [Derrida, 1981; Gross & Mézard, 1984] in the limit of the number of p simultaneously interacting bodies going to infinity. The variance is $\sigma \sim \sqrt{N}$.

1. A system of N particles has 2^N energy levels E_i.

2. The energy levels are distributed according to the Gaussian probability

$$p_{\text{rem}}(E) = \frac{1}{\sqrt{2\pi\sigma^2}} \exp\left\{-\frac{(E-E_0)^2}{2\sigma^2}\right\} \qquad (6.21)$$

   where E_0 is an (arbitrary) energy scale.

3. The energy levels are independent identically distributed (IID) random variables.

The first two requirements are the common features of many spin-glass models, including the p-spin model that works quite well as a mean-field model for structural glasses, despite the lack of microscopic similarity with a glass former.[5] The third property is specific of the present model and simplifies things so much that it can be solved exactly. It leans on the rather strong hypothesis of lack of correlations among energy levels, that are independent, identically distributed (IID), variables.

The REM description can be applied to the PEL formalism (see, e.g., [Keyes, 2000; Sciortino, 2005]) by writing the number of basins at potential energy ϕ as

$$\Omega(\phi)\, d\phi = e^{\alpha N} p_{\text{rem}}(\phi) d\phi \qquad (6.22)$$

The parameter α is the logarithm of the total number of basins of the PEL (divided by the number of particles). It would be the total complexity in an all-temperatures survey of the space of states. The configurational entropy $s_c(T)$, Eq. (6.13), counts the total number of basins available at temperature T, according to (and weighted by) the ϕ-distribution $p(\phi; T)$, Eq. (6.12). The partition function Eq. (6.8) can, then, be exactly computed as

$$Z = \int_{\phi_{\text{m}}}^{\phi_{\text{th}}} d\phi \, \exp\left\{N\left[\alpha - \frac{(\phi-\phi_0)^2}{2\sigma^2} - \beta\phi - \beta f_{\text{vib}}(\phi)\right]\right\} \qquad (6.23)$$

The REM, or Gaussian, approximation in the PEL formalism has been checked in numerical simulations of several different glass former models, e.g., for binary LJ mixtures [Heuer & Büchner, 2000; Sastry, 2001], for the LW model of orthophenyl [La Nave *et al.*, 2002; Mossa *et al.*, 2002], as well as for the SPC/E water model [Starr *et al.*, 2001]. In all these cases one can find a REM-like behavior for the density of states $\Omega(\phi)$, such that

$$\frac{s_c(\phi)}{k_B} = \alpha - \frac{(\phi-\phi_0)^2}{2\sigma^2} \qquad (6.24)$$

---

[5]In the model, groups of p spins (usually spherical [Kirkpatrick & Thirumalai, 1987b; Crisanti & Sommers, 1992] but also Ising [Gross & Mézard, 1984; Gardner, 1985]) interact collectively with a contribution that is a random variable with a Gaussian distribution, quenched with respect to the motion of the spins.

Furthermore, the vibrational entropy (linearly related to the "shape factor" $\mathcal{S}$, Eq. (6.16)) in the harmonic approximation, turns out to be linear in ϕ:

$$\frac{S_{\mathrm{vib}}(\phi)}{k_B} = M - \sum_{i=1}^{M} \log \beta\hbar\omega_i(\phi) \simeq \frac{S_{\mathrm{vib}}(\phi_0)}{k_B} - Nb(\phi - \phi_0) \qquad (6.25)$$

and so is the vibrational free energy $F_{\mathrm{vib}} = U_{\mathrm{vib}} - TS_{\mathrm{vib}}$ (the internal energy U_{vib} contribution amounts to an energy term $k_B T$ for every degree of freedom).

6.2.3 Equation of state

The possibility of working out a universal equation of state for undercooled liquids allows, among other things, for a comparison of numerical simulations, much more easily performed at a fixed volume, and true experiments, more easily carried out at a fixed temperature. The relationship between pressure, volume and temperature has been computed in different models for viscous liquids assuming

1. the REM nature of potential energy levels, Eqs. (6.21) or (6.24)

2. the linearity in ϕ of the vibrational contribution to the free energy (that is numerically established in the case of harmonic vibrations, see above), cf. Eqs. (6.15) and (6.17)

As we already said, these hypotheses turn out to be verified in a number glass former models.

The IS energy in the supercooled state in LJ binary mixtures, and in models for water and orthoterphenyl (OTP), turns out to significantly decrease on cooling, thus strongly hinting that low-lying minima become more and more important. In numerical simulations, though, experiments are performed at constant volume.[6] In order to get more in contact with reality, it becomes rather fundamental to find the relation between pressure and volume.

In Sec. 6.2.1 the formal thermodynamic expression of the pressure in the PEL formalism has been derived, and the so-called inherent structure pressure contribution has been identified. Here, following the case study of La Nave *et al.* [2002] on the LW model for orthoterphenyl, we show how the equation of state looks for the class of models satisfying conditions 1 and 2 above.

Combining Eqs. (6.15) and (6.17) the vibrational free energy is:

$$f_{\mathrm{vib}}(\phi; T, V) = T \sum_i \log \beta\hbar\omega_i(\phi) = T[a(V) + M \log \beta\hbar\omega_0 + b(V)\phi]$$
$$= f_{\mathrm{vib}}(\phi_0; T, V) + Tb(V)(\phi - \phi_0) \qquad (6.26)$$

---

[6]With some exceptions, e.g., the molecular dynamics simulations of undercooled water of Tanaka [1996].

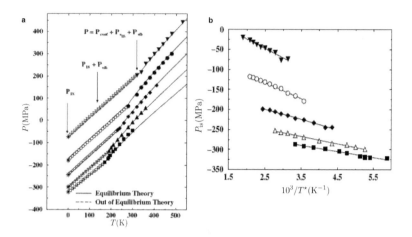

FIGURE 6.5

(a) Equation of state for the LW orthoterphenyl model. The comparison is shown between the molecular dynamic pressure obtained using the virial function (symbols) and the PEL equation of state, Eq. (6.33) (lines). Solid symbols stand for equilibrium values (only reachable at higher T). Open symbols are instead calculated from molecular dynamics data recorded during a constant heating procedure starting from the IS configuration marked by the asterisk. The dashed lines (superimposed to open symbols) are those obtained from Eq. (6.31) for the heating procedure. (b) Inherent structure equation of state: P_{is} versus inverse (effective) temperature. Reprinted figures with permission from [La Nave *et al.*, 2002]. Copyright (2002) by the American Physical Society.

Inserting Eqs. (6.26) and (6.24) into Eq. (6.10) for the mean inherent structure energy, allows for a simple solution in this case:

$$\bar{\phi}(T, V) = \phi_0 - b(V)\sigma^2(V) - \beta\sigma^2(V) \tag{6.27}$$

In this way, one can perform an exact evaluation of the free energy, Eq. (6.9).

Using the free energy, Eq. (6.9), for systems with decoupled IS/vibrational landscape structure, the mechanical definition of pressure can be divided in three contributions, yielding

$$P = -\frac{\partial F}{\partial V} = T\frac{\partial s_c}{\partial V} - \frac{\partial \bar{\phi}}{\partial V} - \frac{\partial f_{\text{vib}}}{\partial V} = P_{\text{conf}} + P_{\bar{\phi}} + P_{\text{vib}} \tag{6.28}$$

where the first two contributions combine in what we have called inherent structure pressure, cf. Eqs. (6.18) and (6.20). In the present subclass of supercooled liquids (random energies, harmonic vibrations and "linear shapes"), one finds the expressions:

$$P_{\text{conf}} = T\frac{\partial S_\infty}{\partial V} \qquad\qquad -\frac{\partial b\sigma^2}{\partial V} - \frac{1}{2T}\frac{\partial \sigma^2}{\partial V} \qquad (6.29)$$

$$P_{\bar\phi} = \qquad\qquad\qquad -\frac{\partial \phi_\infty}{\partial V} + \frac{1}{T}\frac{\partial \sigma^2}{\partial V} \qquad (6.30)$$

$$P_{\text{vib}} = -T\frac{\partial(a+b\phi_\infty)}{\partial V} + \frac{\partial b\sigma^2}{\partial V} \qquad (6.31)$$

$$- - - - - - - - - - - - - - - - - - - \ +$$

$$P = \qquad TP_T \qquad + \quad P_{\text{const}} + \frac{1}{T}P_{1/T} \qquad (6.32)$$

where $S_\infty = \alpha N - b^2\sigma^2/2$ and $\phi_\infty = \phi_0 - b\sigma^2$. The last line enhances the kind of temperature dependence of the various contributions to the pressure. One can further recombine the above terms, eventually obtaining the equation of state:

$$P = T\left(\frac{\partial S_\infty}{\partial V} - \frac{\partial(a+b\phi_\infty)}{\partial V}\right) - \frac{\partial \phi_\infty}{\partial V} + \frac{1}{2T}\frac{\partial \sigma^2}{\partial V} \qquad (6.33)$$

The volume dependence passes through the parameters S_∞, ϕ_∞, a, b and σ. Indeed, their behavior specifies the model [provided Eqs. (6.24)-(6.26) are valid, otherwise the whole formalism has no theoretical support] and has to be deduced each time.

6.2.4   IS equation of state

If we perform the steepest descent procedure at constant volume, starting from equilibrium configurations, the vibrational contribution to the pressure is suppressed and only the IS part, P_{is}, remains, i.e., the sum of the configurational and the mean potential energy contributions. During the descent, the inherent structure pressure keeps the value that it has in the equilibrium configuration at temperature T. Indeed, by definition, the configurational contribution derives from the configurational entropy that counts all the available minima of the PEL, once the steepest descent procedure has been carried out, and the IS energy $\bar\phi$ is the average value of all reachable minima starting from equilibrium at a certain T (i.e., starting from the minima of the FEL).

Moving to such a purely IS description, another equation of state can be derived:

$$P_{\text{is}}(T,V) = -\frac{\partial \phi_0}{\partial V} + T\frac{\partial S_\infty}{\partial V} + \frac{\beta}{2}\frac{\partial \sigma^2}{\partial V} \qquad (6.34)$$

where $\beta = 1/(k_B T)$.

Starting from the purely inherent structure contribution, the total pressure can be seen as P_{is} plus the vibrational contribution arising when the system is "heated" from the bottom of a PEL minimum (that is symbolically at

$T = 0$), at constant volume, performing some kind of "steepest *ascent*." The vibrational contribution is, cf. Eqs. (6.15), (6.17), and (6.28),

$$P_{\text{vib}} = k_B T \frac{\partial \mathcal{S}}{\partial V} \qquad (6.35)$$

In Fig. 6.5 we reproduce the inherent structure equation of state where the prediction of Eq. (6.34) is compared to the data of numerical simulations on the LW model [La Nave *et al.*, 2002].

If dealing with a glass instead of an undercooled liquid, one can try to keep the same formal theoretical structure built up to now, substituting $T \to T^\star$ in Eq. (6.34). The temperature $T^\star$ is an effective temperature that we introduce now for the first time in the PEL formalism. It is the temperature conjugated to the ϕ derivative of the configurational entropy in the out-of-equilibrium analogue of Eq. (6.10):

$$1 + \frac{\partial f_{\text{vib}}(\beta, \phi)}{\partial \phi} - T^\star \frac{\partial s_c(\phi)}{\partial \phi} = 0 \qquad (6.36)$$

If we are dealing with a substance in its viscous liquid regime, $T^\star = T$ and the IS identified by steepest descent are those at equilibrium. Dealing with a vitrified substance, instead, would yield a different effective temperature, as we discussed in the general framework of out-of-equilibrium thermodynamics in Sec. 2.4. We will apply this idea to the PEL theory in the next section. We stress, however, that the effective temperature appearing in Eq. (6.36) is not a uniquely defined expression.

6.3   The solid amorphous phase: thermodynamics out of equilibrium

In a viscous, supercooled liquid the timescale separation between intra-basin and inter-basin dynamics is assumed and expressed in the free energy Eq. (6.9). The equilibration process is complete when the IS energy $\bar{\phi}(T)$, solution of Eq. (6.10), is equal to the mean of all potential energy minima visited by the system in the dynamics (the minima identified by steepest descent).

Lowering the temperature, the kinetic arrest of many degrees of freedom leads to the amorphous phase, and the system falls out of equilibrium. Eq. (6.10) no longer yields anything meaningful, in principle. In an attempt to generalize thermodynamics to include the glass phase in its descriptive frame-work, however, one can define a further temperature-like parameter in the spirit of the effective temperatures widely discussed in the previous chapters.

6.3.1 PEL effective temperature from direct comparison to the aging dynamics

A first way to relate aging dynamic properties of an off-equilibrium system at a given temperature T into a thermodynamic framework is to devise the existence of an equilibrium ensemble at an effective temperature different from T (usually larger, since the glass former relaxation slows down in cooling). This hypothesis has been tested with some success in numerical simulations of, e.g., LJ binary mixture [Kob *et al.*, 2000] and random orthogonal model (ROM) [Crisanti & Ritort, 2000b]. We will refer to it as the "PEL equilibrium matching" effective temperature: T_e^{em}. The operative procedure to define it is the following, as depicted in Fig. 6.6:

1. The average inherent structure $\bar{\phi}$ is computed at different temperatures in systems for which thermalization has occurred (left plot in Fig. 6.6).

2. The value of $\bar{\phi}$ is computed in off-equilibrium systems [7] and is a function of t (right plot in Fig. 6.6). A system that has been quenched from high temperature to a certain low temperature T_f, and has further relaxed towards equilibrium for a time t, displays a time-dependent inherent structure energy $\bar{\phi}(T_f; t)$.

3. If the equivalence between the equilibrium ensemble at T and the off-equilibrium set of states visited at T_f in a time window around t holds, one can connect, through the value of the mean potential energy, the aging system at T_f with an equilibrium system at $T_e^{em}(t)$:

$$\bar{\phi}(T_e^{em}(t)) = \bar{\phi}(T_f, t) \qquad (6.37)$$

Even though some numerical evidence has been provided for the reliability of the definition Eq. (6.37), e.g., in [Kob *et al.*, 2000; Crisanti & Ritort, 2000b], we expect that the thermodynamic behavior of an off-equilibrium observable can be properly described by *adding*, at least, an extra parameter, containing the information relative to the aging dynamic regime, to those used at equilibrium (temperature, pressure, volume, ...). See Chapter 2, Secs. 2.3, 2.4). In the present case, instead, one tries to express the off-equilibrium mean potential energy by the equilibrium $\bar{\phi}$ function of only one temperature. Indeed, instead of considering a further dependence of $\bar{\phi}$ on a time-dependent effective temperature T_e (besides the heat-bath temperature T), one substitutes T with T_e. We will come back to a simple counterexample in Sec. 6.6. We now explore the possibility of other, theoretically more careful, definitions of effective temperatures.

---

[7]In numerical simulations this is obtained, e.g., by considering systems of a larger size than those considered for the equilibrium analysis, or the same size but at lower temperature, or for a shorter time. The average is now taken over different trajectories, starting each time a new simulated dynamics from different initial conditions.

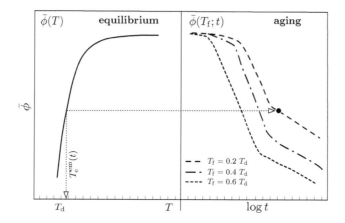

FIGURE 6.6

Pictorial definition of T_e^{em}. The IS energy, $\bar{\phi}(T)$ at equilibrium (left) is compared to the time-dependent $\bar{\phi}$ off equilibrium at three different temperatures T_f. The dynamic glass transition temperature T_d (corresponding to the critical mode coupling temperature, e.g., in [Kob *et al.*, 2000]) is also indicated in the left plot as a reference. The arrow links aging dynamics and equilibrium thermodynamics. Such a connection is encoded in the equilibrium effective temperature T_e^{em}.

6.3.2 PEL effective temperature and pressure in the two temperature thermodynamic framework

As it was presented in Sec. 2.4 (and applied in Sec. 3.3.2) to a specific class of models), one can devise that the entropic contribution of the degrees of freedom that have fallen out of equilibrium (i.e., the configurational entropy s_c), enters the thermodynamic potential with a conjugated temperature field different from the one of the thermal bath. In the latter, the fast, vibrational, degrees of freedom are in equilibrium. What is $T s_c$ for the undercooled liquid (i.e., a liquid below T_d), now becomes $T_e s_c$ for the glass (i.e., for $T < T_g$).

Following the approach of Sciortino & Tartaglia [2001], one starts from the off-equilibrium thermodynamic potential, function of T *and* of a given ϕ value,

$$f(\phi; T, v) = \phi + f_{vib}(\phi; T, v) - T_e^{em} s_c(\phi) \qquad (6.38)$$

Then one looks at the solutions of $\partial f / \partial \phi = 0$, but, at variance with Eq. (6.10), one solves for T_e^{em} at fixed ϕ, rather than for ϕ at fixed T. The expression eventually obtained for the effective "internal" temperature is:

$$T_e^{int}(\phi, T) = \left(1 + \frac{\partial f_{vib}}{\partial \phi}\right)\left(\frac{\partial s_c}{\partial \phi}\right)^{-1} \qquad (6.39)$$

To evaluate the T dependence of the configurational entropy derivative at the denominator, one defines another effective temperature $T_e^{em}(t)$, that is the temperature at which the ensemble of potential energy minima visited in the glassy dynamics on the timescale t, corresponds to the equilibrium ensemble. To be more precise, one supposes that the average value of the set of ϕ-values out of equilibrium is the equilibrium average at a higher temperature: $\bar{\phi} = \bar{\phi}(T_e^{\text{em}})$. The latter can be simply computed from Eq. (6.10) as if at equilibrium, only as a function of a different temperature. The "equilibrium"-like effective temperature is, eventually, the inverse of this solution: $T_e^{\text{em}}(\bar{\phi})$ (see Fig. 6.6 and the previous section).

The vibrational free energy is evaluated at the heat-bath temperature T. For T not too large, the vibrational average contribution to the free energy can be written in the harmonic approximation (Sec. 6.1.5):

$$F_{\text{vib}} = k_B T \sum_{i=1}^{M} \log \beta \hbar \omega_i(\phi)$$

Namely, one neglects any anharmonic contribution to the shape of the minimum of the potential energy and any temperature dependence of the frequencies of the normal modes involved. This implies:

$$\frac{\partial F_{\text{vib}}}{\partial \phi} = k_B T \sum_{i=1}^{M} \frac{\partial}{\partial \phi} \log \omega_i(\phi) \qquad (6.40)$$

$$N \frac{\partial s_c(\phi)}{\partial \phi} = \frac{1}{k_B T_e^{\text{em}}(\phi)} \left[1 + k_B T_e^{\text{em}}(\phi) \sum_{i=1}^{M} \frac{\partial}{\partial \phi} \log \omega_i(\phi) \right] \qquad (6.41)$$

The temperature T_e^{em} is a temperature-like parameter reflecting the slow flow of heat from the system to the thermostat. The quantity defined in Eq. (6.39) should be compared with T_e^{em} but also with an independent definition, e.g., with the fluctuation-dissipation ratio (cf. Sec. 2.8) in order to verify whether such a generalized description can actually hold and incorporate the features of the aging dynamics taking place in amorphous systems. If f_{vib} *does not* depend on the ϕ level, i.e., if all minima yield the same inter-basin contribution to the free energy disregarding their height in the PEL, from Eqs. (6.39)-(6.41) one obtains that $T_e^{\text{em}}(\phi) = T_e^{\text{em}}(\phi)$. In this case the phase space volume of the basins is independent of their depth and one can use a two temperature thermodynamics to describe the off-equilibrium system.[8]

The validity of this theoretical framework has been verified in a binary LJ binary mixture [Sciortino & Tartaglia, 2001] where this last hypothesis of a ϕ-independent vibrational free energy is rejected. Nevertheless, the effective temperature T_e^{FD} measured by the amplitude of the response of the aging system to the external perturbation and the T_e^{em} predicted using the

---

[8]This is a stronger approximation than the harmonic one and is not verified in most of the numerically mentioned models for glass formers addressed in the preceding sections. In particular it is different from what is shown in Fig. 6.2.

PEL formalism are in very good agreement even if this assumption is relaxed. Therefore, T_e^{int} appears to be a definition more consistent than T_e^{em} for the implementation of a two temperature thermodynamics. This observation supports the generalized thermodynamic approach developed in chapter 2 and studied in various models in Chapters 3, 4 and 5.

Following the approach of Sec. 2.8, where T_e^{FD} is defined, the fluctuating variable is now the Fourier transform of the density of the α particles (i.e., those belonging to the A type, cf. Appendix 6.A.2): $\rho_{\boldsymbol{k}}$. The time correlation function is, then, the dynamical structure factor

$$S_k(t, t_w) \equiv \langle \rho_{\boldsymbol{k}}(t) \rho_{\boldsymbol{k}}^\star(t_w) \rangle \tag{6.42}$$

where the brackets refer to the ensemble average. The time t is longer than the waiting time t_w at which measurements begin. The perturbative field, switched on at $t = t_w$, conjugated to the Fourier density is the chemical potential V_0. The integrated response function [cf. Eq. (2.108)] takes, here, the form

$$\chi_{\rho_{\boldsymbol{k}}}(t, t_w) = \langle \rho_{\boldsymbol{k}}(t) \rangle_{t_w} \tag{6.43}$$

where the average is now over the perturbed ensemble. The fluctuation-dissipation ratio, Eq. (2.106), eventually reads:

$$T_e^{\text{FD}} = V_0 \frac{S_k(t_w, t_w) - S_k(t, t_w)}{\chi_{\rho_{\boldsymbol{k}}}(t, t_w)} \tag{6.44}$$

See Fig. 6.7 for the plot of the effective temperatures (internal and FDR) and their comparison.

The same approach has been applied to a binary mixture of soft spheres, by means of Monte Carlo simulations (with different algorithms) confirming the coincidence of T_e^{FD} and T_e^{int} even at low temperature and for long times, in the late aging regime. See Fig. 6.8.

Practically, in the PEL approach, the out-of-equilibrium condition is implemented by imposing the constraint that the inter-basin processes, that allow the system to explore the space of states in the aging regime, are thermalizing in a thermostat different from the one at the heat-bath temperature (at which the intra-basin processes are at equilibrium). One can implement the following generalization of Eq. (6.9) [compare with Eq. (2.50) in Sec. 2.4]

$$f(V, T, T_e) = -T_e s_c(V, \phi) + \phi + f_{\text{vib}}(V, T, \phi) \tag{6.45}$$

where the condition of minimum free energy is yielded by $\phi = \bar{\phi}$, solution of

$$1 + \frac{\partial f_{\text{vib}}}{\partial \phi} - T_e \frac{\partial s_c}{\partial \phi} = 0 \tag{6.46}$$

The free energy contribution f_{vib} is still calculated at the thermostat temperature, whereas the configurational part is weighted by another temperature-like parameter. This separation is based on the freezing in of the slow processes taking place in the, however equilibrated, viscous liquid before it becomes a glass.

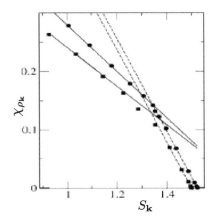

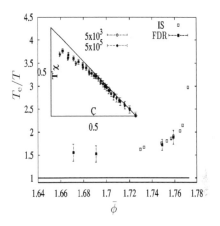

FIGURE 6.7

Fluctuation-dissipation ratio and effective internal temperature (cf. Eq. (6.39)) in a LJBM. Response-correlation plot of $\chi_\rho(S)$ for two very different waiting times: $t_w = 1024$ (circles) and $t_w = 16384$ (squares). Symbols are from direct computation of Eqs. (6.42) and (6.43). Full lines have a slope $V_0/(k_B T_e^{\text{int}})$, where the value of T_e^{int} is derived from Eq. (6.39). Dashed lines have the equilibrium slope $V_0/(k_B T)$. The wave vector magnitude is set equal to $k = 6.7$, the location of the first minimum of S_k^{AA}. Reprinted figure with permission from Sciortino & Tartaglia [2001]. Copyright (2001) by the American Physical Society.

FIGURE 6.8

FDR and T_e^{int}/T as a function of $\bar{\phi}(t_w)$ in a soft sphere binary mixture. The empty squares refer to T_e^{int}, Eq. (6.39), the black squares with error bars to T_e^{FD} of Eq. (2.107). The two sets of data appear to overlap and deviate nontrivially from the value $T_e/T = 1$ indicated by the lower line. In the inset: $\chi(C)$ for $t_w = 5 \cdot 10^3, 5 \cdot 10^5$ MC steps. On the left side, data deviate from the equilibrium prediction (straight line), but they weakly depend on t_w, indentifying a constant effective temperature for the time interval considered. Reprinted with permission from Grigera *et al.* [2004]. Copyright (2004) by American Physical Society.

6.3.3 The pressure in glasses

Generalizing Eq. (6.33) to the two temperature thermodynamic framework, one gets for the pressure in the glassy phase the following expression:

$$P = -\frac{\partial f}{\partial v}\bigg|_{T,T_e} = T_e \frac{\partial s_c}{\partial v}\bigg|_\phi - \frac{\partial f_{\text{vib}}}{\partial v}\bigg|_{T,\phi} \qquad (6.47)$$

where $v = V/N$ is the specific volume. The condition Eq. (6.46) has been employed and, therefore, $\phi = \bar{\phi}(T, T_e, v)$. This would be the equation of state in a generalized parameter space where T_e plays the same role as bath temperature, volume and pressure.

Starting from the above equation, a second expression for the inherent structure pressure, the sum of Eqs. (6.30) and (6.29), can be derived setting $T = 0$ (the quenching equivalent to the steepest descent procedure), $T_e = T_e(T = 0, v)$, $\phi = \bar{\phi}(0, T_e(0, v), v)$ and $f_{\text{vib}} = 0$:

$$P_{\text{is}}(v, \bar{\phi}) \equiv P(0, T_e(v, 0), v) = T_e \frac{\partial s_c}{\partial v}\Big|_\phi \tag{6.48}$$

Using the chain relation

$$\frac{\partial v}{\partial s_c}\Big|_\phi \frac{\partial s_c}{\partial \phi}\Big|_v \frac{\partial \phi}{\partial v}\Big|_{s_c} = -1 \tag{6.49}$$

and Eq. (6.46) evaluated at zero temperature, P_{is} becomes [cf. Eq. (6.20)]

$$P_{\text{is}}(v, \bar{\phi}) = -\frac{\partial \bar{\phi}}{\partial v}\Big|_{s_c} \tag{6.50}$$

What is the physical meaning of such an effective pressure? The mechanical definition of pressure refers to the free energy potential difference due to a differential scaling of configurational coordinates with the volume. The response of the IS energy due to an infinitesimal change of volume does not correspond to such a continuous configurational deformation but, rather, to a change inside the (number-conserving) ensemble of minima considered. In the particular case of a soft sphere model [La Nave *et al.*, 2003b] (see Appendix 6.A) the change in energy associated with the compression of an inherent structure configuration is equal to the change in ϕ. Furthermore, it can be proven that the configurational entropy stays invariant under volume variations. Indeed, the soft sphere potential energy is self-similar, namely

$$V(\lambda \mathbf{r}) = \lambda^{-M} V(\mathbf{r}) \tag{6.51}$$

where M is the total number of degrees of freedom of the N-particle system, and this scaling property implies that the number of IS basins is invariant under any change of real space volume (see Fig. 6.9).

From Eq. (6.51) two main consequences derive:

1. In an isotropic compression a minimum of the potential energy stays a minimum and

2. the IS pressure can be written as

$$P_{\text{IS}}(v, \phi) = -\frac{\partial \phi}{\partial v}\Big|_{s_c}$$

that is Eq. (6.50) as derived earlier from first principle assumptions on the nonequilibrium behavior. The above formula is obtained by realizing that (i) an isotropic change of volume moves the soft sphere system along a path of constant s_c (the number of basins do not change in changing $\phi \to \phi + \Delta\phi$ and $V \to V + \Delta V$) and (ii) the IS pressure is, mechanically speaking, the measure of the potential energy change under compression of an IS configuration.

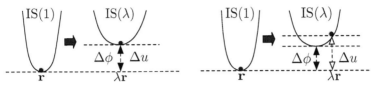

a. Self-similar potential   **b. Generic potential**

$\mathbf{r}$: state point in M-dimensional configurational space

$\bullet$: state point in energy

FIGURE 6.9

The change induced in the energy of a system undergoing homogeneous expansion (or contraction) in configurational space ($\mathbf{r} \rightarrow \lambda\mathbf{r}$, where $\mathbf{r}$ is the M-dimensional system point, $M = DNn$). In the left figure, the special case of the soft sphere model [La Nave *et al.*, 2003b] is shown. In this case the potential energy is self-similar under homogeneous expansion in the configurational space, cf. Eq. (6.51), and the shifts in potential energy ($\Delta\phi$) and in internal energy (Δu) coincide. In the picture on the right, the case of an arbitrary interaction is displayed, where $\Delta\phi \neq \Delta u$.

Provided that the PEL description holds (see discussion in Sec. 6.1.3 and cf. Fig. 6.1) and that the features of the models with a self-similar interaction potential can be exported to more complicated materials, the effective temperature $T_{\mathrm{e}}^{\mathrm{em}}$ and the inherent pressure P_{is} allow for an implementation of the out-of-equilibrium thermodynamics in terms of an extra temperature and an extra effective field, as considered in Chapters 2 (Sec. 2.3) and 3 (Sec. 3.3.1).

6.4   Fragility in the PEL

The "fragility" of a glass-forming liquid is a quantity introduced by Angell [1985, 1991] to measure the rapidity with which the properties of a liquid change as the glassy state is approached in cooling. The mainly measured property is the viscosity and looking at its variation around the (thermal) glass transition, a fragility index can be defined (Sec. 1.6). Its origin has been discussed in Secs. 1.2 and 1.6 and we are now interested in understanding how the fragility is related to the properties of the PEL.

The connection between the degree of fragility of a glass former and the properties of the PEL has been first established by Sastry [2001] on a LJ binary mixture model and later confirmed by other studies, e.g., [Martinez

& Angell, 2001; Ruocco *et al.*, 2004]. Fragility depends both on changes of the vibrational properties of individual potential energy minima, on their total number (i.e., the total configurational entropy) and, furthermore, on the distribution of the number of distinct minima in potential energy.

We recall two of the definitions for a fragility index, introduced in Sec. 1.6 (*global* definitions). The first one comes from the assumption of a functional Vogel-Fulcher dependence for $\eta(T)$:

$$K_\eta \equiv \frac{1}{T/T_0 - 1} \frac{1}{\log[\eta(T)/\eta_0]} \tag{6.52}$$

Notice that this definition involves the logarithm with base 10. The second definition is based on the linearity of $Ts_c(T)$ near the Kauzmann temperature (that is the key assumption connecting the Adam-Gibbs relation to the VF phenomenological law, cf. Sec. 1.5):

$$K_{\text{ag}} \equiv \frac{TS_c(T)}{T - T_K} \tag{6.53}$$

The Adam-Gibbs relation seems to hold for LJ binary mixtures (see, e.g., the recent simulations at varying density by Sastry [2001]). The configurational entropy at a certain temperature, in the PEL formalism, is determined by the number of potential energy minima sampled by the liquid at that temperature [according to Eq. (6.12)] and it is equal to the difference between the total entropy and the entropy of the typical basin (due to the T-dependent vibrational contribution): $S_c = S_{\text{tot}} - S_{\text{vib}}$.

When the harmonic approximation is assumed to hold and the density of states at given ϕ level is assumed to be Gaussian (i.e., when Eqs. (6.24)-(6.26) properly describe the rugged landscape representing the many-molecules glass former), the vibrational entropy is [Eq. (6.25)]

$$S_{\text{vib}}(\phi) \equiv \sum_{i=1}^{M} [1 - \log \beta \hbar \omega_i(\phi)]$$

and the difference between the basin entropy at the generic ϕ and the one at $\phi = \phi_0$ is

$$\frac{\Delta S_{\text{vib}}}{N} = \frac{S_{\text{vib}}(\phi) - S_{\text{vib}}(\phi_0)}{N} = -b(\phi - \phi_0) \tag{6.54}$$

One can, thus, derive for the fragility parameter the following expression [Sastry, 2001]:

$$K_{\text{ag}}^{\text{PEL}}(T) = \left(\frac{\sigma \sqrt{\alpha}}{2} - \frac{\sigma^2 b}{4Nk_b} \right) \left(1 + \frac{T_K}{T} \right) + \frac{\sigma^2 b}{2Nk_B} = \frac{\sigma^2}{4Nk_B} \left(\frac{1}{T_K} + \frac{1}{T} + 2b \right) \tag{6.55}$$

where

$$T_K = \frac{\sigma}{2Nk_B\sqrt{\alpha} - \sigma b} \tag{6.56}$$

In this, apparently simple, case, K_{ag} is not constant in T and, therefore, the Adam-Gibbs relation does not exactly lead to a VF-like relaxation law. However, since in viscous liquids it always holds that $T > T_g > T_K$, the $1/T$ term in Eq. (6.55) yields a second order contribution that seems to decrease rapidly in the LJ case study. This explains why numerically the two definitions, VF and AG, almost always coincide.

Eq. (6.55) is an explicit example of the connection between the parameter expressing the rapidity of increase of the viscosity as temperature is decreased (as well as the sharpness of the jump in specific heat at T_g) and the parameters describing the main features of the PEL: the total number of minima (α), the spread of the distribution of potential energy minima (σ), and the slope of the vibrational entropy decrease in ϕ (the parameter b). In particular, it shows that the fragility grows as the square root of the total number of minima of the PEL at any temperature ($\alpha = \max_\phi s_c(\phi)$), yielding theoretical support to the heuristic belief that fragile systems have a large number of basins, in comparison with strong glass formers [Angell, 1995, 1997; Debendetti & Stillinger, 2001].[9]

6.5 PEL approach to the random orthogonal model

We present the PEL analysis of the random orthogonal model (ROM) [Marinari *et al.*, 1994a,b], i.e., a mean-field glassy model belonging to the class of "one step of replica symmetry breaking" spin-glass models with quenched, random, infinite range interactions (cf., e.g., [Kirkpatrick & Thirumalai, 1987b; Crisanti & Sommers, 1992]). These are models reproducing all the basic properties of a glass, although with some prescription due to their mean-field nature.

The model Hamiltonian is

$$\mathcal{H} = -2 \sum_{i,j}^{1,N} J_{ij} s_i s_j \qquad (6.57)$$

where $s_i = \pm 1$ are N Ising spins and J_{ij} is a random symmetric orthogonal matrix whose diagonal elements are equal to zero. In the thermodynamic limit, the dynamic glass transition occurs at $T_d = 0.536$, at which the highest

---

[9]In the example brought about by Sastry [2001] the volume (or the density) is kept constant, different from experiments usually performed at constant pressure. This generates a counterintuitive artifact. Indeed, it makes the vibrational entropy contribution of a basin, $S_{vib}(\phi)$, decrease and the normal mode frequencies $\omega_i(\phi)$ increase as ϕ increases, implying that the basins at higher potential energy, that are those visited by the system at higher temperature, are *narrower* than those visited at low temperatures.

configurational entropy is found at energy $\phi_{th} = -1.87$. The Kauzmann temperature is $T_K = 0.25$, for which the lowest energy value is reached, at $\phi_m = -1.936$.[10] The study of the FEL exhibits a very large number of basins, growing exponentially with N [Parisi & Potters, 1995].

In the PEL formalism, ϕ_m and ϕ_{th} are the lowest and the highest values between which the many minima are found. As $N \to \infty$, no activated processes are allowed in the ROM dynamics, since the barriers among metastable glassy states are infinite. Cooling down the temperature from $T > T_d$ to $T = 0$, therefore, traps the system in one of the most probable basins, that is those whose configurational entropy is maximal, those at ϕ_{th}.

Performing numerical simulations at finite N, though, allows for the exploration of minima below the threshold value. Actually, for finite size systems, stationary points with $\phi > \phi_{th}$ also can be visited. Though they should be saddles instead of minima [Cavagna *et al.*, 1998], the finiteness of the degrees of freedom when N is not too large can stabilize some of them [Crisanti & Ritort, 2000b]. Indeed, finite size effects can help to introduce activated processes, and, therefore the slow structural relaxation, in otherwise dynamically stuck mean-field models, retaining, however, the solubility and the cleaner physical description of the latter.

6.5.1   Effective temperature in the ROM

Crisanti & Ritort [2000c,a,b] study the relaxational dynamics out of equilibrium of the ROM by means of Monte Carlo numerical simulations. The thermal glass temperature is not defined analytically in mean-field systems (dynamics is stuck as soon as $T < T_d$ because thermodynamic fluctuations are neglected), however, at finite size N it will depend both on N and on the largest timescale obtainable for which equilibrium is reached, mimicking in this way the experimental timescale (cf. Sec. 1.1). For $N = 300$ Crisanti & Ritort [2000c] find a $T_g \simeq 0.5$. They, then, perform a fast cooling from high temperature down to $T < T_g$ and look for one- and two-time dynamics.

Considering an inherent structure spin configuration $\{s_i^{is}(t)\}$ at time t and the external small perturbing local field $\boldsymbol{h} = \{h_i\}$, one can define correlation and response functions:

$$C(t, t_w) = \frac{1}{N} \sum_{i=1}^{N} s_i^{is}(t) s_i^{is}(t_w) \tag{6.58}$$

$$G(t, t_w) = \frac{1}{N} \sum_{i=1}^{N} \frac{\delta s_i^{is}(t)}{\delta h_i(t_w)}, \qquad t > t_w \tag{6.59}$$

$$\chi(t, t_w) = \int_{t_w}^{t} dt' G(t, t') \tag{6.60}$$

---

[10]These are the same properties of the p-spin spherical model [Crisanti & Sommers, 1992].

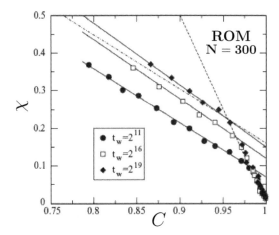

FIGURE 6.10

Susceptibility versus correlation. The slope of the dashed line is the inverse of the heat-bath temperature ($T = 0.2$). The points derive from Monte Carlo simulation data, plotting parametrically χ and C at fixed waiting times. The full lines are the prediction $1/T_e^{\text{em}}$, Eq. (6.39). The dashed line has a slope $1/T$. The dot-dashed line has a slope $1/T_e^{\text{FD}}$, obtained from Eq. (6.61) at $t_w = 2^{11}$. It appears to underestimate the data points. Reprinted figure with permission from [Crisanti & Ritort, 2000a]. Copyright (2000) by the European Physical Society.

As the quench is performed to low temperature, both functions display the aging phenomenon and the $\chi(C)$ parametric plot, Fig. 6.10 is qualitatively similar to the one for LJ binary mixture (Fig. 6.7).

The FDR effective temperature has been computed using definition Eq. (2.107) that can be rewritten as

$$T_e^{\text{FD}} = \frac{\partial_{t_w} C(t, t_w)}{G(t, t_w)} = -\left(\frac{\partial \chi}{\partial C}\right)^{-1} \qquad (6.61)$$

The PEL internal effective temperature, Eq. (6.39) has been computed as well. Practically no ϕ dependence has been detected for the vibrational free energy contribution ($\partial f_{\text{vib}}/\partial \phi = 0$) implying a coincidence of T_e^{em} with the rougher T_e^{em} defined in Sec. 6.3.1. We recall that numerical agreement between the internal effective temperature and the FDR one was also found in the LJ binary mixture (see Sec. 6.3 and [Sciortino & Tartaglia, 2001]), even though in that case $\partial f_{\text{vib}}/\partial \phi \neq 0$.

The outcome is shown in Fig. 6.10 where the different effective temperatures are compared. At given t_w the system is in the stationary regime for larger C (i.e., shorter times), where it satisfies the FDT, and then it falls out of equilibrium as the correlation decreases.

6.6   The PEL approach to the harmonic oscillator models

In Chapter 3 we introduced a class of kinetic models, the HOSS models, and we gave the description of their statics and of their Monte Carlo dynamics. Here we present how the PEL approach can be applied to the dynamic and thermodynamic analysis of such models [Leuzzi & Nieuwenhuizen, 2001b].

The characteristics of a glassy system can be represented by means of a multidimensional potential energy function with a complex topography. The spatial patterns of atoms in crystals and in amorphous systems, at low temperature, represent minima in the potential energy function describing the interactions. In the case of the HOSS model (3.50) all the complex chemical properties of real glass formers are not represented, nevertheless the system exhibits several aspects of their complex features, indicating that the model is complicated enough for what concerns the description and the comprehension of the basic long time properties of a glass.

We have seen that, in the PEL approach, cf. Sec. 6.1, α processes are represented by escape processes from one deep minimum within a large scale valley to another valley, whereas β processes are related to simpler relaxations between neighboring minima. Note that in our HOSS models we allow all kinds of β processes in our short timescale, since their timescales are in any case much shorter than the observation time considered, so they are just thermally equilibrated.

As we will see, the HOSS model is built in such a way that every $\{x_i\}$ configuration is an inherent structure. Indeed, at a given $\{x_i\}$ configuration at finite T, the $\{S_i\}$ are fast variables and they contribute to the energy and to the other observables as a noise depending on temperature. If we take away this contribution we do not actually change the configurations of the minima of the slow variables. In the case of the system without constraint on the configuration space, nor contrived dynamics [see Sec. (3.2.3)], any $\{x_i\}$ configuration is an inherent structure. For what concerns the constrained model (where $(1/N)\sum x_i^2 - [(1/N)\sum x_i]^2 \geq m_0$), instead, certain configurations are not allowed (Sec. 3.2.4). Moreover, the presence of the constraint (3.65) produces (entropic) barriers higher than in the case without it to get from a certain IS to a different one. This just models the well-known fact that the dynamics through the inherent structures is even slower in the fragile glass case than in the strong glass case.

First of all, we have to define the steepest descent procedure for the model. For that purpose we will minimize the Hamiltonian, Eq. (3.50),

$$\mathcal{H}[\{x_i\}, \{S_i\}] = \sum_{i=1}^{N} \left(\frac{1}{2}Kx_i^2 - H\,x_i - J\,x_iS_i - L\,S_i \right)$$

with respect to the spins. Indeed, that corresponds to the zero temperature limit of the thermodynamic potential F given by Eq. (3.53). Such a correspondence boils down to looking for the minimum of

$$\Phi \equiv \min_{\{S_i\}} \left[\mathcal{H} + \lambda \sum_{i=1}^{N} S_i^2 - \lambda N \right] \tag{6.62}$$

with respect to the spins, in order to get rid of their contribution, i.e., to get rid of the fast modes (we implemented the spherical constraint $\sum_i S_i^2 = N$ by using the Lagrange multiplier λ). We get $S_i^{(\min)} = (J x_i + L)/(2\lambda)$, $\forall i$. Solving the spherical condition $\sum_{i=1}^{N} \left(S_i^{(\min)} \right)^2 = N$ for λ, we find $\lambda = w_{is}/2$, where

$$w_{is} \equiv \sqrt{J^2 m_2 + 2J L m_1 + L^2} \tag{6.63}$$

This has to be compared with w from Eq. (3.52) as $T \to 0$. The variables $m_{1,2}$ are equal to $\sum_i x_i^{1,2}/N$. The minimum $\{S_i\}$ configuration for a given set of $\{x_i\}$ is, thus, given by

$$S_i^{(\min)} = \frac{J x_i + L}{w_{is}}, \qquad \forall i \tag{6.64}$$

Eventually, Eq. (6.62) becomes

$$\Phi = N \left[\frac{K}{2} m_2 - H m_1 - w_{is} \right] \tag{6.65}$$

that is the energy function of the inherent structures. Consequently, the partition sum over inherent structures is defined by

$$Z_{is} = \int \mathcal{D}x \exp\left[-\beta \Phi(\{x_i\}) \right] = \int dm_1 dm_2 \exp\left[S_c(m_{1,2}) - \beta \Phi(m_{1,2}) \right] \tag{6.66}$$

Due to the minimization, any explicit dependence on T in the effective Hamiltonian disappears [compare Eq. (3.53) and Eq. (6.65)]: the minimization with respect to the spherical "fast" $\{S_i\}$ variables is equivalent to taking the $T \to 0$ limit for them.[11]

The configurational entropy for ISs comes from the Jacobian of the transformation of variables $\mathcal{D}x = e^{S_c} dm_1 \, dm_2$, where

$$s_c = \frac{S_c}{N} = \frac{1}{2} \left[1 + \log\left(m_2 - m_1^2 \right) \right]$$

---

[11] In Eq. (3.59), we integrated over the spins, instead of minimizing with respect to them, and therefore we also had an entropic term $T S_{ep}$ for the fast processes, with S_{ep} given by Eq. (3.56), and a slightly different internal energy [Nw instead of $N w_{is}$, with w given in Eq. (3.52) and w_{is} in Eq. (6.63)]. In the inherent structure approach, instead, carrying out steepest descent makes the entropic term vanish (only the minimal configuration is taken into account) and the effective Hamiltonian, given in Eq. (6.65), has no explicit dependence on the temperature.

It is the same as in the finite T case, since any allowed configuration $\{x_i\}$ is also an IS. The static average of Φ/N is given by

$$\bar{\phi}(T) = \frac{1}{N}\Phi(\bar{m}^{\text{is}}_{1,2}(T)) \tag{6.67}$$

where $\bar{m}^{\text{is}}_{1,2}(T)$ are the solutions of the saddle point equations relative to Eq. (6.66):

$$\bar{m}^{\text{is}}_1 = \frac{\tilde{H}_{\text{is}}}{\tilde{K}_{\text{is}}}; \qquad \bar{m}^{\text{is}}_2 - (\bar{m}^{\text{is}}_1)^2 = \frac{T}{\tilde{K}_{is}} \tag{6.68}$$

and we have defined

$$\tilde{H}_{\text{is}} \equiv H + \frac{JL}{w_{\text{is}}}; \qquad \tilde{K}_{\text{is}} \equiv K - \frac{J^2}{w_{\text{is}}} \tag{6.69}$$

with w_{is} from Eq. (6.63). The combination $\tilde{H}_{\text{is}}J + \tilde{K}_{\text{is}}L = HJ + KL = D$ is simply a constant, as in Eq. (3.58).

In the case at finite T the full static partition function (3.59) was

$$Z = \int dm_1 dm_2 \exp\left(S_c - \beta \mathcal{H}_{\text{eff}}\right)$$

with $\mathcal{H}_{\text{eff}}$ defined in Eq. (3.53) and s_c in Eq. (3.60). The saddle point equations (6.68) are different from Eqs. (3.62) and (3.63) valid in the realistic case, thus yielding different results: $\bar{m}^{\text{is}}_{1,2} \neq \bar{m}_{1,2}$. We stress, however, that $\bar{m}^{\text{is}}_{1,2}$ depend on T also in the IS case.

The HOSS free energy, Eq. (3.61), has to be compared in the PEL formalism [cf. Eq. (6.9)] with the IS contribution to the free energy deriving from Eq. (6.66) plus the vibrational contribution F_{vib}:

$$F = \mathcal{H}_{\text{eff}}(m_{1,2}) - TS_c(m_{1,2}) = \Phi(m_{1,2}) + F_{\text{vib}}(m_{1,2}) - TS_c(m_{1,2}) \tag{6.70}$$

As we mentioned above, the configurational entropy of the inherent structures is the same as for the finite temperature case, Eq. (3.60), implying

$$f_{\text{vib}} = \frac{F_{\text{vib}}}{N} = \frac{T}{2}\log\left(\frac{w + T/2}{T}\right) - w + w_{\text{is}} \tag{6.71}$$

where w is defined in Eq. (3.52) and w_{is} in Eq. (6.63). Notice that f_{vib} explicitly depends on the parameters m_1 and m_2 of the IS and is thus not a constant (cf. Secs. 6.3.1, 6.5 and [Stillinger & Weber, 1982, 1984; Sciortino et al., 1999; Kob et al., 2000; Crisanti & Ritort, 2000b]).[12]

---

[12]Notice that for the HO model of Sec. 3.1, where no spins (that is to say, no fast processes) are present, Φ is trivially equal to the Hamiltonian (3.1), that is the potential energy of a set of N uncoupled harmonic oscillators linearly coupled to an external field.

6.6.1 PEL effective temperature in the HOSS model

In Chapter 2 we presented different ways in which the effective (or fictive) temperature can be defined. All those definitions should be equivalent if the quantity called effective temperature were actually a temperature. In Chapter 3 we tested, therefore, those (and other) definitions on the HOSS model in order to verify whether, at least in this solvable dynamically facilitated model, the thermodynamic picture with two temperatures (T and T_e) can consistently describe the behavior of (a class of) slow relaxing, aging, glassy systems.

Further possible ways of defining an effective temperature exist in the PEL formalism, e.g., the one described in Sec. 6.3.1 obtained by comparing the time-dependent out-of-equilibrium IS energy at temperature T with the equilibrium IS energy expression at a temperature $T_e \neq T$. The out of equilibrium average of ϕ is built up in practice by taking the dynamics of a system out-of-equilibrium at temperature T and repeating it many times starting from different initial conditions. A statistical ensemble of trajectories is constructed in this way. The configurations that each sample is visiting are determined at any given time t.

In the HOSS model, the energy ϕ, averaged over the ensemble of different trajectories, reads

$$\bar{\phi}(T,t) = \frac{K}{2}m_2(t) - Hm_1(t) - \sqrt{J^2 m_2(t) + 2JLm_1(t) + L^2}$$

$$\simeq \frac{K}{2}\bar{m}_2^{is} - H\bar{m}_1^{is} - \sqrt{J^2 \bar{m}_2^{is} + 2JL\bar{m}_1^{is} + L^2}$$

$$+ \tilde{K}_{is}(\bar{m}_1^{is}, \bar{m}_2^{is})\delta\mu_2(t) + c(\bar{m}_1^{is}, \bar{m}_2^{is})\delta\mu_2(t)^2$$

where $\delta\mu_2(t) \equiv \mu_2(t) - \bar{\mu}_2(T)$ is reported in Eq. (3.89) for the Arrhenius case and in Eq. (3.119) for the Vogel-Fulcher case. The asymptotic value $\bar{\mu}_2$ is obtained by Eq. (3.67). The static counterpart of the IS energy is expressed by

$$\bar{\phi}(T) = \frac{K}{2}\bar{m}_2^{is} - H\bar{m}_1^{is} - \sqrt{J^2 \bar{m}_2^{is} + 2JL\bar{m}_1^{is} + L^2} \qquad (6.72)$$

The equilibrium IS energy $\bar{\phi}(T)$ will be a different function of temperature in the strong and in the fragile versions of the model. Furthermore, the second order expansion for long times of the off-equilibrium $\bar{\phi}$ will be needed only for the strong glass case, where

$$c(\bar{m}_1^{is}, \bar{m}_2^{is}) = \frac{DJ^4 K^2}{8(D+J^2)^4} \qquad (6.73)$$

Fixing the time t, we define the effective temperature as the temperature at which the system at equilibrium would visit the same configurations visited by the system out of equilibrium at temperature T, with the same frequency (cf. Fig. 6.6): the one such that

$$\bar{\phi}\left(T_e^{em}(t)\right) = \bar{\phi}(T,t) \qquad (6.74)$$

For the HOSS model it is possible to work out an analytic expression for such a temperature-like parameter, both in the strong and in the fragile glass case.

T_e^{em} in the fragile HOSS model

In the version of the HOSS model representing fragile glasses, linearizing in $T - T_0$, we get a time dependence for the effective parameter that is different from the thermodynamic effective temperature T_e, that we obtained from several different approaches (including the fluctuation-dissipation ratio) in Sec. 3.3, cf. Eqs. (3.130, 3.141, 3.158, 3.188).

Away from the Kauzmann temperature we are not able to derive any simple expression of the IS energy (6.72), but, in any case, we can numerically solve Eq. (6.74) exactly. The results are shown in Figs. 6.11-6.12 for a given choice of the values of the interaction parameters and of the Vogel-Fulcher exponent γ of the model. As one can see, $T_e^{em}(t)$ turns out to be different from $T_e(t)$ at any time decade. As a matter of fact, what we are comparing now with the off-equilibrium $\bar{\phi}(T, t)$ is a function $\bar{\phi}\left(T_e^{em}(t)\right)$ of the effective temperature alone, while we know that out of equilibrium any proper thermodynamic function cannot simply depend on just one temperature as the thermodynamic function of equilibrium systems does (see Sec. 2.4). It is not surprising, thus, that the two effective temperatures do not coincide.

T_e^{em} in the strong HOSS model

For the strong glass case the analytic treatment is by far easier and, expanding Eqs. (6.72) and (6.72) near zero temperature up to second order in T, we can work out a simple analytic expression for the equilibrium-like effective temperature:

$$T_e^{em} \simeq T + \frac{KD}{D + J^2}\delta\mu_2(t) + \frac{J^4 K^2}{2(D + J^2)^3}T\delta\mu_2(t) + \mathcal{O}(T^3) + \mathcal{O}\left((\delta\mu_2(t))^3\right) \quad (6.75)$$

Here above terms of $\mathcal{O}(T^2)$ and $\mathcal{O}(\delta\mu_2(t)^2)$ cancel out. This effective temperature is obtained from Eq. (6.74) with

$$\bar{\phi}(T, t) = -\frac{(H + J)^2}{2K} - L + \frac{T}{2} - \frac{J^4 K}{8D(D + J^2)^2}T^2 + \frac{KD}{2(D + J^2)}\delta\mu_2(t)$$

$$+ \frac{J^4 K^2}{8D(D + J^2)^2}T\delta\mu_2(t) + \frac{DJ^4 K^3}{8(D + J^2)^4}\delta\mu_2(t)^2 \qquad (6.76)$$

$$+ \mathcal{O}\left(T^3\right) + \mathcal{O}\left(T^2\delta\mu_2(t)\right) + \mathcal{O}\left(T\delta\mu_2(t)^2\right) + \mathcal{O}\left(\delta\mu_2(t)^3\right)$$

Comparing with the effective temperature of the HOSS model obtained considering the finite temperature states, if we expand Eq. (3.130) at the same order as above we find:

$$T_e = \tilde{K}(m_2 - m_1^2) = T + \tilde{K}\delta\mu_2(t) \qquad (6.77)$$

$$\simeq T + \frac{KD}{D + J^2}\delta\mu_2(t) + \frac{T}{2}\left(\frac{JK}{D + J^2}\right)^2\delta\mu_2(t) + \frac{DJ^4 K^3}{2(D + J^2)^4}\delta\mu_2(t)^2$$

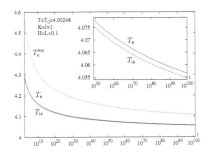

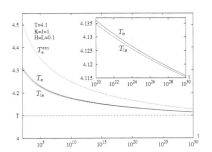

FIGURE 6.11

Effective temperatures vs. t at $T = T_K = 4.00248$. In Eq. (3.50) parameters are set as $K = J = 1$, $H = L = 0.1$. The constraint constant is $m_0 = 5$ and $\gamma = 2$. The upper curve shows the effective temperature obtained by matching out of equilibrium and equilibrium $\bar{\phi}$. The one in the middle is the behavior of Eq. (3.130), for systems at finite T, and the lowest one is the IS quasi-static effective temperature, Eq. (6.81). The inset exposes the difference between the lower two curves. Reprinted figure with permission from [Leuzzi & Nieuwenhuizen, 2001b]. Copyright (2001) by the American Physical Society.

FIGURE 6.12

The same effective temperatures, for the same choice of parameters as before, are plotted at a heat-bath temperature $T = 4.1$, above the Kauzmann temperature. Reprinted figure with permission from [Leuzzi & Nieuwenhuizen, 2001b].

As we see from the formulas above and from Figs. 6.13 and 6.14, in the case with Arrhenius relaxation, T_e and T_e^{em} are very similar. Their difference is one order of magnitude less than in the model with contrived dynamics (cf. Figs. 6.11, 6.12).

6.6.2 Quasi-static definition of IS effective temperature

Here we propose an alternative way to identify an effective temperature that maps the dynamics between inherent structures into a thermodynamic quantity. We follow a quasi-static approach using a partition sum, just as we did in the finite T case, in Sec. 3.3. Following exactly the same approach we used in Sec. 3.3.1, including the substitution of the real external field H with an effective one, H_{is}, we compute the partition function counting all the macroscopically equivalent ISs, through which the system is evolving in this

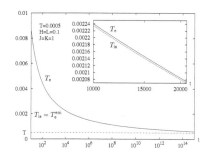

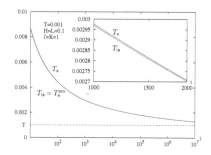

FIGURE 6.13

Time evolution of the effective temperatures at the heat-bath temperature $T = 0.0005$ in the model with Arrhenius relaxation. Constants in Eq. (3.50) are set to: $K = J = 1$, $H = L = 0.1$, while $m_0 = 0$. The lower curve shows T_e^{em}, Eq. (6.75), obtained by matching the out-of-equilibrium and equilibrium IS internal energies. To order $\delta\mu_2$ it coincides analytically with the IS effective temperature, Eq. (6.86). Second order differences are too small to appear in the plot. The upper curve is the behavior of Eq. (3.130), at finite T. The inset magnifies a part of the plot. Reprinted figure with permission from [Leuzzi & Nieuwenhuizen, 2001b].

FIGURE 6.14

The same effective temperatures, for the same choice of parameters as before are plotted for a different heat-bath temperature: $T = 0.001$. Comparing the timescales of the two plots, we can clearly observe the decrease of the Arrhenius relaxation time τ_{em} that takes places upon rising of temperature. Reprinted figure with permission from [Leuzzi & Nieuwenhuizen, 2001b].

symbolic dynamics, at a given time t:

$$Z_e^{\text{is}}(m_1, m_2) = \int \mathcal{D}x \exp\left[-\Phi\left(\{x_i\}; T, H_{\text{is}}\right)/T_{\text{is}}\right]$$

$$\times \delta\left(Nm_1 - \sum_i x_i\right) \delta\left(Nm_2 - \sum_i x_i^2\right)$$

$$= \exp\left\{-N\left[\frac{K}{2}m_2 - H_{\text{is}}m_1 - \bar{w}_{\text{is}} - \frac{T_{\text{is}}}{2}\log\left(m_2 - m_1^2\right)\right]/T_{\text{is}}\right\}$$

$$\simeq \exp\left\{-\left[\Phi\left(m_1, m_2; T, H_{\text{is}}\right) - T_{\text{is}}S_c\left(m_1, m_2\right)\right]/T_{\text{is}}\right\} \qquad (6.78)$$

The parameters T_{is} and H_{is} describe the behavior of the system evolving exclusively through ISs. Minimizing the free energy $F_e^{\text{is}} \equiv -T_{\text{is}}\log Z_e^{\text{is}}$ with respect to $m_{1,2}$ we obtain:

$$T_{\text{is}} = \tilde{K}_{\text{is}}\left(m_1, m_2\right)\left[m_2 - m_1^2\right] \qquad (6.79)$$

$$H_{\text{is}} = H - \tilde{K}_{\text{is}}\left(m_1, m_2\right)\mu_1 \qquad (6.80)$$

where $\tilde{K}_{is}$ was defined in Eq. (6.69). By inserting the time-dependent values of m_1 and m_2 we now look at the time evolution of the effective temperature (6.79) for long times, in the aging regime, and we compare it with the behavior of the thermodynamic effective temperature (3.130).

Fragile HOSS T_{is}

For the dynamically constrained model, as time goes to infinity, $T_{is} \rightarrow T$ (if $T > T_0$). When $t_0 \ll t < \infty$, however, the way the effective temperature approaches the heat-bath temperature is different from the behavior of T_e found in Sec. 3.3. For a comparison, their first order expansions are:

$$T_{is} \simeq T + \left(1 + \frac{P_\infty^{is}}{1 + Q_\infty^{is} D}\right) \tilde{K}_\infty^{is}(T) \, \delta\mu_2(t) \qquad (6.81)$$

$$T_e \simeq T + \left(1 + \frac{P_\infty}{1 + Q_\infty D}\right) \tilde{K}_\infty(T) \, \delta\mu_2(t) \qquad (6.82)$$

with

$$\tilde{K}_\infty^{is}(T) = \lim_{t\to\infty} \tilde{K}_{is}\left(m_1(t), m_2(t); T\right) \qquad (6.83)$$

$$Q_\infty^{is} = \lim_{t\to\infty} \frac{J^2 D}{\tilde{K}_{is}^3 w_{is}^3} \qquad (6.84)$$

$$P_\infty^{is} = \lim_{t\to\infty} \frac{J^4(m_2 - m_1^2)}{2\tilde{K}_{is}^2 w_{is}^3} \qquad (6.85)$$

The functions $\tilde{K}$, Q_∞ and P_∞ defined in Eqs. (3.57), (3.79) and (3.80), respectively. The dynamic variable $\delta\mu_2(t)$ is the same in both cases (apart from the parameter t_0 influencing only the short times) while the coefficients in front of it are different at any temperature, including T_0.

In the fragile case this second IS effective temperature does not coincide with T_e^{em} and it is much more similar to Eq. (6.82), at any time. However, even if this T_{is} is conceptually more properly chosen, we still do not get exactly the same parameter describing the finite T dynamics in a thermodynamic frame. The inherent structure approach yields an actually very good approximation but is, nevertheless - and not surprisingly - never analytically correct in describing the real temperature dynamics, not even in the extremely simplified description provided by the present model. To show how good this approximation is, we can take, as an instance, a certain realization of the model with given values of the "fields" and "coupling constants." We plot in Figs. 6.11 and 6.12 the behavior of $T_e^{em}(t)$, $T_{is}(t)$ and $T_e(t)$ at heat-bath temperatures equal to and just above the Kauzmann temperature.

Strong HOSS T_{is}

For the strong glass case we also expand for temperatures near to zero and for long times, up to second order, in T and $\delta\mu_2(t)$, yielding:

$$T_{is}(t) = T + \tilde{K}_{is}\delta\mu_2(t) = T + \frac{KD}{D + J^2}\delta\mu_2(t) \qquad (6.86)$$

$$+ \frac{T}{2}\frac{J^4 K^2}{(D + J^2)^3}\delta\mu_2(t) + \frac{DJ^4 K^3}{2(D + J^2)^4}\delta\mu_2(t)^2$$

$$= T_e(t) - \frac{DJ^2 K^2}{2(D + J^2)^3}T\delta\mu_2(t)$$

$$= T_e^{em}(t) + \frac{DJ^4 K^3}{2(D + J^2)^4}\delta\mu_2(t)^2$$

where terms of $\mathcal{O}(T^3), \mathcal{O}(T^2\delta\mu_2(t)), \mathcal{O}(T\delta\mu_2(t)^2)$ and $\mathcal{O}\left(\delta\mu_2(t)^3\right)$ have been neglected. The effective temperature T_e mapping the dynamics of the system evolving at finite temperature T have the same behavior of T_{is} in approaching the heat-bath temperature up to the order $T\delta\mu_2(t)$ where they start deviating one from the other. For a quenching to zero temperature the two effective temperatures coincide. Moreover, in this case the IS effective temperature T_{is} is equal to T_e^{em} given in Eq. (6.75) up to the order $\delta\mu_2(t)$ in time and up to order T^2 in temperature. See Figs. 6.13, 6.14 for a plot of their behavior at two different temperatures near zero.

Though not exact, the practical match between the IS and the thermodynamic effective temperatures obtained with the quasi-static approach works rather well, confirming the validity of the approximated PEL description.

6.A   Many-body glassy models

We report, very shortly, some of the computer models of simple liquids undergoing a glass transition, when cooled down or compressed, that have been studied starting from the 1970s. We do not pretend, by any means, to give a comprehensive list, but we only refer to those models mentioned in the book, mainly in the present chapter, as a guide for the reader. We give a sketch of the soft sphere model [Cape & Woodcock, 1980], of the soft sphere binary mixture [Bernu et al., 1985], of the monoatomic Lennard-Jones (LJ) model [Angell et al., 1977], of the Lennard-Jones binary mixture [Kob & Andersen, 1994], and of the LJ-based models for water [the simple point charge extended (SPC/E) model [Berendsen et al., 1987]] and for orthoterphenyl [the Lewis-Wahnström model (LW) [Lewis & Wahnström, 1993]].

6.A.1   Soft spheres

Monoatomic system

The model consists of idealized, spherically symmetric particles interacting in pairs through repulsive forces that are not strong enough to prevent the partial overlap of two spheres, therefore they are called soft [Cape & Woodcock, 1980]. The soft sphere potential is:

$$\Phi(\mathbf{r}) = \epsilon \sum_{i,j} \left(\frac{\sigma}{|\mathbf{r}_i - \mathbf{r}_j|} \right)^n \qquad (6.A.1)$$

where ϵ and σ fix the energy and length-scales respectively. The $M = DN$ dimensional vector $\mathbf{r}$ represents the configuration of all the particles of the system. We will call $r_{ij} = |\mathbf{r}_i - \mathbf{r}_j|$ the inter-particle distance. The only independent parameter is $\epsilon\sigma^n$, where the exponent n is usually taken equal to 12, making the potential quite short ranged. The reason for its introduction was, mainly, that the short range nature of the interaction allowed for accurate molecular dynamics simulations of systems larger with respect to those with longer range. Moreover, a unique scale is present, allowing for clearer comparison with experiments.

    Besides the repulsive term, a background potential field can be added, to describe attractive interactions, as, e.g., in [Shell *et al.*, 2003]. Further applications and analysis based on soft-sphere monoatomic compounds can be found, among others, in [Hansen & McDonald, 2006; Debenedetti *et al.*, 1999; Stillinger *et al.*, 2001; Hall & Wolynes, 2003]

Binary mixture

The binary mixture of spheres interacting via a soft potential (SSBM) was introduced by Bernu *et al.* [1985, 1987] in order to more easily prevent the nucleation of the crystal state in cooling. The softsphere binary mixture has been a very useful benchmark for the study of the glass transition and of the aging, off-equilibrium properties of amorphous systems (cf., e.g., [Barrat *et al.*, 1990; Hansen & Yip, 1995; Parisi, 1997b,a,c; Coluzzi & Parisi, 1998; Coluzzi *et al.*, 1999]).

    Two kinds of spherical particles are involved, distinguished by having different diameters, with a prescribed ratio, as shown in Table 6.1 (the parameter for type B is chosen equal to unity, without loss of generality).

TABLE 6.1
Soft sphere potential excluded volume parameters σ in units of σ_{BB}.

σ	A	B
A	1.2	1.1
B	1.1	1

For these values and concentrations 50%-50%, the crystallization is strongly inhibited and clear and interesting signs both of off-equilibrium dynamics [Parisi, 1997b,c] and a possible thermodynamic glass transition (to the ideal glass state) have been found at about $T_K \simeq (\rho/1.65)^4$ [Coluzzi et al., 1999]. The dynamical transition temperature depends on the density as $T_d = (\rho/1.45)^4$ [Barrat et al., 1990].

6.A.2   Lennard-Jones many-body interaction potential

Monoatomic system

An argon-like model of monoatomic molecules interacting via a Lennard-Jones (LJ) potential has the potential energy [Angell et al., 1977; Stillinger & Weber, 1982, 1983]:

$$v(r_{ij}) = 4\epsilon \left[\left(\frac{\sigma}{r_{ij}} \right)^{12} - \left(\frac{\sigma}{r_{ij}} \right)^6 \right] \tag{6.A.2}$$

Here σ is the excluded volume and ϵ the energy scale.

It is sometimes approximated by the auxiliary (finite range) interaction potential:

$$v_p(r_{ij}) = A \left[\left(\frac{\sigma}{r_{ij}} \right)^{12} - \left(\frac{\sigma}{r_{ij}} \right)^6 \right] \exp\left\{ \frac{\sigma}{r_{ij} - a} \right\} \tag{6.A.3}$$

Numerically, one sees that the vast majority of inherent structures are amorphous packings. An example of a study of a monoatomic LJ system undergoing a glass transition under pressure variation and displaying an effective temperature for the off-equilibrium state is the one of di Leonardo et al. [2000] and an example of a PEL analysis of the aging dynamics can be found in [Angelani et al., 2001].

Binary mixture

The LJ binary mixture model was originally proposed by Weber & Stillinger [1985] as a model for $Ni_{80}P_{20}$ and then applied to glass physics by Kob & Andersen [1994]. Besides the references mentioned in the chapter, LJ binary mixtures in the context of glass-forming liquids have been widely studied in the last years, e.g., by Kob & Andersen [1995a,b]; Vollmayr et al. [1996]; Sastry et al. [1998]; Sciortino et al. [1999]; Coluzzi et al. [2000a,b]; Kob et al. [2000]; Broderix et al. [2000]; Angelani et al. [2000]; Sastry [2000], to mention a few.

The mixture consists of two kinds of particles having the same mass but different diameters. Type A particles are larger and type B are smaller. They interact via a Lennard-Jones potential of the form

$$V_{\alpha\beta}(r_{ij}) = 4\epsilon_{\alpha\beta} \left[\left(\frac{\sigma_{\alpha\beta}}{r_{ij}} \right)^{12} - \left(\frac{\sigma_{\alpha\beta}}{r_{ij}} \right)^6 \right] \qquad \alpha, \beta = A, B \tag{6.A.4}$$

TABLE 6.2
Values of the LJ parameters. σ_{AA} is the length unit and ϵ_{AA}/k_B the temperature unit.

σ	A	B	ϵ	A	B
A	1	0.8		1	1.5
B	0.8	0.88		1.5	0.5

where r_i denotes the position of particles $i = 1, \ldots, N$ and $r_{ij} \equiv |r_i - r_j|$. The parameters are reported in Table 6.2.

The binary mixture is usually composed by 80% of A particles and 20% B particles. At these concentrations and with the values of the interaction parameters reported in Table 6.2 crystallization is prevented, as well as phase separation. Sometimes a cutoff for the potential is used at some interparticle distance r_{cut} (e.g., in [Sastry *et al.*, 1998; Kob *et al.*, 2000; Broderix *et al.*, 2000]).

A very much used model is the 80:20 LJ mixture with the NVT (Nose-Hoover) thermostat at the density of $\rho = 1.2$. (This thermostat is a specific modeling of the bath). In this case the dynamical glass transition occurs for $T = T_{\text{mc}} = T_{\text{d}} = 0.435$, whereas the Kauzmann point is $T_K = 0.32$ [Coluzzi *et al.*, 2000a]. In the PEL formalism, for $T < 0.8$, it turns out that $f_{\text{vib}}(T, \phi) = f_{\text{vib}}(T)$ is independent of ϕ, and the system can be considered as composed of two independent subsystems, respectively described by the IS and by the vibrational part. For $T < 0.5$ the average potential energy $\bar{\phi}$ turns out to be very near to the value $(3/2)k_B T$ supporting the hypothesis that the system mainly probes a harmonic potential landscape.

6.A.3 Lewis-Wahnström model for orthoterphenyl

With respect to the central pair potentials considered so far, [Lewis & Wahnström, 1993] improved the model description of physical interaction including rotational degrees of freedom (see also [Wahnström & Lewis, 1993; Lewis & Wahnström, 1994a,b]). The model they introduced consists of a three site LJ potential that is able, provided the parameter values are tuned appropriately, to mimic the behavior of OTP, a nonpolar organic liquid consisting of three molecular units (three connected benzene rings). The intermolecular interactions are short range, van der Waals-like and the model shows little tendency to enucleate a crystal phase. It is one of the most studied fragile glasses in the last forty years.

In the LW model, the OTP molecule is represented by three sites (one for each benzene ring). The molecule is supposed rigid, neglecting internal degrees of freedom, and planar: an isosceles triangle with two short sides. To each molecule a fixed bond length is assigned, equal to the length of a short side and corresponding to an exluded volume σ of a LJ potential. Also the angle θ between short sides is fixed, at 75°. The interaction between two sites on different molecules is of the LJ type (reproducing in molecular dynamics simulations the van der Waals forces), cf. 6.A.2. Nine site-to-site interactions have, thus, to be evaluated for each pair of molecules, of the form:

$$V(r_{i_a j_b}) = 4\epsilon \left[\left(\frac{\sigma}{r_{i_a j_b}} \right)^{12} \left(-\frac{\sigma}{r_{i_a j_b}} \right)^{6} \right] + \lambda_1 + \lambda_2 r_{i_a j_b} \qquad (6.A.5)$$

where $r_{i_a j_b}$ is the distance between the site $a = 1, 2, 3$ of the molecule $i = 1, \ldots, N$ and the site b of the molecule j. The LJ parameters are $\sigma = 0.483$ nm and $\epsilon = 5.276$ KJ/mol. The values of the extra parameters are $\lambda_1 = 0.461$ kJ/mol and $\lambda_2 = -0.313$ kJ/(mol nm) and are chosen in such a way that $V(r)$ and $V'(r)$ are both zero at $r = 1.2616$ nm. The PEL anaylis of the LW model has been carried out by Mossa *et al.* [2002]; La Nave *et al.* [2002].

6.A.4 Simple point charge extended model for water

The SPC/E potential has been introduced to model the behavior of under-cooled liquid and amorphous solid water [Berendsen *et al.*, 1987]. In particular, it reproduces the experimentally measured property that the diffusion constant, as a function of the pressure or the density at constant temperature, displays a maximum, that becomes more and more evident as temperature is decreased. The model involves orienting electrostatic effects (on the plane), cf. Fig. 6.15, and Lennard-Jones interactions, cf. Eq. (6.A.2), whose σ parameter accounts for the size of the molecules.

At short distances repulsion occurs, ensuring that the structure cannot collapse due to the electrostatic interactions. At intermediate distances, instead, the interaction is significantly attractive but nondirectional and it competes with the directional attractive electrostatic interactions.

Examples related to the PEL approach can be found in [Roberts *et al.*, 1999; Scala *et al.*, 2000].

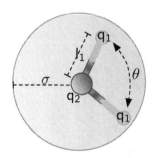

SPC/E

$l_1 = 1$ Å $q_1(e) = 0.4238$
$\theta = 109.47°$ $q_2(e) = -0.8476$

Lennard-Jones parameters
$\sigma = 3.166$ Å $\epsilon = 0.65$ kJ mol^{-1}

FIGURE 6.15

Parameters for the water molecule H_2O are represented in the SPC/E model. The LJ parameters for the molecule-molecule interaction are also reported.

7

Theories of the glassy state

When the moon is not full,
the stars shine more brightly

Bahamian proverb

Thermodynamics acts as a law of large numbers in constraining the energy exchange in macroscopic systems. In previous chapters we have developed a scenario for its application to the glassy state and discussed its limits. A different fundamental question is which mechanism lies at the origin of the slow dynamics that leads to glassy behavior. Obviously, this has a strong system-dependent component, though certain types of universality are expected. A number of scenarios that are known in literature will be discussed in the present chapter. In particular, we will give a short presentation of the mode-coupling theory and of the mean-field replica theory applied to models related to structural glasses, both with and without *ad hoc* quenched disorder. We will, then, dedicate some more space to the state of the art for what concerns the avoided critical point theory and the random first order transition theory for the *mosaic state*. Some theories have been widely referred to in the previous chapters and we report them for self-consistency and to provide up-to-date bibliography for interested readers. Some other theories are, in our opinion, interesting recent developments, whose properties we critically analyze and discuss, trying, as well, to compare different points of view in the literature. Many other approaches, among which we mention those of Schulz [1998] and Franz [2005, 2006], fall outside the scope of this book.

7.1   Mode-coupling theory

Mode-coupling theory (MCT) is a mean field-type approach valid above and near the onset of glassy behavior, that is to say, in the region where the α and β relaxation times start to deviate from each other. This occurs at a "mode-coupling" temperature T_{mc}, corresponding to what we called the dynamical transition temperature T_{d} [Götze, 1984; Bengtzelius *et al.*, 1984;

Leutheusser, 1984]. This temperature lies far above the glass temperature with its typical timescale of hours. For glass-forming liquids, the relevant dynamics is typically in the 100 nanosecond regime, so the theory can be tested by spectroscopic tools [Götze, 1985, 1991; Götze & Sjögren, 1992] . A modern application is soft systems, such as Laponite solutions in water, where the intrinsic timescale is much larger, of the order of hours or more, so that the mode-coupling regime is the only relevant regime for laboratory experiments [Kroon *et al.*, 1996].

One of the most important features of MCT is that just knowing the static structure factor, one can derive the whole dynamics. In its simplest form, one starts with the density fluctuations of the N molecules of the fluid at wave vector $\mathbf{q}$,

$$\delta\rho(\mathbf{q}, t) = \sum_{j=1}^{N} e^{i\mathbf{q}\cdot\mathbf{r}_j(t)} \tag{7.1}$$

The coherent intermediate scattering function is

$$F(q, t) = \frac{1}{N} \langle \delta\rho(\mathbf{q}, t)\delta\rho^*(\mathbf{q}, t) \rangle \tag{7.2}$$

and the static structure factor is

$$S(q) = F(q, 0) \tag{7.3}$$

where $q = |\mathbf{q}|$. Normalizing Eq. (7.2)

$$\Phi(q, t) = \frac{F(q, t)}{F(q, 0)} , \tag{7.4}$$

one may write down the evolution equation in the form

$$\ddot{\Phi}(q, t) + \Omega_q^2 \Phi(q, t) + \Omega_q^2 \int_0^t dt' M(q, t - t')\dot{\Phi}(q, t') \tag{7.5}$$

where

$$\Omega_q^2 = \frac{q^2 kT}{mS(q)} \tag{7.6}$$

with m the mass of the particle of the fluid. All the difficulty of the problem has now been hidden in the memory kernel $M(q, t)$. It has a regular fast part M^{reg}, the short time memory kernel, related to the short time behavior of the liquid, which is a difficult kernel, but unrelated to glassiness. The rest, $M - M^{\text{reg}}$, is slow. Within the Mori-Zwanzig projection operator technique, which itself is exact, a leading order truncation has been made, which yields

$$M(q, t) - M^{\text{reg}}(q, t) = \frac{1}{2(2\pi)^3} \int d^3k \, V^{(2)}(q, k, |\mathbf{q}-\mathbf{k}|)\Phi(k, t)\Phi(|\mathbf{q}-\mathbf{k}|, t) \tag{7.7}$$

The approximated vertex function can be expressed as

$$V^{(2)}(q, k, |\mathbf{q} - \mathbf{k}|) = \frac{n}{q^2} S(q) S(k) S(|\mathbf{q} - \mathbf{k}|) \left(\frac{\mathbf{q}}{q} \cdot [\mathbf{k}c(k) + (\mathbf{q} - \mathbf{k})c(|\mathbf{q} - \mathbf{k}|)] \right)^2$$
(7.8)

Here $n = N/V$ is the density and $c(k) = n/[1 - 1/S(k)]$ the direct correlation function. The whole theory is, thus, determined by the static structure factor $S(q)$.

Another approach, called "fluctuating hydrodynamics," leads to the very same equations [Das *et al.*, 1985; Das & Mazenko, 1986; Schmitz, 1988]. After the proper identifications have been made, these equations also coincide with the dynamic equation of a subclass of spin-glass models in their paramagnetic phase [Bouchaud *et al.*, 1996]. So they appear to bear some universality and, as a consequence, have some universal scope of application.

The above equations are still very complicated. In a numerical approach, Bengtzelius *et al.* [1984] noticed that the main contribution comes from wave vectors near the peak q_0 of the static structure factor $S(q)$. A not unreasonable simplification is then to assume that only the peak is relevant, viz., $S(q) \sim \delta(q - q_0)$. This leads to the so-called schematic theory, in our situation the Leutheusser [1984] model. The mode-coupling equations reduce to

$$\ddot{\phi}(t) + \Omega^2 \phi(t) = -c_2 \Omega^2 \int_0^t dt' \phi^2(t - t') \dot{\phi}(t')$$
(7.9)

where $\phi(t) = \Phi(q_0, t)$, c_2 results from the weight of the peak and the regular part of M has been dropped since it does not enter the long time dynamics. This simplification is not crucial for the rest of our discussion.

To get a feel for the structure of the theory, let us analyze this schematic equation in some detail. Neglecting the second derivative, we write it as

$$\phi(t) + c_2 \frac{d}{dt} \int_0^t dt' \phi^2(t') \phi(t - t') = c_2 \phi^2(t)$$
(7.10)

We set, then,

$$\phi(t) = f + G(t)$$
(7.11)

where f, called the non-ergodicity parameter, is a plateau value of ϕ, where the system will stay some while (for $T > T_{mc}$) or will relax for long times (for $T < T_{mc}$). It holds that $G(0) = 1 - f$. Expanding up to order G^2 we have

$$\frac{1}{2} \left(\frac{1}{c_2 f} - 2 + 3f \right) G(t) + \frac{d}{dt} \int_0^t ds \, G(s) G(t - s) = \sigma + \frac{1 - f}{2f} G^2(t)$$
(7.12)

The linear term in G vanishes if we choose

$$f = \frac{1}{3} \left(1 + \sqrt{1 - \frac{3}{c_2}} \right).$$
(7.13)

The constant term

$$\sigma \equiv \frac{f}{2}\left(1 - f - \frac{1}{c_2 f}\right) = \frac{1}{6 - c_2 + c_2\sqrt{1 - 3/c_2}}\frac{c_2 - 4}{2c_2} \tag{7.14}$$

clearly vanishes for $c_2 \searrow c_2^* = 4$. At that point, $f = f^* \equiv \frac{1}{2}$. As c_2 will be temperature dependent, the point $c_2 = c_2^*$ will correspond to the mode-coupling temperature T_{mc}, and σ is proportional to the deviation from it,

$$\sigma = C\frac{T - T_{mc}}{T_{mc}} \tag{7.15}$$

where C is a system parameter. After these steps, Eq. (7.12) reads

$$\frac{d}{dt}\int_0^t ds\, G(s)G(t - s) = \sigma + \lambda G^2(t) \tag{7.16}$$

with, in our case, $\lambda = f^*/[2(1 - f^*)] = \frac{1}{2}$. In general, however, f^* and λ can take model dependent values, $f^* > 0, \lambda < 1$.

For times in the β-regime, $\tau_0 \ll t \ll \tau_\beta$, we will have a decay towards the plateau of the form

$$G(t) = A\left(\frac{t}{\tau_\beta}\right)^{-a} \tag{7.17}$$

Taking Laplace transforms, one finds the condition

$$\frac{\Gamma^2(1 - a)}{\Gamma(1 - 2a)} = \lambda \tag{7.18}$$

for the exponent a, while, generally, the amplitude A is of the order of unity, provided we match in the non-asymptotic regime with the σ term, thus choosing

$$\tau_\beta \sim \tau_0\sigma^{-1/(2a)} \tag{7.19}$$

where τ_0 is the microscopic attempt time. In the ergodic phase ($\sigma > 0$) this decay will go to the long time value set by Eq. (7.16),

$$G(\infty) = \sqrt{\frac{\sigma}{1 - \lambda}} \tag{7.20}$$

The total non-ergodicity parameter $f + G(\infty)$ thus exhibits a square root signature on the approach to the mode-coupling transition, which is well visible on top of linear terms in σ.

Above the mode-coupling transition, the system is ergodic and the correlator will decay from the plateau on longer times, for $\tau_\beta \ll t \ll \tau_\alpha$. We may set

$$g = -B\left(\frac{t}{\tau_\alpha}\right)^b \tag{7.21}$$

which is called "von Schweidler's law" [Götze, 1985]. This will bring us to an exponent relation similar to Eq. (7.18),

$$\frac{\Gamma^2(1+b)}{\Gamma(1+2b)} = \lambda \tag{7.22}$$

The matching of the amplitudes for $t \sim \tau_\beta$ implies $\tau_\beta/\tau_\alpha \sim \sigma^{1/(2b)}$. The α processes are, thus, predicted to have a power law divergent relaxation time,

$$\tau_\alpha = C_\tau \tau_0 \sigma^{-\gamma} = C'_\tau \tau_0 \left(\frac{T_{\mathrm{mc}}}{|T - T_{\mathrm{mc}}|}\right)^\gamma \tag{7.23}$$

where γ reads

$$\gamma = \frac{1}{2a} + \frac{1}{2b} \tag{7.24}$$

Both a and b are set by λ, cf. Eqs. (7.18), (7.22), so also γ may vary from system to system. In practice, λ can be derived from a, b or γ.

In general, beyond the schematic approximation, every correlator can be decomposed according to the "factorization property"

$$\phi(t) = f + hG(t) \tag{7.25}$$

where f and h are constants that depend on the considered dynamical variable of which ϕ is the correlation function (for instance, when considering $\rho_q(t)$, they may depend on q), but G is a universal function:

$$G(t) = \sqrt{|\sigma|}\, g_\pm\left(\frac{t}{t_\beta}\right) \tag{7.26}$$

Here g_+ is a scaling function applying to $T > T_{\mathrm{mc}}$ while g_- applies to $T < T_{\mathrm{mc}}$. The functions $g_\pm(s)$ satisfy the equations

$$\frac{\mathrm{d}}{\mathrm{d}s}\int_0^s \mathrm{d}s'\, g_\pm(s - s')g_\pm(s') = \pm 1 + \lambda g_\pm^2(s) \tag{7.27}$$

and $G = \sqrt{|\sigma|}\, g_\pm$ has the above-discussed asymptotics. Notice also that, for a given observable, time enters in a scaling form t/τ_β, which allows us to present data at different T in a master curve after determining τ_β.

An impressive confirmation of MCT has been achieved in the temperature-frequency plane [Götze, 1991]. When one comes closer to T_{mc}, the theory, however, fails [Schmitz *et al.*, 1993]. This is because it assumes a divergence of the α timescale, which is not present in reality. Though some improvements of the approach can be made, it has remained difficult to capture the relevant physics below T_{mc}, that is the domain of hopping dynamics, or intermittency, due to trapping in mildly long-lived states, or, equivalently, of a timescale separation related to mildly metastable states.

More recent applications of MCT include aging dynamics of soft systems such as Laponite solutions in water. Here the intrinsic timescale is much longer, and experimental timescales can, for Laponite concentrations of the order of a few percent, range up to 200 hours [Kroon *et al.*, 1996]. Probably just for this reason, dynamics will mostly stick in the initial time regime, where mode-coupling applies. In agreement with this, microrheology experiments show that the FDT is satisfied [Jabbari-Farouji *et al.*, 2007].

The reason for the success of MCT may have been uncovered recently. It was pointed out by Biroli *et al.* [2006] that in MCT a divergent length-scale occurs, not in the two point correlation function, as happens in standard phase transitions, but in a four-point correlation function, in a manner in which it typically happens in spin-glasses. The presence of the underlying cooperative mechanism explains why this theory has ubiquitous applications, even if the cooperation is an effect of the initial stage, that gets leveled off beyond a certain scale, and no true divergence happens at T_{mc} (and, in fact, not even at T_g).

For recent overviews treating extensively the MCT, the reader can consult, e.g., the reviews of Götze [1999]; Cummins [1999]; Das [2004] and the book of Binder & Kob [2005].

7.2 Replica theory for glasses with quenched disorder

Many spin-glass models have been studied sharing features similar to those occurring in real glass formers. Most of them are built with a disordered interaction between the dynamic variables that is *quenched*, by this meaning that, once prepared, couplings never change during the rest of the system's history. In the majority of cases, the dynamic variables are magnetic spins and the interaction is magnetic (mixing both ferro- and antiferromagnetic bounds). The liquid phase is, thus, represented by a paramagnet, whereas the glass phase is represented by a particular spin-glass phase, computed by means of the so-called one step replica symmetry breaking (1RSB) Parisi Ansatz.[1] The quenched randomness is a rather good tool to *ad hoc* implementation of the feature of spatial disorder of the molecules in the glass, and it turns out that models with built-in disorder do not yield different behaviors from those with self-induced disorder (like in real systems). The subject is widespread and here we just want to briefly recall the main results and link them to the actual behavior observed in real glasses. Also, we will pay attention to the order parameters arising in the description of the frozen phase. Many reviews, books and lectures have been dedicated to mean-field spin-glass models yielding the behavior of structural glasses. We mention, e.g., the review of Bouchaud

---

[1]This is not the spin-glass phase actually reproducing the behavior of true amorphous magnets, where the replica symmetry must be broken infinite times in order to yield a thermodynamically stable phase [Mézard *et al.*, 1987].

et al. [1998] and the Les Houches lectures of Parisi [2003] for a comprehensive survey of the static and dynamic properties, the pedagogic review of Cavagna & Castellani [2005], dedicated to the special case of the spherical *p*-spin model, and the recent books of Binder & Kob [2005] and De Dominicis & Giardina [2006].

7.2.1 The random energy model

The simplest model for glassy systems, the REM [Derrida, 1980, 1981], has been already introduced in Sec. 6.2.2, where it was applied to the PEL formalism. We now analyze its properties a little bit further.

The REM was initially devised as a limit case of a set of N Ising spins $s_i = \pm 1$ exchanging disordered interaction. The model consists of 2^N IID energy levels, whose probability distribution is

$$p(E) = \frac{1}{\sqrt{2\pi N J^2}} \exp\left(-\frac{E^2}{2\pi N J^2}\right) \tag{7.28}$$

Computing the thermodynamics, one observes that, coming from high temperature, the energy per spin changes behavior in T at a critical temperature T_s: from $e = -1/T$ at high temperature, it becomes constant at $e = -1/T_s$. At that temperature, the entropy goes to zero, with a discontinuous first derivative. At high temperature the relevant configurations are, therefore, infinite (as $N \to \infty$) and we have the usual paramagnetic (fluid) phase, whereas at low temperature only a subextensive number of configurations dominate. To represent this, an order parameter can be defined in terms of the original Ising spin variables, the *overlap*

$$q_{ab} \equiv \frac{1}{N} \sum_i s_i^a s_i^b \tag{7.29}$$

where a and b label two configurations of spins, $\mathbf{s}_{a,b} = \{s_i^{a,b}\}$. The overlap q_{ab} is equal to one if the two configurations a and b are equal ($s_i^a = s_i^b$), it is zero if the two configurations are orthogonal ($\mathbf{s}_a \cdot \mathbf{s}_b = 0$), and it is -1 if they are opposed ($s_i^a = -s_i^b$).

For $T > T_s$ it turns out to be $q_{ab} = 0$, $\forall (a, b)$, whereas for $T \leq T_s$ the overlap can take value zero with probability m and one with probability $1 - m$, where

$$m = T/T_s \tag{7.30}$$

As $T < T_s$ the latter parameter is smaller than one and the probability of finding two equal configurations is finite. Formally, the probability distribution of the overlap for the REM is

$$P(q) = (1 - m)\,\delta(q - 1) + m\,\delta(q) \tag{7.31}$$

In the transition at T_s, no latent heat is involved and the specific heat has a discontinuity. The specific heat, contrary to standard continuous transitions but similar to the (thermal) glass transition, cf. Sec. 1.1, jumps from a larger value at high temperature to a lower value (zero) below T_s, i.e., some part of the phase space is hidden. This corresponds to the fact that the free energy of the paramagnetic phase extrapolated at low temperature is *lower* than the free energy of the frozen phase, contrary to what happens in standard statistical mechanics. Apart from the inversion displayed by the specific heat, according to these elements, the REM transition would seem a second order phase transition. However, no diverging susceptibility is found and, furthermore, the order parameter q jumps discontinuously from zero to one as the system undergoes the critical temperature. This kind of transition is also found in other, more complicated, mean-field models, that we will consider below. It is called either discontinuous phase transition [Bouchaud *et al.*, 1998] or random first order transition (RFOT) [Kirkpatrick *et al.*, 1989] to distinguish it from the standard first order phase transition between liquid and crystalline solid. It might be argued that it is of order "one and half," sharing some properties both of first and second order phase transitions [Parisi, 2003].

7.2.2　　The p-spin model

The model of which the REM is a limiting case is the many-body interacting, known as the Ising p-spin model [Derrida, 1980; Gross & Mézard, 1984; Gardner, 1985], whose Hamiltonian is

$$\mathcal{H} = -\sum_{i_1 < i_2 < \ldots < i_p} J_{i_1 i_2 \ldots i_p} s_{i_1} \ldots s_{i_p} \tag{7.32}$$

The couplings $J_{i_1 i_2 \ldots i_p}$ are quenched random variables with Gaussian distribution

$$P(J_{i_1 i_2 \ldots i_p}) = \sqrt{\frac{N^{p-1}}{p!\pi J^2}} \exp\left\{ -\frac{J_i^2 N^{p-1}}{p! J^2} \right\} \tag{7.33}$$

In this model, the energies of different configurations are no more uncorrelated. If one takes the limit $p \to \infty$, however, the correlation vanishes and one finds the REM model back. The p-spin model displays a dynamic transition at some temperature T_d, below which the thermodynamic solution is still paramagnetic, but the free energy landscape becomes so corrugated and its valleys so numerous and deep that the dynamics gets stuck in one of them without being able to ever escape. This is an artifact of the mean-field nature of the model causing the barriers to grow like N. At a lower temperature, T_s, a thermodynamic phase transition occurs to a frozen glassy phase that is a mean-field description for the possible ideal glass phase. Indeed, the transition temperature is identified with the Kauzmann temperature (see Sec. 1.4), $T_s = T_K$. At a still lower temperature, the Ising p-spin undergoes a second thermodynamic phase transition to a proper spin-glass phase.

Since, in this book, we are focusing on the glass transition, we will now shortly report the features of a simplified version of the model where the spins are spherical ($\sum_i s_i^2 = N$) instead of Ising and where only a dynamic and a static random first order transition occur, between a fluid (paramagnet) and a phase sharing many properties of the glass phase. Because of its simplicity and because of the similarities with glasses (in the behavior, not in the microscopic description) the spherical p-spin model has been widely studied in literature since its introduction [Crisanti & Sommers, 1992], see, e.g., [Crisanti *et al.*, 1993; Kurchan *et al.*, 1993; Cugliandolo & Kurchan, 1993, 1994; Nieuwenhuizen, 1995; Cavagna *et al.*, 1998, 1999; Crisanti *et al.*, 2003], just to mention a few.[2]

Looking at the Hamiltonian, Eq. (7.32), it can be seen that there is no spatial structure in the model, since every spin can interact with any other: the model is mean-field.

The thermodynamic of the glassy phase can be computed by averaging over the quenched disorder the logarithm of the sample dependent partition function

$$Z_J = \sum_{\{s_{i_1} \dots s_{i_p}\}} e^{-\beta \mathcal{H}[\{s\}]} \tag{7.34}$$

This is performed by means of the replica trick, that is, by making n identical copies of the original model, whose collective partition function reads

$$Z_J^n = \sum_{s_{i_1}^a \dots s_{i_p}^a} e^{-\beta \sum_{a=1}^n \mathcal{H}[\{s^a\}]} \tag{7.35}$$

The replicated, or *total*, free energy is, then, computed considering the analytic continuation of the above function of n for non-integer values and taking the limit for $n \to 0$:

$$-\beta F_{\text{tot}} = \overline{\log Z_J} = \lim_{n \to 0} \frac{\overline{Z_J^n} - 1}{n} \tag{7.36}$$

The total free energy per spin of the spherical p-spin model is

$$\beta f_{\text{tot}} = -\lim_{n \to 0} \left[\frac{\beta^2}{4} \sum_{ab} q_{ab}^p + \log \det \hat{q} \right] \tag{7.37}$$

where $\hat{q}$ is the overlap matrix,[3] i.e., the order parameter. In the paramagnetic phase all elements are zero. Below the static phase transition the stable phase is obtained by breaking the original symmetry of the replicas and imposing that the elements q_{ab} of $\hat{q}$ are not all equal to each other, as one would expect

---

[2]The similarity between the p-spin model and the glass was initially put forward by Kirkpatrick & Thirumalai [1987b,a]. They studied yet another version of the model, with continuous spins and $\mathcal{H}_{\text{soft}} = \mathcal{H} + \sum_i (\lambda s_i^2 + u s_i^4)$. Another model with random interaction, the Potts glass with $Q > 4$, also displays a discontinuous dynamic transition, as well as an underlying static random first order transition and similar glassy properties [Gross *et al.*, 1985; Kirkpatrick & Wolynes, 1987b].

[3]Not to be confused with the wave vector $\mathbf{q}$ of the mode-coupling theory, see Sec. 7.1.

Thermodynamics of the glassy state

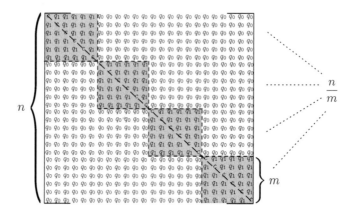

FIGURE 7.1

The $n \times n$ matrix $\hat{\mathbf{q}}$ in the generic 1RSB Ansatz. The diagonal elements are irrelevant constants (c in the figure). The diagonal block elements (in grey, $m \times (m-1)$ in each block) are equal to the self-overlap q_1, the Edwards-Anderson order parameter. The $n/m(n/m-1)$ elements of the off-diagonal blocks are equal to the minimum overlap q_0. In absence of an external magnetic field $q_0 = 0$.

if all replicas were equivalent. The 1RSB Ansatz put forward in the matrix shown in Fig. 7.1 [Parisi, 1980], turns out to yield the stable, consistent, low temperature phase [Crisanti & Sommers, 1992]. The elements can take two values, q_1 and $q_0 < q_1$, and the matrix is organized in diagonal and off-diagonal blocks. Formally, this can be written as

$$q_{ab} = (1 - q_1)\, \delta_{ab} + (q_1 - q_0)\, \epsilon_{ab} + q_0 \tag{7.38}$$

where $\epsilon_{ab} = 1$ if a and b belong to a diagonal block, zero otherwise.

The 1RSB free energy reads

$$f_{\text{tot}} = -\frac{\beta}{4}\left[1 - (1-m)q_1^p - mq_0^p\right] - \frac{1-m}{m}\log\tilde{\chi}(q_1) + \frac{1}{m}\log\tilde{\chi}(q_0) + \frac{q_0}{\tilde{\chi}(q_0)} \tag{7.39}$$

where $\tilde{\chi}(q) \equiv 1 - q_1 + m(q_1 - q)$. The self-consistency equations for the order parameters q_0, q_1 and m are (in the zero external field)

$$q_0 = 0 \tag{7.40}$$

$$(1 - m)\left[\frac{\beta^2 p}{2}q_1^{p-1} - \frac{q_1}{\tilde{\chi}(q_1)\tilde{\chi}(0)}\right] = 0 \tag{7.41}$$

$$\frac{\beta}{2}q_1^p + \frac{1}{m^2}\log\left(\frac{\tilde{\chi}(q_1)}{\tilde{\chi}(0)}\right) + \frac{q_1}{m\tilde{\chi}(0)} = 0 \tag{7.42}$$

At a high temperature, the stable solution is $q_1 = 0$ and one has the paramagnetic phase (m is undetermined and irrelevant). As T equals some critical value, Eqs. (7.41)-(7.42) are solved for $m = 1$ and $q_1 \neq 0$.

Introducing the universal Crisanti-Sommers [1992] function

$$z(y) \equiv -2y\frac{1 - y + \log y}{(1 - y)^2} \tag{7.43}$$

where

$$y \equiv \frac{\tilde{\chi}(q_1)}{\tilde{\chi}(0)} \tag{7.44}$$

the static transition temperature can be written as [Kurchan *et al.*, 1993]

$$T_K = \sqrt{\frac{py^\star}{2}}(1 - y^\star)^{(p-2)/p} \tag{7.45}$$

where $y^\star$ is the solution of $z(y) = 2/p$. For $p = 3$, e.g., $y^\star = 0.35499$ and $T_K = 0.58605$. Eq. (7.44), computed at $y = y^\star$ and $m = 1$, yields the value of $q_1 = 1 - y^\star$, that discontinuously takes a nonzero value at the transition, hinting that the states constituting the frozen phase are already formed.

In terms of probability distribution of the overlap values, the framework is the same as for the REM model, cf. Eq. (7.31), i.e., $P(q) = m\,\delta(q) + (1 - m)\delta(q - q_1)$. At $T = T_K$, $m = 1$ and the probability of states with overlap $q - 1$ is zero: even though they are there, they have no thermodynamic weight. As $T < T_K$, m decreases from 1 and they progressively acquire a nonzero weight.

Even though at higher temperature the stable phase is paramagnetic, from the study of the dynamics, in particular of the equation of motion of the time correlation functions, one sees that, at some temperature, the correlation functions starts developing a plateau (see, e.g., Fig. 1.2) that, further lowering the temperature, becomes persistent to infinite times at

$$T_d = \sqrt{\frac{p(p - 2)^{p-2}}{2(p - 1)^{p-1}}} \tag{7.46}$$

that is larger than T_K. The plateau sets in at the value of the correlation equal to

$$q_d = (p - 2)/(p - 1) \tag{7.47}$$

At $p = 3$, for instance, $T_d = 0.61237$ and $q_d = 0.5 < q_1 = 0.645$.

7.2.3 Complexity

The entropy is the logarithm of the number of configurations of the system, $S = \log \mathcal{N}$.[4] In standard statistical mechanics it can be regarded, e.g., in

---

[4]We set the Boltzmann constant equal to one.

the canonical ensemble at constant volume, as the Legendre transform of the free energy, $F = U - TS$, where the inverse temperature β and the internal energy U are conjugated variables related by $\beta = \partial S / \partial U$ (a Maxwell relation, cf. Sec 2.1.3). In systems displaying a huge number of metastable states, organized in such a way that some kind of ergodicity breaking occurs (weak ergodicity breaking, cf. Sec. 1.1.1), the lost part of the phase space is coded into the configurational entropy (Sec. 1.4), called the *complexity* function in the framework of spin-glasses. It is computed as the average logarithm of the number of the minima of the free energy landscape for a given realization of the disorder: $S_c = \overline{\log \mathcal{N}_J}$. The function whose minima are counted, is generally the one devised following the approach of Thouless *et al.* [1977] (TAP), that is, the free energy functional reproducing the right mean-field equations for average magnetizations in the frozen phase.

Crisanti & Sommers [1995] first applied the TAP approach to the spherical p-spin model and the complexity of the spherical p-spin reads (see, e.g., [Crisanti *et al.*, 2003])

$$S_c(z) = \frac{1}{2} \left[\frac{2-p}{p} - \log \frac{p\hat{z}^2}{2} \frac{p-1}{p} \hat{z}^2 - \frac{2}{p^2 \hat{z}^2} \right] \qquad (7.48)$$

where $\hat{z}$ turns out to be a function of the potential energy values in the minima of the TAP free energy, that is, of the values of the Hamiltonian function computed at its minimal configurations $\{s_i^{\min}\}$, $e = \mathcal{H}(\{s_i^{\min}\})/N$:

$$\hat{z} = \frac{p}{2} \left(e + \sqrt{e^2 - e_{\text{th}}^2} \right) ; \qquad e_{\text{th}} = -\sqrt{\frac{2(p-1)}{2}} \qquad (7.49)$$

The complexity is defined as long as $\hat{z}$ is real, i.e., $e < e_{\text{th}}$ and for $e > e_0$ where it becomes negative, that is, where the number of states becomes exponentially small with the size of the system. The energy e_0 is the ground state of the p-spin model. This comes about because the TAP mean-field equations for the average site magnetizations of the spherical p-spin model do not depend on temperature and they formally coincide with the equation of minimization of the Hamiltonian (see [Crisanti & Sommers, 1995; Cavagna & Castellani, 2005]). In other words, in this very special case, there is an exact one-to-one correspondence between minima of the FEL and minima of the PEL (cf. the discussion in Sec. 6.1).

The overlap, as well as the internal energy $u = U/N$ and free energy $f = F/N$ per particle do, instead, depend on the temperature (and $u(e, T = 0) = e$). In particular, the Edwards-Anderson parameter q_1 at the threshold is obtained by solving

$$(p-1)^2 q_1^{p-2} (1 - q_1)^2 = T^2 e_{\text{th}}^2 \qquad (7.50)$$

At the dynamic transition temperature, the threshold overlap coincides with q_d [cf. Eq. (7.47)], signaling that the states visited at the dynamic transition are the threshold ones (i.e., the most numerous ones and those at the highest potential energy): there the dynamics is trapped and the divergence of the height of the barriers in the thermodynamic limit prevents the system from reaching the (paramagnetic) equilibrium state.

Knowing that many states are present in an energy domain between e_0 and $e_{\rm th}$, the partition function can be written as

$$Z \equiv \sum_\gamma e^{-\beta N f_\gamma} = \int de \; e^{S_c} e^{-\beta N f(e,T)} = \int de \; e^{-\beta N f_{\rm tot}(e,T)} \qquad (7.51)$$

where

$$f_{\rm tot}(e,T) \equiv f(e,T) - T s_c(e,T) \qquad (7.52)$$

The above function can be regarded as a Legendre transform of the complexity with conjugated variables f and

$$T = \frac{\partial s_c}{\partial f} \qquad (7.53)$$

Its minimum value is displayed at the average equilibrium value of the energy at temperature T, $\bar{e}(T)$ [compare with $\bar{\phi}$ in Eq. (6.9)] and allows us to evaluate the integral in Eq. (7.51) by the saddle point method. Eventually,

$$-\beta F_{\rm tot}(\bar{e}(T),T) = -\beta N f_{\rm tot}(\bar{e}(T),T) = \log \sum_\gamma e^{-\beta N f_\gamma}$$

7.2.4 Mean-field scenario

Summarizing, the single state free energy per particle is $f = u - Ts$ (s is the entropy of a state), whereas the total free energy per particle is

$$f_{\rm tot} = f - T s_c = u - T(s + s_c) \qquad (7.54)$$

This is the same decomposition carried out in Sec. 2.5, cf. the free energy expressed by Eq. (2.66), where, however, equilibrium is supposed to hold and, therefore $T_e = T$.[5]

The temperature at which $\bar{e}(T)$ equals the ground state is, not surprisingly, the thermodynamic transition temperature T_K, for which $s_c(\bar{e}(T_K)) = 0$ and $f_{\rm tot}(T_K) = f(T_K)$. Below T_K, equilibrium is dominated by stable states (with sub-extensive complexity).

Lowering the temperature of a liquid system, at some point $T_{\rm d}$, dynamic

---

[5]This depends on the fact that the breaking parameter m is equal to one above $T_s = T_K$. In other words, the statics is the one of the liquid phase and we are adopting a static, equilibrium mean-field description. As one goes below T_K, or the dynamics is considered, the factor in front of the complexity can become $T_e = T/m$, with $m < 1$, see, e.g., [Kirkpatrick et al., 1989; Cugliandolo & Kurchan, 1993; Nieuwenhuizen, 1998a; Mézard & Parisi, 1999a].

arrest occurs. The ergodicity of the liquid phase is broken and the system is stuck into one metastable vitreous state. The liquid phase is still the stable thermodynamic phase but the barriers to overcome in order to end up in this state are infinite in the thermodynamic limit, so that the equilibration time is infinite.

Below $T_{\rm d}$, metastable glassy states are very many. Their number increases exponentially with the number of particles of the system, $\sim e^{N s_c}$. The liquid states are of a negligible number in comparison (their entropic contribution is subextensive). Therefore, e.g., in a cooling experiment, it is statistically impossible to reach a liquid phase: the system cooled down from an arbitrary high temperature will almost certainly find itself in a glassy metastable state and will be forever unable to leave it.

A glassy state will appear basically as a liquid phase for what concerns the topological ordering but will display a nonzero Debye-Waller factor (solid-like). In the MCT language this amounts to saying that, below $T_{\rm d}$, the non-ergodicity parameter is finite. The *dynamic temperature* $T_{\rm d}$ is, indeed, the mode-coupling temperature $T_{\rm mc}$. The transition out of a glassy state is a non-perturbative phenomenon, thus inexistent in the mean-field approximation. From the point of view of the disordered systems description by means of the overlap order parameter [Mézard *et al.*, 1987], the self-overlap q_1 of each metastable state is finite, but the overlap q_0 between two different states is zero: the metastable glassy states are incongruent, or completely decorrelated. The liquid state underlying has, instead, also a zero q_1.

Given this picture of ergodically disconnected sectors (i.e., metastable states with infinite lifetimes) we have defined two free energy functions: the logarithm of the partition functions of the *states* $f_{\rm tot}$, Eqs. (7.51)-(7.52), and the average state free energy:

$$f = \bar{f} \equiv \frac{1}{N} \sum_{\gamma} P_{\gamma} f_{\gamma}; \qquad P_{\gamma} \equiv \frac{e^{-\beta N f_{\gamma}}}{Z} \qquad (7.55)$$

where $f_{\gamma} = -1/(\beta N) \log \sum_{c \in \gamma} \exp\left(-\beta \mathcal{H}[c]\right)$. The entropic contribution of all states γ is in all respects the configurational entropy

$$S_c \equiv -\sum_{\gamma} P_{\gamma} \log P_{\gamma} \qquad (7.56)$$

In the mean-field case here considered, the configurational entropy is mathematically well defined even for infinite timescales but physically a bit weird. Indeed, it counts the entropic contribution of all the states that will *never* be reached by the system.

When $T_K < T < T_{\rm d}$, the configurational entropy $T S_c = F - F_{\rm tot}$ is extensive. The total free energy in the presence of exponentially many metastable states is equal to the free energy of the underlying liquid. It would be the free energy of the liquid phase if the dynamic arrest would not prevent the system from exploring the whole phase space.

As the configurational entropy goes to zero decreasing the temperature we can define the zero point as the Kauzmann temperature of a hypothetic thermodynamic phase transition to a stable glassy phase. At that point the average free energy becomes equal to the liquid one: the glassy phase becomes the stable one. This kind of transition is signaled by the jump of the order parameter of the stable phase (the global minimum of the thermodynamic potential) from 0 (liquid, $T = T_K^+$) to a finite q_1 (*ideal* glass, $T = T_K^-$).

7.3 Glass models without quenched disorder: clone theory

If ordinary phase transition theory would hold, to study the equilibrium thermodynamics of the ideal glass state below T_K one should apply an external weak field, selecting one particular state across the transition point and eventually send it to zero (working in the thermodynamic limit). Unfortunately, due to the intrinsic frustration of the system, that is, to the inability of single particles to contemporarily minimize each of their couplings, to build such a *pinning* field would already require the knowledge of all states, clearly an unfeasible task.

A way to overcome the problem is to consider two (or more) copies of the same glassy system coupled in such a way that they are constrained to be in the same state [Franz & Parisi, 1995]. The coupling must be weak (proportional to $\epsilon \to 0$) and short range in order not to modify the FEL, and has to be sent to zero at the end of the selection procedure, just like the pinning field in ordinary phase transitions. By replicating the system, an order parameter can be devised for the glass state, with the property of jumping discontinuously from zero as the transition occurs, like the overlap of mean-field models with quenched disorder. An example is a sort of replicated pair correlation function of the distance r between the position $r_i^{(1)}$ of the molecule i in the (real) replica 1 and the position $r_j^{(2)}$ of the molecule j in the (real) replica 2:

$$g_{12}(r) = \lim_{\epsilon \to 0} \lim_{N \to \infty} \frac{1}{\rho N} \sum_{ij} \left\langle \delta(r_i^{(1)} - r_j^{(2)} - r) \right\rangle \qquad (7.57)$$

where ρ is the density of the particles.

7.3.1 Equilibrium thermodynamics of the cloned m-liquid

To study the ideal solid glass phase at $T < T_K$, it is necessary to use an arbitrary number m of replicas [Mézard & Parisi, 1996; Mézard & Parisi, 1999b; Mézard, 1999; Mézard & Parisi, 1999a, 2000; Parisi, 2003]. The replicated partition sum is

$$Z_m = \int_{f_0}^{f_{\text{th}}} df \ e^{S_c(f,T)} e^{-\beta m N f} \equiv e^{-\beta m F_{\text{tot}}} \tag{7.58}$$

where the total free energy per particle of the m replicas,

$$m \frac{F_{\text{tot}}}{N} = m f_{\text{tot}} = m f - T s_c(f, T) \tag{7.59}$$

is the Legendre transform of the configurational entropy. This also means that, allowing, as was done in Sec. 7.2.2 for the parameter n, for analytic continuation of $m f_{\text{tot}}$ to non integer values of m, the dependence of f from m, and vice versa, can be obtained from the relations $\beta m = \partial s_c / \partial f$ or $f = \partial(m\phi)/\partial m$. The latter, together with Eq. (7.58), yields the useful formula for the configurational entropy,[6]

$$s_c(m) = \beta m^2 \frac{\partial f_{\text{tot}}(m)}{\partial m} \tag{7.60}$$

The derivative of the configurational entropy as a function of the free energy f, computed at the lowest value, $f_0(T)$, for which $s_c \searrow 0$, describes how the system approaches the Kauzmann transition (the total configurational entropy, counting contributions from all f-levels, is zero at the transition, by definition). In the non-replicated system ($m = 1$), T_K is yielded by

$$1 = T_K \frac{\partial s_c}{\partial f} \bigg|_{f_0(T_K)} \tag{7.61}$$

In the replicated system, instead, the temperature $T_K^{(m)}$ at which $s_c \to 0$ can be obtained solving

$$m = T_K^{(m)} \frac{\partial s_c}{\partial f} \bigg|_{f_0(T_K^{(m)})} \tag{7.62}$$

If one carries out the analytic continuation of Eq. (7.59) for continuous $m < 1$, one can, thus, explore the region $T < T_K$ employing a liquid made of molecules containing m particles bounded together.[7]

The ideal glass phase cannot be probed coming from the fluid phase as $T \to T_K$ and extrapolating the thermodynamic properties below it, because of

---

[6] The m replicas introduced in this section play a different role than those introduced in the replica approach of Sec. 7.2 for the computation of the average of the free energy of a system with a quenched disorder, where, beyond the analytic continuation, the zero limit was taken. Here, replicas are introduced to describe the (ideal) amorphous state as an equilibrium state in statistical physics. To stigmatize this difference, we often use the word *clone* instead of replica.

[7] The small attractive coupling of strength ϵ is, actually, not sufficient to yield a molecular liquid at any m, but, depending on the temperature, this only holds in a particular interval of m values for which the total free energy is minimized.

the phase transition. However, one can still probe the glass phase by means of the cloned molecular liquid just defined. Indeed, we first observe that, in such a fluid, for $T < T_K^{(m)}$ the configurational entropy is zero. That is, $f_{\text{tot}} = f$, independently of m. In particular, below T_K, $f_{\text{tot}}(1, T) = f_{\text{tot}}(m^\star, T) = f$. At a given temperature, the total free energy grows with m, starting from small values, up to the value $f_{\text{tot}}(m^\star, T) = f$, where $m^\star$ is determined solving Eq. (7.62).

Before showing explicit applications of the cloned liquid method we observe that at $T < T_K$, for $m \geq m^\star(T)$, $f_{\text{tot}}(m) = f$ is larger than the continuation of $f_{\text{tot}}(m)$ of the liquid. In the latter case, indeed, $\partial f_{\text{tot}}/\partial m < 0$ and, therefore, $s_c < 0$. The fact that $f_{\text{tot}}(m^\star)$ (stable solid) is larger than $f_{\text{tot}}(m > m^\star)$ (metastable liquid) is the source of the inverted jump in specific heat that decreases at the thermodynamic glass transition as temperature is decreased. Indeed, this is the opposite of what happens in ordinary second order phase transitions, but it is the right static counterpart of what happens in off-equilibrium real glasses at the calorimetric glass transition temperature $T_g > T_K$ (Sec. 1.1).

7.3.2 Analytic tools and specific behaviors in cloned glasses

We now consider a generic glass former of N particles interacting by pair potential $\mathcal{V}(r)$ in d-dimensional space, whose Hamiltonian is

$$\mathcal{H}[\mathbf{r}] = \sum_{i<j}^{1,N} \mathcal{V}(\mathbf{r}_i - \mathbf{r}_j) \tag{7.63}$$

where $\mathbf{r}_i$ is the position vector of the atom i. The partition function is, then,

$$Z = \frac{1}{N!} \int \prod_{i=1}^{N} d\mathbf{r}_i e^{-\beta \mathcal{H}[\mathbf{r}]} \tag{7.64}$$

Replicating m times the system and introducing a weak coupling among replicas, the total Hamiltonian reads

$$\mathcal{H}_m = \sum_{a=1}^{m} \mathcal{H}[\mathbf{r}^a] + \epsilon \sum_{i,j}^{1,N} \sum_{a<b}^{1,m} W\left(\mathbf{r}_i^a - \mathbf{r}_j^b\right) \tag{7.65}$$

where the replica coupling potential can be of any shape provided it is short range, e.g., one can take $W(r) \sim 1/(1 + (r/c)^2)^6$, where $c \simeq 0.2a$ (a is the intermolecular distance).

The partition function of the "cloned" glass former is

$$Z_m = \frac{1}{N!^m} \int \prod_{i=1}^{N} \prod_{a=1}^{m} d\mathbf{r}_i^a e^{-\beta \mathcal{H}_m[\{\mathbf{r}^a\}]} \tag{7.66}$$

As an order parameter, the inter-clone cross correlation is defined:

$$\rho(\{\mathbf{r}_0^a\}) = \frac{1}{N} \sum_{\{i_a\}} \left\langle \prod_{a=1}^{m} \delta(\mathbf{r}_{i_a}^a - \mathbf{r}_0^a) \right\rangle \tag{7.67}$$

At low temperatures, where thermal fluctuations are relatively small with respect to the typical distance a between atoms and where diffusion processes can be neglected, one can identify the single molecule positions in the amorphous phase, as is usually possible in a crystal lattice. The W coupling takes care of the fact that where a system exhibits a particle, any other of its clones also displays a particle very nearby. It is, then, possible to relabel the particles in every clone, so that nearby particles are labeled by the same index ($\mathbf{r}_j^a \simeq \mathbf{r}_j^b$, $\forall(a,b)$). All labeling permutations yield a factor $N!^{m-1}$. We are, then, left with a system of N molecules (indices i,j), each of them formed by m atoms (indices a,b). In terms of centers of mass $\mathbf{R}$ and relative coordinates $\mathbf{u}$, we can write $\mathbf{r}_i^a = \mathbf{R}_i + \mathbf{u}_i^a$ ($\sum_a \mathbf{u}_i^a = 0$) and the partition function reads

$$Z_m = \frac{1}{N!} \int \prod_{i=1}^{N} d\mathbf{R}_i \prod_{i=1}^{N} \prod_{a=1}^{m} d\mathbf{u}_i^a \prod_{i=1}^{N} \left(m^d \delta^{(d)} \left(\sum_{a=1}^{m} \mathbf{u}_i^a \right) \right) \tag{7.68}$$

$$\exp \left\{ -\beta \sum_{i<j}^{1,N} \sum_{a=1}^{m} \mathcal{V}(\mathbf{R}_i - \mathbf{R}_j + \mathbf{u}_i^a - \mathbf{u}_i^b) - \beta\epsilon \sum_{i=1}^{N} \sum_{a<b}^{1,m} W(\mathbf{u}_i^a - \mathbf{u}_i^b) \right\}$$

In order to perform explicit computation, different approaches, yielding differently refined approximations, have been developed in the literature.

Harmonic resummation (HR). One approach consists in expanding $\mathcal{V}(\Delta\mathbf{R} + \Delta\mathbf{u})$ in the exponent of Eq. (7.68) up to the second order in $\Delta\mathbf{u}$ and performing the Gaussian integrations in $\mathbf{u}_i^a$ building an effective potential for the center of mass variables.

Small cage expansion (SCE). A complementary approach is to expand the exponential of Eq. (7.68) in powers of $\mathbf{u}$, keeping only the quadratic term in W. Integrating in $\mathbf{u}$ yields, in this case, an expansion of the free energy in powers of $1/\epsilon$, whose Legendre transform is a generalized thermodynamic potential function of the "cage radius" $A = 1/[2dm(m-1)] \sum_{ab} \langle |\mathbf{u}_i^a - \mathbf{u}_i^b|^2 \rangle$, conjugated to $1/\epsilon$. Since at low T the cages are small, one eventually performs an expansion in A.

Molecular HNC free energy. Yet another computational scheme consists in defining the functional Legendre transform of the free energy functional

$$F_{\text{tot}}[W] = -\frac{T}{m} \log Z_m[W]$$

cf. Eqs. (7.65)-(7.66), with respect to the generalized inter-clone correlation $\rho(r_1, \ldots, r_m)$, cf. Eq. (7.67), whose conjugated function is the clone-clone interaction W. The Hyper-netted chain (HNC) approximation [Hansen &

TABLE 7.1

Kauzmann temperature T_K as obtained by analytic approaches in the framework of the cloned molecular liquid theory in the literature for different glass models and techniques. HR stays for harmonic resummation and SCE for small cage expansion. For binary mixtures also the dynamic temperatures are reported.

Model (technique)	T_K	T_d
Soft Spheres, $\rho = 1$ (HR)	0.182	
	[Coluzzi et al., 1999]	
Soft Spheres, $\rho = 1$ (SCE)	0.203	
	[Coluzzi et al., 1999]	
SSBM, $\rho = 1$ (HR and SCE)	0.135	0.226
	[Coluzzi et al., 1999]	[Hansen & Yip, 1995]
LJBM, $\rho = 1.2$	0.32	0.435
	[Coluzzi et al., 2000a]	[Kob & Andersen, 1994]

McDonald, 2006] is then applied to the cloned free energy, taking into account only the molecular density $\rho(r)$ and the two point correlation $g^{(2)}(r, r')$. The trial molecular density is, then, expressed as a function of a single variational parameter, the cage size A. In this framework, an expansion for small cage sizes is, eventually, once again performed to yield thermodynamic results.

These methods are complementary and they have been applied to models for soft spheres (SS) [Mézard & Parisi, 1999b,a; Coluzzi et al., 1999], to soft spheres binary mixtures (SSBM) [Coluzzi et al., 1999] and to Lennard-Jones binary mixtures (LJBM) [Coluzzi et al., 2000a,b] (for a description of the models see Appendix 6.A). In Table 7.1 the obtained values of T_K are reported, as well as the known values for T_d of those models. In Fig. 7.2 the behavior of the specific heat is compared for SS, SSBM and LJBM models. The jump occurs at T_K and it is downward lowering the temperature. The (ideal) glass specific heat $C = 3/2$ in each case, is nothing other than the Dulong-Petit law: the specific heat of the glass is equal to the one of the crystal, in good agreement with experimental data.

7.3.3 Effective temperature for the cloned molecular liquid

Finally, a short remark on the ratio T/m^*: it stays relatively constant as T varies below T_K, being almost always close to the latter, independently of T. It is, therefore, $m^* \simeq T/T_K$. In Fig. 7.3 we show, for the models and the techniques considered in literature, the behavior of m^* versus temperature. This linear behavior is similar to the one found for the statics of many discontinuous spin-glass models below T_K (e.g., in the REM, Eq. (7.30), or the Ising p-spin model [Gross & Mézard, 1984; Crisanti et al., 2005]), as well as for dynamics below T_d in the spherical p-spin model, where T/m^* turns out to be the FDR and $m^* \simeq T/T_d$.

Interpreting the ratio T/m^* as an effective temperature T_e, Eq. (7.59) can be rewritten as

$$F_{\text{tot}} = F - T_e S_c \tag{7.69}$$

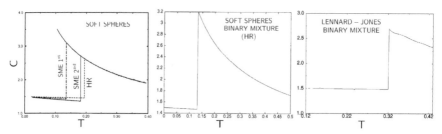

FIGURE 7.2

Specific heat across the thermodynamic glass transition for (left) the soft spheres (Appendix 6.A.1), (center) the 50%-50% soft spheres binary mixture with $\sigma_{AA}/\sigma_{BB} = 1.2$ (Appendix 6.A.1) and (right) the 80%-20% Lennard-Jones binary mixture (Appendix 6.A.2). Reprinted figures with permission from [Mézard & Parisi, 1999a; Coluzzi *et al.*, 1999, 2000a], left to right respectively. Copyright (1999,2000) by the American Institute of Physics.

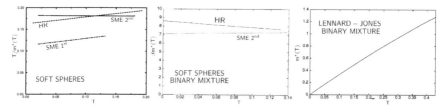

FIGURE 7.3

The temperature behavior of $m^\star$: (left) $T/m^\star(T)$ for the soft spheres model, (center) $\beta m^\star(T)$ in the soft spheres binary mixture, (right) $m^\star(T)$ for the Lennard-Jones binary mixture. Reprinted figures with permission from [Mézard & Parisi, 1999a; Coluzzi *et al.*, 1999, 2000a], left to right respectively. Copyright (1999, 2000) by the American Institute of Physics.

Considering that the single state free energy is $F = U - TS$, where U is the internal energy and S is the entropy of the single state, this is identical to the generalized Helmholtz free energy introduced in Sec. 2.5 in the framework of the two temperature thermodynamics, cf. Eq. (2.66) (where F_{tot} was called F). By analogy with statistical mechanics of non-frustrated systems (and by direct computation), one realizes that the entropy counting the minima of the free energy landscape can be expressed as the Legendre transform of the total free energy, Eq. (7.39), as

$$S_{\text{c}} = -\beta_{\text{e}} F_{\text{tot}} + \beta_{\text{e}} F \qquad (7.70)$$

where the inverse effective temperature $\beta_{\text{e}} = 1/T_{\text{e}} = m/T$ and the free energy of the single state f are conjugated by

$$\beta_{\text{e}} = \frac{\partial S_{\text{c}}}{\partial F} \qquad (7.71)$$

We notice that this is equivalent to Eq. (3.143), (where the typical state free energy F was called $\bar{F}$) and, indeed, the parameter β_e is precisely the inverse of the effective temperature in a generalized two temperature thermodynamic description of the glass state.

7.4 Frustration limited domain theory

Though MCT works reasonably well near the mode-coupling temperature and in soft matter systems like Laponite solutions, the crucial property of a glass is the enormous enhancement of the viscosity or of the equilibration timescale. Indeed, an enhancement of 15 orders of magnitude may occur over a reasonably small temperature interval, say some $50K$. Such a huge amplification of timescales is very atypical for other solids, but more usual in geology and astronomy where, indeed, glassy behavior occurs.

The dynamical arrest occurring in glass-forming liquids may have a common cause. Indeed, in many systems one observes that the viscosity increase is enhanced with respect to a simple Arrhenius law, Eq. (1.2), $\eta = \eta_0 e^{A_0/kT}$. In realistic cases, this enhancement will apply to the infinite frequency limit of the frequency-dependent viscosity. One may define the effective free energy barrier as

$$A(T) = kT \log \frac{\eta}{\eta_0} \tag{7.72}$$

Fig. 7.4 shows that, for moderately high temperatures, A is constant (on the left in the figure), while it presents a rapid increase below some crossover temperature T_*. The ubiquity of such an enhancement in various glass-forming liquids and polymer mixtures suggests that some universal mechanism is underlying here. If a Vogel-Fulcher (VF) law $\eta = \eta_0 \exp[B/(T - T_0)]$ were present, one would have the effective barrier

$$A(T) = \frac{BT}{T - T_0} \tag{7.73}$$

The aim of the theory presented in this section is to derive an enhancement like this, probably having a different analytical shape, from a microscopic theory.

7.4.1 Geometric frustration

In crystalline structures, local order can be extended as far as one wishes: more material leads to a larger crystal. Subtle cases are quasicrystals, where it is still possible to tile the entire space, but only in a manner without translational symmetry [Steinhardt & Ostlund, 1987]. Geometric frustration applies

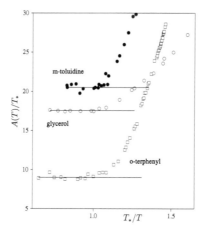

FIGURE 7.4

Crossover from Arrhenius to enhanced Arrhenius behavior for three glass-forming liquids. Both effective Arrhenius free energy A and temperature T are scaled with respect to the crossover temperature T_*, where A starts to deviate from a constant. Reprinted figure with permission from [Tarjus *et al.*, 2005]. Copyright (2005) Institute of Physics Publishing.

to situations where a local tendency to order is not completely fitting.[8] It can, then, be extended only up to a certain finite length-scale, beyond which the collective mismatch effects hinder further ordering. In that situation, one expects a mosaic pattern of locally ordered "droplets."

In metallic spin-glasses a related phenomenon exists, known as mictomagnetism. Consider, for example, the gold-iron mixture $Au_{1-c}Fe_c$, at concentration $c = 30\%$. Local Fe regions can contain as much as 1000 atoms, and, thus, 1000 ferromagnetically ordered spins [Mydosh, 1993]. These regions, sometimes called "fat spins", are easily recordable in magnetic experiments. Basically, the system behaves as a set of weakly coupled fat spins. A similar pattern of locally ordered regions may underlie the glass state. Whereas in metallic spin-glasses the cause of the domains is the substitutional disorder, which in combination with the oscillating Rudermann-Kittel-Kasuya-Yosida (RKKY) interaction cannot build large ferromagnetic regions, in glass-forming liquids the finiteness of the domains may be supposed to originate from a geometric reason, namely the impossibility to tile space beyond a certain length-scale. Likewise, this frustration will lead to finite droplets of the ordered phase, that together make up a mosaic structure.

In solid state physics it is often possible to go to a field theoretic description

---

[8]Frustration, a concept introduced by Toulouse [1977] in the field of spin-glasses, expresses that different contributions to the Hamiltonian cannot be simultaneously optimal. This leads to the field of complex systems, where the system may for some time have benefit from some part of the interaction couplings, at the cost of other ones, and change these parts in time.

of the system, the so-called Ginzburg-Landau approach. The focus is on the slow modes of the system, that must be properly identified, and the fast modes are integrated out. This approach, then, many times allows us to perform a renormalization group approach in the neighborhood of a critical point, where the slow modes become very slow. For the glass transition, various scenarios for such an approach have been attempted with varying levels of success, see the recent overview of Tarjus *et al.* [2005].

7.4.2 Avoided critical point

The notion of an avoided critical point is widespread. If in an Ising magnet, for some reason, there exists a small external field, then the zero field phase transition gets smeared, it is "avoided." This holds, in particular, when the field is random in sign. More generally, a modest amount of frustration can wipe out phase transitions, such that on lowering the temperature the behavior first looks like a phase transition, but on approaching the presumed transition more and more, it drifts off from a real transition. This scenario is known, for instance, for transitions that would exist in the mean-field, but do not occur in reality, since the dimension of the system is below the lower critical dimension. That case, where thermal fluctuations themselves pose a large frustration, will not be considered here.

An interesting approach is the one founded on the concept of the avoided critical point [Chayes *et al.*, 1996; Kivelson *et al.*, 1997; Tarjus *et al.*, 1997] and brought on in the late 1990s. In this theory the frustration, i.e., the principal cause of glass formation, is said to act as a source of strain free energy opposing the spatial expansion of locally preferred structures. As the system size increases, this strain intensifies, causing the breaking up of the liquid into domains. As the temperature decreases, the size and growth of these domains are limited by the frustration, preventing any stable, globally ordered organization of the molecules. This is called the *frustration-limited domain theory* [Kivelson *et al.*, 1995; Tarjus & Kivelson, 1995] and provides an explanation for the onset of super-Arrhenius behavior of the structural relaxation time and the viscosity.

The question whether glasses are connected with avoided critical points, finds motivation in the presumed existence of a Kauzmann transition where the configurational entropy would vanish. Though, by construction, this does happen in the HOSS model discussed in Chapter 3, in practice, this transition has never been confirmed in real glasses. Likewise, the VF law for the relaxation time $\tau = \tau_0 \exp[B/(T - T_0)]$ exposes a blocking of dynamics below T_0, but practical fits to this shape will always remain stuck at the level of a "reasonable fit," never becoming a "convincing fit." This could all be related to avoidance of a true critical point.

Several contributions have been made to a frustration-based field theoretic approach of supercooled liquids and the glass transition. Here we wish to mention one scaling approach, which presents an alternative for the VF law

[Kivelson *et al.*, 1995, 1996]. Let us consider a critical theory, to which small frustration is added. At temperatures somewhat below T_*, the critical temperature of the theory without frustration, two length-scales appear. The first one is the correlation length of the would-be ordered low temperature phase,

$$\xi_0 \sim a\epsilon^{-\nu}, \qquad \epsilon = \frac{T_* - T}{T_*} \tag{7.74}$$

where a is the lattice spacing and ν a critical exponent.

The effect of weak, long range frustration is to destabilize this ordering at a larger scale $R_D \gg \xi$. In some models (one of which we discuss in the next subsection), a scaling approach brings the form

$$R_D \sim \frac{a^2}{\xi_0 \sqrt{K}} \tag{7.75}$$

where K is the strength of frustration and a is the lattice constant.

In the correlation functions of some order parameter O of the would-be solid, these two scales would show up as [Kivelson *et al.*, 1995, 1996]

$$\langle O(r)O(0)\rangle = \begin{cases} m^2 + c_1 \frac{a}{r} e^{-r/\xi}, & \xi \ll r \ll R_D \\ c_2 m^2 \frac{R_D}{r} e^{-r/R_D}, & r \gg R_D \end{cases} \tag{7.76}$$

where $m = \langle O \rangle$, vanishing at the critical point $\varepsilon \to 0$, is the order parameter in the absence of frustration, and we have assumed the Ornstein-Zernicke decay for the connected correlation functions. At the scale R_D, fragmentation then leads to a mosaic structure of ordered "droplets" of typical size R_D.

From finite size studies of ordinary systems below their critical temperatures, Kivelson *et al.* [1995] recall that there is a timescale τ associated with the relaxation of the order parameter, which proceeds via nucleation and motion of a domain wall and that the divergence of τ in the thermodynamic limit is the signal of a broken symmetry state. For system size $L \gg \xi_0$ one expects $\log \tau$ to be proportional to $\sigma L^2/T$, where the domain wall surface tension scales as

$$\sigma \sim \frac{T_*}{\xi_0^2} \tag{7.77}$$

It is next assumed that this may be applied to domain wall sizes up to R_D, leading to

$$\log \frac{\tau}{\tau_0 \exp(A_0/T)} \sim \frac{\Delta F}{T}, \qquad \Delta F \sim \sigma R_D^2 \sim T_* \frac{R_D^2}{\xi_0^2} \tag{7.78}$$

where τ_0 is the microscopic collision timescale of the order of 10^{-15} s and A_0 is the energy barrier for local changes of the conformation. Combining this with Eqs. (7.75) and (7.74) brings a free energy cost

$$\Delta F = \frac{T_*}{K} \frac{a^4}{\xi_0^4} = \frac{T_*}{K} \left(\frac{T_*}{T_* - T}\right)^{4\nu} = BT_* \left(\frac{T_*}{T_* - T}\right)^{\psi} \tag{7.79}$$

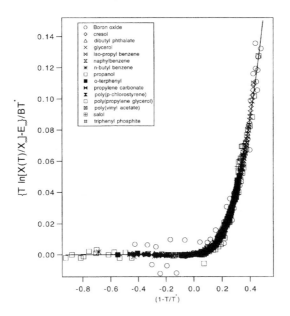

FIGURE 7.5

Reduced activation free energy, $y = \{T\log[X(T)/X_0] - A_0\}/(BT_*)$, versus reduced temperature $x = 1 - T/T_*$, where $X = \eta$ or τ, for 14 liquids listed in the inset. The parameters A_0, X_0, B and T_* are fitted for each liquid. The solid line presents the theoretical curve, the pure Arrhenius shape $y = 0$ for $x \le 0$ and the avoided criticality shape $y = x^{8/3}$ for $x > 0$. Reprint with permission from [Kivelson *et al.*, 1996]. Copyright (1996) by the American Physical Society.

where B is a system parameter and $\psi = 4\nu$ is an exponent that lies between 7/3 and 3. For ordinary three dimensional critical phenomena in the absence of disorder, one has $\nu \approx 2/3$, which brings the best fit

$$\psi = \frac{8}{3} \qquad (7.80)$$

This free energy barrier will modify the energy barrier A_0 in the Arrhenius law, leading to the modified shape

$$\tau = \tau_0 e^{\beta A(T)}, \qquad A(T) = A_0 + \Delta F = A_0 + BT_*(1 - T/T_*)^\psi \qquad (7.81)$$

Notice that ΔF sets in *below* some crossover temperature T_*. Empirical evidence for such a scenario was presented in Fig. 7.4. As shown in Fig. 7.5, for a variety of undercooled liquids this modification of the Arrhenius law below T_*, appears to present a good fit to the data for all these liquids in the whole temperature regime.

Thus one may conclude that the VF law is not the only realistic description of the glass physics, moreover its gradual onset from high temperatures is challenged by the present approach, where the onset occurs below some T_*.

7.4.3 Critical assessment of the approach

Let us reconsider the model employed by Kivelson *et al.* [1995], a spin model with short range ferromagnetic interactions and long range, Coulomb anti-ferromagnetic interactions, that we put on a simple cubic lattice with lattice parameter a. The Hamiltonian reads

$$H = -J \sum_{\langle i,j \rangle} \mathbf{S}_i \cdot \mathbf{S}_j + \frac{KJa}{8\pi} \sum_{i \neq j} \frac{\mathbf{S}_i \cdot \mathbf{S}_j}{|\mathbf{R}_i - \mathbf{R}_j|} \qquad (7.82)$$

where the $\mathbf{S}_i$ are Heisenberg spins and where the first sum is over nearest neighbor pairs only. The interaction parameters J and $K \ll 1$ are both chosen to be positive.

For simplicity, we use classical spins and consider the model in the spherical approximation, where we relax the constraint on the length of the individual spins, and instead require

$$\sum_{i=1}^{N} \mathbf{S}_i^2 = \sum_{i=1}^{N} (S_{i,x}^2 + S_{i,y}^2 + S_{i,z}^2) = N\sigma \qquad (7.83)$$

The partition sum now becomes

$$Z = \int \prod_{i=1}^{N} \mathrm{d}\mathbf{S}_i \, e^{-\beta H} \delta \left(\sum_{i=1}^{N} \mathbf{S}_i^2 - N\sigma \right)$$

$$= \beta \int_{-i\infty}^{i\infty} \frac{\mathrm{d}\mu}{2\pi i} \int \prod_{i=1}^{N} \mathrm{d}\mathbf{S}_i \, e^{-\beta H} e^{-\frac{1}{2}\beta\mu(\sum_{i=1}^{N} \mathbf{S}_i^2 - N\sigma)} \qquad (7.84)$$

where we inserted a plane wave representation of the δ-function. The integrals are now Gaussian, and yield

$$\beta F = \frac{3N}{2} \frac{a^3}{(2\pi)^3} \int \mathrm{d}^3 q \, \log \left\{ \beta \left[\mu - J(q) + \frac{KJ}{q^2 a^2} \right] \right\} - \frac{1}{2} N\beta\mu\sigma \qquad (7.85)$$

As in Bose-Einstein condensation, the parameter μ should be taken at its saddle point, which is real, and set by the "gap equation"

$$\frac{3a^3}{2(2\pi)^3} \int \frac{\mathrm{d}^3 q}{\mu - J(q) + KJ(qa)^{-2}} = \frac{1}{2}\beta\sigma \qquad (7.86)$$

It appears that the problem can be solved in the low wavelength limit. For a simple cubic lattice with lattice constant a one has

$$J(q) = J \sum_{\rho} e^{iq\rho} = 2J(\cos q_x a + \cos q_y a + \cos q_z a) \approx 6J - Jq^2 a^2 \qquad (7.87)$$

Decomposing μ as

$$\mu = (6 - 2\sqrt{K} + 4\nu^2 a_0^2)J \tag{7.88}$$

we may write the gap equation in the form

$$\frac{3a}{2(2\pi)^3} \int \frac{d^3q}{4\nu^2 + (q - \sqrt{K}a^{-2}q^{-1})^2} = \frac{1}{2}\beta J\sigma \tag{7.89}$$

Since for nonzero K the integral diverges as $1/\nu$ for small ν, it is seen that the ferromagnetic phase transition is suppressed by the Coulomb-type long range antiferromagnetic interaction. But the spin density wave at wave vector

$$q_* = \frac{K^{1/4}}{a} \tag{7.90}$$

only orders itself at zero temperature.

The spin-spin correlator in the long wavelength limit reads

$$C(r) = \langle \mathbf{S}(0) \cdot \mathbf{S}(r) \rangle = \frac{3aT}{2(2\pi)^3 J} \int d^3q \frac{e^{i\mathbf{q}\cdot\mathbf{r}}}{4\nu^2 + (q - q_*^2/q)^2} \tag{7.91}$$

$$= \frac{3aT}{(2\pi)^2 J} \int_0^\infty dq \frac{q^2 \sin qr}{qr[4\nu^2 + (q - q_*^2/q)^2]}$$

This can be solved explicitly. First one takes the integral as half the one from $-\infty$ to ∞, and next one separates the poles,

$$C(r) = \frac{3aT}{4(2\pi)^2 J\nu r} \Im \int_{-\infty}^\infty dq \frac{q^2 \sin qr}{q^2 - q_*^2 - 2i\nu q}$$

$$= \frac{3aT}{4(2\pi)^2 J\nu r} \Im \int_{-\infty}^\infty dq \frac{q^2}{(q - i\kappa_+)(q - i\kappa_-)} \frac{e^{iqr} - e^{-iqr}}{2i} \tag{7.92}$$

where

$$\kappa_\pm = \nu \pm \sqrt{\nu^2 - q_*^2} \tag{7.93}$$

both have a positive real part. Therefore, we obtain

$$C(r) = \frac{3aT}{16\pi J\nu} \Re \left(\frac{\kappa_+^2}{\kappa_+ - \kappa_-} \frac{e^{-\kappa_+ r}}{r} - \frac{\kappa_-^2}{\kappa_+ - \kappa_-} \frac{e^{-\kappa_- r}}{r} \right) \tag{7.94}$$

We are interested in the region not far below the would-be ferromagnetic ordering temperature and weak long range interaction, i.e., where $\nu \gg q_*$ and $\kappa_\pm$ are real. Let us define the correlation length of the would-be ferromagnetic phase,

$$\xi = \frac{1}{\kappa_+} = \frac{1}{\nu + \sqrt{\nu^2 - q_*^2}} \tag{7.95}$$

It deviates from the pure value

$$\xi_0 \equiv \frac{1}{2\nu} \tag{7.96}$$

but the difference is small because we assume a weak disordering interaction, leading to $q_* \ll \nu$. The second length-scale in Eq. (7.94) is the disorder length

$$R_D = \frac{1}{\kappa_-} = \frac{\nu + \sqrt{\nu^2 - q_*^2}}{q_*^2} = \frac{1}{q_*^2 \xi} = \frac{a^2}{\sqrt{K} \, \xi} \tag{7.97}$$

In terms of these variables the correlator, Eq. (7.94), reads

$$C(r) = \frac{3aT}{16\pi J \nu \xi (1 - \xi/R_D)} \left(\frac{e^{-r/\xi}}{r} - \frac{\xi^2}{R_D^2} \frac{e^{-r/R_D}}{r} \right) \tag{7.98}$$

From this we may deduce a few aspects:

- There is no steady ($r = \infty$) term, since the spins are not condensed globally. Locally, there would be ferromagnetic ordering up to a length ξ, but, due to directional randomness, this averages out globally.

- The first term describes ferromagnetic clustering up to the length $\xi_* \simeq 2\xi \log(R_D/\xi)$, beyond which the second term takes over and makes C negative.

- The second term describes antiferromagnetic clustering, as expressed by its negative value. It is much weaker, since $\xi \ll R_D$, but extends up to the much larger scale R_D.

- Our Eq. (7.97) for the disorder length coincides with the scaling Eq. (7.75).

- Most aspects of the present analysis agree with conclusions drawn from dimensional analysis of the first and second order contributions to the free energy in powers of K [Kivelson *et al.*, 1995].

Let us recall an important point: the ferromagnetic clustering expressed by Eq. (7.98) extends up to a size ξ_* where $C(\xi_*) = 0$, that is, up to

$$\xi_* \simeq 2\xi \log \frac{R_D}{\xi} \tag{7.99}$$

This size, a logarithmic enhancement of ξ_0, is far less than the supposed droplet size R_D of the previous section, proving that the domain of validity of the behaviors stated in Eq. (7.76) is incorrect.[9] The positive part of the correlator brings a size ξ_* for the maximal ferromagnetic droplets, which would not bring R_D but ξ_* as the relevant scale in Eq. (7.78). This would

---

[9] Also other points are in conflict with our exact result, Eq. (7.98): we find that though $m = 0$ in the first domain, the decay in the second domain does not vanish, but has a negative sign, reflecting the antiferromagnetic ordering tendency.

end up in a barrier ΔF from which powers of ξ drop out, leaving only the square of a logarithm,

$$\Delta F \sim \sigma \xi_*^2 \sim \frac{T_*}{\xi_0^2} \xi_*^2 \sim T_* \ln^2 \frac{R_D}{\xi} \tag{7.100}$$

This could in no way describe the strong enhancement exposed in Fig 7.5.

As a technical remark, let us notice that the limit $R_D \to \infty$ in our Eq. (7.98) does not reproduce the m^2 term of the pure system's correlation function, as it is denoted in the first line of Eq. (7.76). This happens because we have taken the thermodynamic limit first, when we replaced sums over momenta by integrals. To accommodate the steady magnetization, one should split off the $k = 0$ term as in Bose-Einstein condensation. The m^2 term would then reappear when R_D becomes of the order of the system size.

The argument above exposed seems fairly general and probably applies beyond the employed spherical approximation. Inspecting the crossover between the short and long distance behaviors in Eq. (7.76), however, one can, then argue that the range of the first behavior of Eq. (7.76) must be of the order ξ rather than R_D, because when the two behaviors have the same order of magnitude, due to the exponentials and the inequality $R_D \gg \xi$, this can only happen for some not-too-large value of r/ξ. This analysis, therefore, appears to invalidate the picture of a long range disordering field as a cause of slow glassy behavior, undermining the derivation of the avoided critical point theory prediction for the relaxation time, Eq. (7.81).

If one replaces the long range disordering field by a short range random field, one arrives at the random field Heisenberg model. This leads to a different mechanism to disorder the ferromagnetic state, though, which goes beyond the philosophy of Kivelson *et al.* [1997]; Tarjus *et al.* [1997].

7.4.4 Heuristic scaling arguments

In their recent review paper, Tarjus *et al.* [2005] present a heuristic scaling argument. As a first step towards a scaling analysis, one can include the effect of aborted nucleation of the ideal ordered phase in the liquid phase. At temperatures sufficiently below some ordering temperature T_*, the free energy of the ideal phase in a disordered liquid surrounding can be written

$$F = -\phi(T)L^3 + \sigma(T)L^\theta + s(T)L^5 \tag{7.101}$$

As usual, the first term is the gain in bulk free energy, the second the interface free energy cost, and one takes $\theta = 2$. The last term represents strain free energy due to frustration, that grows as L^{d+2} in d-dimensions. We may, indeed, expect that the Coulomb propagator $1/q^2$ of the Hamiltonian (7.82) brings an extra factor L^2 to the volume factor L^d for an excitation of size L.

Tarjus *et al.* [2005] consider the free energy density $\Phi = F/L^3$, having the form

$$\Phi = -\phi(T) + \frac{\sigma(T)}{L} + s(T)L^2 \tag{7.102}$$

This expression is then optimized in L, bringing $L_* = (\sigma/s)^{1/3}$ and then an interface free energy $\Delta F \sim \sigma L_*^2 \sim \sigma^{5/3} s^{-2/3} \sim \xi^{-10/3} s^{-2/3} \sim \tau^{10\nu/3} s^{-2/3}$. For $\nu = 2/3$ this argument leads to $\Delta F \sim \tau^{\psi}$ with $\psi = 20/9 = 2.222$, not very far from the previous estimate $\psi = 8/3 = 2.666$.

However, the free energy density should not be optimized, but rather the free energy itself. Then, the surface term and the Coulomb term add up, rather than compensate for each other, and together compensate for the bulk term, which leads to different results.

7.5   Random first order transition theory

In the mean-field case it is quite clear what the glass state is (see Sec. 7.2), but the mean-field scenario is, in many respects, different from the actual phenomenological behavior of the glass. One fundamental feature lacking is, e.g., the existence of the calorimetric temperature T_g at which the undercooled liquid falls out of equilibrium and vitrifies into an amorphous solid.

To move on in the quest for a comprehensive microscopic theory of glasses, one has, then, to relax the mean-field hypothesis, allowing for thermodynamic fluctuation and considering the case where interactions are finite ranged. In doing so, it is very important to clearly understand what will remain of the mean-field scenario and, bearing the latter in mind, how its properties will be related to the physics of real glassy systems. First of all, referring to Secs. 7.2, 7.3 and to Chapter 1, we set a correspondence between transition points in the mean-field approximation and in the real world.

- What does T_d become?

  The temperature T_d, in the mean-field theory, is defined as the temperature at which the free energy landscape develops high energy minima whose barriers grow like N, i.e., they go to infinite in the thermodynamic limit. Since these higher lying excited amorphous states are more numerous than any less excited state, the dynamics will get stuck there. These states are metastable, because they are not at the lowest available free energy for the system, but their lifetime is infinite because of the lack of fluctuations and activated processes. T_d corresponds to the mode-coupling temperature at which the relaxation time of liquids algebraically diverges and the time-dependent structure factor develops a persistent plateau.

  In real systems, cf. Sec. 1.1.1, we know that T_d is a crossover temperature between two different behaviors of the slowest possible processes taking place in the liquid. Cooling down a liquid glass former, at some temperature higher than T_d (we called it T_{cage} in Fig. 1.1), some of the molecules will start to hinder each other's way, creating "cages" among themselves with the effect of obliging small groups of particles to stay close together for times longer than the simple collisional time. The eventual diffusion across a cage is called an activated process and takes longer to occur than standard collisions. As $T \rightarrow T_d$, the activated processes become dominant with respect to the collisional ones and a bifurcation of timescales of slower (α) and faster (β) processes occurs upon further lowering the temperature. A structural step in relaxation will, now, be feasible only if a relatively large number of particles are contemporarily rearranging themselves, and this kind of rearrangement corresponds to an α process. Other, smaller rearrangements of the molecule packings are, however, possible and they correspond to the β processes.[10]

• What does the static transition at $T = T_K$ become?

The temperature $T_K \simeq T_0$ is out of experimental reach. Therefore, it is not defined operatively in a strict way. It can be interpreted as the fitting temperature in the Vogel-Fulcher (VF) relaxation law $\tau \sim \exp\left(A/(T - T_0)\right)$, cf. Sec. 1.2, or as the point at which the excess entropy can be extrapolated to zero, cf. Sec. 1.4.1. One can, thus, infer, that at that temperature, where the relaxation time diverges and the entropy of the amorphous phase undergoes the one of the crystal, a phase transition to an *ideal glass* phase takes place. If, at T_K, a thermodynamic phase transition is taking place, some growing correlation distance is also expected. Other effects could, however, take place in the actual experiment preventing the realization of the ideal glass phase. The subject is still controversial. We notice, however, that, even if a true thermodynamic phase transition would not occur, the results borrowed from mean-field theory might still adequately describe the behavior of real glass formers at higher temperatures, above and around vitrification.

Assuming that a discontinuous (otherwise named random first order) transition occurs also in short range systems, we have, then, to tackle other, verifiable, issues.

What will a mean-field state become in the presence of fluctuations?

Will the viscous glass former be in a homogeneous phase or will it be in a spatially heterogeneous phase?

Will the standard theory of critical phenomena also be applicable in the case of amorphous solids?

---

[10]Notice that the latter, even though so fast to be able to reach equilibrium inside a solid glass, are yet slower than the cage rattling occurring by thermal oscillations. A complementary classification of rearranging processes between α and β has been described in the context of the PEL approach, in Chapter 6, as inter-basin and intra-basin transitions, respectively.

Before attempting to present a theory that can, at least partially, answer these questions, for the time being, we notice that a metastable state will now be truly metastable, i.e., its mean lifetime will be finite, though potentially amazingly long in good glassy materials at low temperatures. Its definition depends, indeed, on setting a timescale for the observation of the system (our $t_{\exp}$). We can, then, call a *glass state* whatever construction survives on the timescale of our experiment. Therefore, the configurational entropy, that counts the metastable states, also will now become a purely dynamic observable at a given temperature. It still decreases on lowering the temperature, but - at fixed T - also decreases on increasing the observation time. See Sec. 1.4 for a detailed discussion on the definition(s) of configurational entropy.

The configurational entropy will play the key role in the theory we are going to expose. To warm up, let us start with a critical revisiting of the first entropic approach that we ran into, in Sec. 1.5.

7.5.1   Adam-Gibbs theory, revisited

One of the first formulations of a microscopic theory leading to physical predictions such as the VF law for the viscosity and the relaxation time was provided by Adam & Gibbs [1965] (AG), cf. Sec. 1.5. We shortly recall their argument, explicitly introducing the linear length-scale ξ of a droplet $\mathcal{C}(n)$ of n particles (dynamically speaking, a CRR). Let us call ω the number of preferred configurations in which the n particles inside a droplet can arrange themselves. The configurations counted by ω correspond to different minima of a landscape, separated by energy barriers. Its logarithm yields the droplet configurational entropy of the supercooled glass former:

$$s_c^\star(n, T) = k_B \log \omega \qquad (7.103)$$

The total configurational entropy of the whole macroscopic system can be written as, cf. Eq. (1.13), $S_c \simeq \mathcal{N}(n) s_c^\star(n) = V \tilde{s}_c$, where $\tilde{s}_c$ is the entropy per unit of volume, $s_c^\star$ the entropy per rearranging droplet and $\mathcal{N}(n)$ is the number of droplets containing n particles. In the AG theory, $\omega = e^{s_c^\star/k_B}$ is not supposed to depend on n, provided that $n \geq n^\star$, and, hence, on the linear size ξ of the droplet [assumption I], defined as $\xi = (Vn/N)^{1/3}$. Adam and Gibbs took $\omega = 2$. Eventually, one has

$$V\tilde{s}_c = \frac{V}{\xi^3} k_B \log \omega \qquad (7.104)$$

We look, then, at the barrier that a CRR must overcome in order to rearrange itself into a different packing. According to Eq. (1.12), the free energy barrier scales like

$$\Delta F \simeq n^\star \Delta\mu \simeq \Delta\mu \left(\frac{\xi}{a}\right)^3 \qquad (7.105)$$

where a is the average particle distance in the amorphous packing. That is, the barrier is assumed to scale with the volume [assumption II].

Eqs. (7.104)-(7.105) imply, for the relaxation time (i.e., the characteristic time needed to escape a metastable state),

$$\frac{\tau_{eq}}{\tau_0} \equiv \exp\left(\beta\Delta F\right) = \exp\left(\frac{V s_c^\star \Delta\mu}{a^3 k_B T S_c(T)}\right) = \exp\left(\frac{C}{T S_c(T)}\right) \qquad (7.106)$$

i.e., the AG Eq. (1.14) that accounts very well for fitting and comparing experimental data.

The two assumptions behind this prediction are, however, not very intuitive. First, it is unlikely that a finite number of "preferred configurations" (each representing some kind of pure metastable state) in a volume ξ^3 is enough to let the system relax. It seems, actually, rather odd that the number of metastable states *does not* increase as the size of the region increases. The work of Johari [2000], indeed, shows that hypothesizing the value of $s_c^\star = k_B \log 2$, the typical size $n^\star$ of a CRR is even *less than one* for many materials (including glucose and glycerol), according to the data table provided by Adam & Gibbs [1965] themselves at T_g. Even relaxing that hypothesis, though, and estimating $s_c^\star$ by independent experimental data, a survey of 33 glass formers further shows that $n^\star \lesssim \mathcal{O}(10)$.

Furthermore, the fact that the barrier grows like the volume of $\mathcal{C}(\xi)$ implies that, in order to have a cooperative rearranging process, a finite fraction of the total number of molecules must be involved. This is clearly unfeasible at large ξ, besides being incompatible with a small, finite $n^\star$.

We, then, have to switch to a theory that reproduces the AG relation (and the related VF law) starting from physically sound hypotheses.

7.5.2 Entropic driven "nucleation" and mosaic state

In the real world, a glassy system can leave a metastable state (however defined) in a finite time. Where to? Possibly to another one of the exponentially many states that are statistically equivalent (i.e., having more or less the same free energy) and incongruent (i.e., with no similarity in shape with the initial one). As we already said, the transition occurs when, by fluctuation, a CRR of molecules succeeds in transforming itself into a configuration belonging to a different metastable state.

Let us take a droplet of typical length-scale ξ: $\mathcal{C}(\xi)$. Contrary to the nucleation of the crystal in a liquid, the driving force pushing the change is not the free energy gain, since all metastable states have the same free energy. Instead, the transition can start right because of the huge amount of equivalent alternatives available. The driving is, then, said to be *entropic*. Inspired by mean-field theory, Kirkpatrick, Thirumalai and Wolynes [Kirkpatrick *et al.*, 1989] hypothesized a driving force equal to $-T s_c(T)\xi^d$ in a d-dimensional droplet, contrasted by the energy increase due to the mismatch of bounds on the boundary of $\mathcal{C}(\xi)$ between the generating droplet of the new state and the rest of the system still in the initial uniform metastable state. The latter interaction can be generically written as a surface tension-like term

$\Upsilon \xi^\theta$, where the exponent θ has to be determined according to the microscopic mechanisms accompanying the rearrangement of C. Generally speaking, we can only expect that, being connected at most with a surface effect, it cannot be larger than $d - 1$. We can finally define a droplet activation free energy expression as a function of the radius of the droplet, much as in conventional crystal nucleation:[11]

$$F^\ddagger(\xi) \sim -Ts_c \xi^d + \Upsilon \xi^\theta \qquad (7.107)$$

We can observe that, for small lengths, the restoring force dominates, whereas for large ξ the CRR escapes the metastable state. In order to overcome the opposing force, the CRR must be larger than the distance maximizing Eq. (7.107):

$$\xi^\ddagger \sim \left(\frac{\Upsilon}{Ts_c(T)}\right)^{\frac{1}{d-\theta}} \qquad (7.108)$$

In other words, it has to overcome the barrier around the metastable state, consequently scaling as

$$\Delta F^\ddagger \sim \Upsilon \left(\frac{\Upsilon}{Ts_c(T)}\right)^{\frac{\theta}{d-\theta}} \qquad (7.109)$$

Accordingly, the characteristic time to leave the state is related to the configurational entropy by a generalized AG relation

$$\frac{\tau(\xi^\ddagger)}{\tau_0} = \exp\left\{\left(\frac{\Upsilon}{Ts_c(T)}\right)^{\frac{\theta}{d-\theta}}\right\} \qquad (7.110)$$

At low enough temperatures, deep in the viscous regime ($T \ll T_d$) and near to T_K, the configurational entropy can be linearly expanded as

$$s_c(T) \sim \frac{T - T_K}{T_K} \equiv \bar{t} \qquad (7.111)$$

Substituting this into Eq. (7.110), we find the VF law with an exponent $\gamma = \theta/(d - \theta)$.

To determine θ, one has to make some crucial assumptions. Kirkpatrick *et al.* [1989] initially used the theory of critical phenomena, conjecturing that

1. the droplet is large enough to be considered a thermodynamically independent system

2. the ideal glass transition exists at T_K

---

[11] If we consider transitions between the amorphous and uniform liquid the same prescription holds. Indeed, from the mean-field theory we know that the difference between the free energy averaged over the metastable states and the free energy of the uniform liquid is $\bar{F} - F_{\text{unif}} = TS_c$.

3. scaling laws hold around such a transition, as in ordinary critical phenomena

Under these assumptions, the correlation length ξ of the new phase scales, in reduced temperature, as $\bar{t}^{-\nu}$. Therefore, the droplet free energy scales as $F(\bar{t}) \sim -\bar{t}^{1-d\nu} + \bar{t}^{-\theta\nu}$. In order for the CRR to expand and drive the transition out of the state, the first term must scale faster than the second one: $1 - d\nu \leq -\theta\nu$, or

$$\theta \leq \frac{d\nu - 1}{\nu} \tag{7.112}$$

To estimate ν, the fluctuation formula is used

$$(\delta T)^2 = \frac{k_B T^2}{nC} \tag{7.113}$$

where C is the specific heat (at this point the difference between constant pressure and volume is not influent). Notice that here equilibrium is assumed to hold since we are considering the case in which the thermodynamic transition at T_K is reachable, avoiding the purely kinetic, off-equilibrium vitrification that occurs in real systems (the thermal glass transition at T_g). The number n of molecules is proportional to the volume ξ^d and, according to critical scaling formulation, $C \sim \bar{t}^{-\alpha}$. Setting T near enough to T_K in order to expand the configurational entropy, but not too close, so that the thermal fluctuations do not lead below T_K ($\delta T \lesssim T - T_K = \bar{t}T_K$), Eq. (7.113) implies

$$\alpha + d\nu \geq 2 \tag{7.114}$$

If the equality holds we have the usual hyper-scaling relation.

Assuming, further, that the ideal glass transition is qualitatively similar to the thermal one (cf. Figs. 1.4 and 7.2), one sets for the discontinuous specific heat $\alpha = 0$. In this way $1/\nu \gtrsim d/2$. If the limiting values are employed, $\theta \sim d/2 \sim 1/\nu$ and $\gamma \sim 1$, a VF exponent equal to one is obtained. This is just a choice, with no preferential motivation, apart from the fact that it yields back the VF law.

About this derivation of the scaling laws, Eqs. (7.108)-(7.111) and of the exponents, Eqs. (7.112)-(7.114), we comment that none of the above assumptions 1-3 has been proved true. Experimentally, one cannot know the size of the CRR around T_K, because the material is vitrified at T_g and there, as we will see, the correlation length turns out to be only of the order of 5 molecular interspacing distance units, too small to consider those regions as independent systems. Furthermore, no experimental evidence exists for the Kauzmann transition and no explanation is provided for the weird mixing of ordinary algebraic divergences at the candidate critical point (as in standard critical phenomena) and exponential divergences in viscosity and relaxation time. Moreover, equilibrium can only be assumed in an idealized experiment for which the Kauzmann transition is actually reachable.

In order to clarify, or constructively criticize, the idea of nucleation of a truly metastable glass state, as well as to yield experimentally verifiable predictions and put forward reasonable values for the exponent of the mismatch energy scaling, the above presented theory, called RFOT theory has been recently revised. We are going to analyze the improvements in the following.

Before that, we summarize the features of the proposed state for the viscous liquid: the mosaic state. If a CRR of critical length larger than $\xi^{\ddagger}$ is created, then the system leaves the metastable state in which it was initially prepared. This is not, however, a true nucleation of a new supercooled liquid state. Indeed the droplet, once it has grown beyond a typical volume $(\xi^{\star})^{d}$ [only slightly larger than $(\xi^{\ddagger})^{d}$],[12] does not immediately expand over the whole system as in the liquid-crystal phase transition. Once the observation time is longer than $\tau(\xi^{\ddagger})$, the kind of entropically driven transition we are speaking about can occur continuously on this timescale because there will always be a huge number of metastable states reachable by fluctuation. Different regions of the space can start to rearrange next to each other and/or a new "minimal" packing can start growing inside an already expanded droplet. All of them are energetically degenerate, so that none of them will be preferred. The idea is that the system, after a transient time $\tau(\xi^{\ddagger})$, cf. Eqs. (7.106), (7.110), ends up relaxing into a mosaic state composed by dynamically heterogeneous tiles, corresponding to finite subsets of different metastable states. This is called a viscous liquid - different from a (warm) uniform liquid, that is in a unique, homogeneous, phase. The length $\xi^{\ddagger}(T)$ grows, as the temperature decreases towards T_K, as a power law, cf. Eqs. (7.108) and (7.111). The dynamics for times longer than $\tau(\xi^{\ddagger})$ is characterized by *entropically driven* activated processes, i.e., by creation (and destruction) of droplets of a size larger than $\xi^{\ddagger}$. To compare with mean-field models, the viscous liquid free energy is what is called the total free energy [cf. Eqs. (7.51)-(7.52)]

$$F_{\text{tot}} \simeq -\frac{1}{\beta} \log \sum_{\tilde{\gamma}} e^{-\beta F_{\tilde{\gamma}}} \qquad (7.115)$$

where we are summing over subsets $\tilde{\gamma}$, conceptually related to metastable states but finite in space, and not over ergodically separated metastable states as in the mean-field case.

Eventually, we can schematize the dynamics leading to the viscous, heterogeneous liquid phase, in terms of the typical size of the mosaic constituents:

1. $\xi < \xi^{\star}$. The system stays in the initial uniform glassy state, like in the mean-field scenario.

2. $\xi = \xi^{\star}$. The uniform glassy state is fragmented.

3. $\xi > \xi^{\star}$. The viscous liquid is in a mosaic state.

---

[12]The difference between $\xi^{\ddagger}$ and $\xi^{\star}$ will be clarified later, but it will turn out to be irrelevant as far as qualitative scaling laws are involved.

7.5.3 Density functional for the RFOT theory

Assuming a harmonic behavior for the individual oscillations of the particles in the cages formed as the temperature of the liquid decreases below T_{cage}, the emergence of amorphous packings, or *aperiodic crystals*, can be expressed by adopting a functional density theory constituted by an entropic localization term for ideal gas plus an interaction term expanded to the second order around a uniform liquid state: the Ramakrishnan & Yussouff [1979] density functional [Oxtoby, 1991],

$$\beta F[\rho(\mathbf{r})] = \int d^3\mathbf{r}\,\rho(\mathbf{r})\left[\log \rho(\mathbf{r}) - 1\right] \tag{7.116}$$
$$+\frac{1}{2}\int d^3\mathbf{r}\int d^3\mathbf{r}'\left[\rho(\mathbf{r}) - \rho_0\right]c(\mathbf{r},\mathbf{r}';\rho_0)\left[\rho(\mathbf{r}') - \rho_0\right] + \beta F_{\text{unif}}$$

where $\rho(\mathbf{r})$ is a nonuniform density distribution, ρ_0 its mean value and $c(\mathbf{r},\mathbf{r}';\rho_0)$ is the correlation function of the fluid evaluated at the mean density ρ_0, i.e., a renormalized form of the bare two body interaction potential. F_{unif} is the free energy of the uniform liquid.

Singh *et al.* [1985], looked at the approximately harmonic motion in the cages centered around the particles pinned at $\mathbf{r}_i$, choosing as a trial density function the sum of Gaussian distributions:

$$\rho(\mathbf{r}) = \rho(\mathbf{r},\{\mathbf{r}_i\}) = \sum_{i=1}^{N}\left(\frac{\alpha}{\pi}\right)^{3/2}e^{-\alpha(\mathbf{r}-\mathbf{r}_i)^2} \tag{7.117}$$

where the inverse variance α is a variational parameter that works as a measure of the localization degree yielded by the cages. If it is zero, we are in the completely delocalized case of the uniform liquid. On the contrary, $\alpha > 0$ characterizes a localized environment of quasi-harmonic processes. Indeed, α can be used as an order parameter to signal a discontinuous phase transition to the viscous liquid, see Fig. 7.6.

In mean-field models we have seen, in Sec. 7.2, that $F - F_{\text{tot}} = TS_{\text{c}}$, cf. Eqs. (7.51), (7.52), (7.56), for temperatures below the dynamic transition (and above the static one). In the present case, for $T_K < T < T_{\text{d}}$, F_{tot} is equal to the uniform fluid free energy F_{unif}, whereas the role of F is played by the heterogeneous free energy functional, Eq. (7.116), yielding, together with Eq. (7.117), $F[\rho(\alpha)] - F_{\text{unif}} = TS_{\text{c}}$.

For large α and in the assumption that the interfaces between a localized packing and the uniform liquid are thin, from Eqs. (7.116)-(7.117), the localization term can be expressed as a surface tension energy contribution $\sigma(r) \times$ surface [Xia & Wolynes, 2000], and, for low enough temperature, far below T_{d}, the surface tension can be approximated by

$$\sigma(r) \simeq \sigma_0 = \frac{3}{4\beta a^2}\log\frac{\alpha_{\text{loc}}a^2}{\pi\,e} \tag{7.118}$$

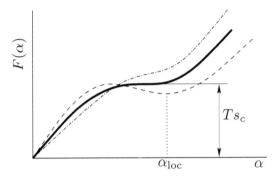

FIGURE 7.6

Free energy density functional of the (amorphous) undercooled liquid versus the inverse of the mean square displacement α of the particles around the aperiodic lattice site. This is also to be considered as the effective spring constant of the quasi-harmonic motion of the particles inside the cages. $\alpha = 0$ represents the uniform liquid. At $T = T_{\mathrm{d}}$, $F(\alpha)$ develops a spinodal point leading to a secondary minimum for $T < T_{\mathrm{d}}$. This corresponds to the presence of metastable localized nonuniform amorphous packings. Reprinted figure with permission from [Xia & Wolynes, 2000]. Copyright (2000) by the American Physical Society.

where a is the average inter-particle distance in the amorphous lattice and α_{loc} is the local spring constant for the localized, heterogeneous, liquid phase. The latter also represents the inverse of the mean square displacement of a localized particle from its lattice site $\mathbf{r}_i$. If the lattice would be regular and we had a crystal, $1/\sqrt{\alpha_{\mathrm{loc}}}$ would be the Lindemann length d_L, measuring the vibrational motion extent at the edge of mechanical stability. In the amorphous case, it plays a similar role, even though we are not, exclusively, considering amorphous solids but we are also dealing with undercooled liquids and for "mechanical stability" we have to understand "cages dominance," a far less precise concept. In any case, as $a/d_L \simeq 10$ for all crystals, also in the amorphous case such a constant value appears to be approximately verified (since it enters in a log expression, corrections will be quite likely negligible, anyway), implying $\beta \sigma_0 a^2 \simeq 1.8453$.

If we assume that the region considered, belonging to a minimum of the FEL, is a spherical droplet of radius r, its activation free energy will, then, be written as

$$F^{\ddagger}(r) = -\frac{4}{3}\pi \left(\frac{r}{a}\right)^3 T s_{\mathrm{c}} + 4\pi r^2 \sigma(r) \qquad (7.119)$$

where s_{c} is the configurational entropy per "mobile unit", S_{c}/N, and $F^{\ddagger}(r)$ is the free energy variation above the minimum value of the packing of the droplet. The typical radius will, therefore, be given by equating $F^{\ddagger}(r) = 0$, that is when the region, in spite of the mismatch at the boundary, cooperatively rearranges itself into a configuration corresponding to a new local minimum of the FEL. The above expression resembles conventional nucleation, but the driving force is substituted by an entropic term.

We spoke about entropy per mobile unit of the molecules forming the liquid [Schulz, 1998], also called "beads" [Angell & Smith, 1982], implying that N here counts those, instead of the molecules directly. Indeed, molecular liquids, above all polymers, will display more mobile units per molecule, depending on their chemical composition. Typical beads can be side chains in polymers, benzene rings, chemical groups generically oscillating independently from the rest of the molecule. A systematic treatment of the identification of beads in specific compounds, both strong and fragile, can be found in the works of Lubchenko & Wolynes [2003] and Stevenson & Wolynes [2005].

In order for a CRR to move from one amorphous minimum to another one, it has to overcome an energy barrier, in other words, to increase beyond a given critical radius $r^{\ddagger}$. This is computed by looking at the maximum of Eq. (7.119) and strongly depends on the behavior of the surface tension. Naively taking $\sigma(r) = \sigma_0$ one obtains

$$\frac{r^{\ddagger}}{a} = \frac{2\sigma_0 a^2}{T s_c} \tag{7.120}$$

$$\Delta F^{\ddagger} = \frac{16}{3}\pi \frac{(\sigma_0 a^2)^3}{(T s_c)^2} \tag{7.121}$$

Can this be correct? We know that the structural relaxation time is the time needed for a CRR to overcome the barrier dividing the initial packing from analogous, energetically degenerate, amorphous packings belonging to different "states," $\tau_{eq} = \tau_0 \exp \beta \Delta F^{\ddagger}$, implying the AG-like relation

$$\log \frac{\tau_{eq}}{\tau_0} \sim (T s_c)^{-2} \tag{7.122}$$

that is qualitatively different from the original AG relation, cf. Eq. (1.14), usually recognized as a very good phenomenological law for data fit. It cannot be excluded *a priori* that a quadratic law would do the job as well (only experimental probes might confirm or rule it out) but it is, anyway, worth seeing whether the present argument can be improved considering a more realistic surface tension term and how predictions change. Xia & Wolynes [2000] take into account the fact that, since the amorphous optimal packings are many and degenerate, a very large set of packings with any kind of surface mismatch is available. When a CRR tries to move into another state, growing in size, inside a given initial state, it has to compete with other CRRs developing around the interface and *wetting* it. Applying the wetting argument first developed for the random field Ising model[13] by Villain [1985], the surface

---

[13] For the interested reader we mention the very recent textbook of De Dominicis & Giardina [2006] about the random field Ising model.

tension at low temperature ($\gtrsim T_K$) is estimated as

$$\sigma(r) = \sigma_0 \sqrt{\frac{a}{r}} \qquad (7.123)$$

With this correction one obtains, in place of Eqs. (7.120), (7.121),

$$\frac{r^\ddagger}{a} = \left(\frac{3}{2}\frac{\sigma_0 a^2}{T s_c}\right)^{2/3} \qquad (7.124)$$

$$\Delta F^\ddagger = 3\pi \frac{(\sigma_0 a^2)^2}{(T s_c)} \qquad (7.125)$$

i.e., the original AG relation, Eq. (1.14), is reproduced with $C = 3\pi(\sigma_0 a^2)^2 N$.
   The typical radius of the droplet is, instead, the one for which the activation $F^\ddagger(r^\star)$ is zero (i.e., the free energy is minimal),

$$\frac{r^\star}{a} = 2^{2/3}\frac{r^\ddagger}{a} \qquad (7.126)$$

and the typical number of molecules composing the rearranged region is, thus, $n^\star = 4\pi/3(r^\star/a)^3$. Relaxing the spherical constraint, one can define a typical length-scale for the CRR, $(\xi^\star/a)^3 \simeq n^\star$, such that

$$\frac{\xi^\star}{a} = \left(\frac{16\pi}{3}\right)^{1/3}\frac{r^\ddagger}{a} \qquad (7.127)$$

   Universal laws can be constructed by means of the following relationships and checked by comparison with available experimental data. For instance, from Eqs. (7.118), (7.125), the configurational entropy turns out to depend exclusively on the timescales over which the glass former is analyzed:

$$\frac{s_c}{k_B} \simeq \frac{32}{\log \tau_{eq}/\tau_0} \qquad (7.128)$$

where we recall that s_c is the configurational entropy per bead (not per molecule) [Xia & Wolynes, 2000]. By definition, at the glass temperature T_g, $\tau_{eq}/\tau_0 \simeq 10^{17}$, cf. Fig. 1.1, and the configurational entropy *per bead* turns out to be $s_c \simeq 0.82 k_B$ for all substances.
   The typical length, Eq. (7.127), of the mosaic components is computed, using Eq. (7.124), as

$$\frac{\xi^\star}{a} \simeq 4\left(\frac{\log \tau_{eq}/\tau_0}{3\sqrt{3\pi \log \frac{(a/d_L)^2}{\pi\, e}}}\right)^{2/3} \simeq 0.4995\left(\log \frac{\tau_{eq}}{\tau_0}\right)^{2/3} \qquad (7.129)$$

implying, always at T_g, $\xi^\star \simeq 5.1a$ and $n^\star \simeq 140$ beads. A "state portion" is, thus, composed by a relatively small number of molecules, even though this estimate is far larger than the one based on the (already a bit improved) AG theory, for which $n^\star \lesssim 10$ [Johari, 2000]. Looking at the whole system, this turns out to be heterogeneous: a mosaic built by similar tridimensional tiles belonging to different local minima of the FEL, whose interfaces display surface tension.

Looking at experiments, detecting heterogeneity in glass formers and measuring the size of the heterogeneous regions, the estimate above seems to be approximately confirmed. Experimental observation on polyvinylacetate, indeed, report a typical correlation length of 2-4nm, corresponding to $n^\star \simeq 25$-180 beads (the inter-particle distance is $a = 0.7$nm) at a temperature ten degrees above T_g [Tracht *et al.*, 1998], confirmed by probes slightly below T_g, for which $n^\star \simeq 30 - 90$. In colloids, Weeks *et al.* [2000] estimated, for high critical densities, a number of beads $n^\star \simeq 60$.

Other predictions can be devised, that can be experimentally tested. For instance, assuming the usual linear behavior for the configurational entropy, $s_c = \Delta \tilde{c}_p (1 - T_K/T)$, where $\Delta \tilde{c}_p$ is the specific heat jump at the glass transition computed *per bead*, Xia & Wolynes [2000], using Eqs. (7.118) and (7.128), determine the fragility expression[14]

$$\frac{1}{K} \simeq \frac{32 k_B}{\Delta \tilde{c}_p} \qquad (7.130)$$

where the global fragility is defined in Eq. (1.20).

Another expression can be derived adopting the local, or kinetic, definition of the fragility, Eq. (1.21). Using Eq. (7.128) Lubchenko & Wolynes [2003] derive

$$K_{\text{loc}} \simeq 34.7 \frac{T_m}{\Delta H_m} \Delta c_p \qquad (7.131)$$

where the fusion latent heat ΔH_m and the specific heat are now computed per mole of molecules and not per mole of beads.[15] This has to be compared with the empirical law devised by Wang & Angell [2003]:

$$K_{\text{loc}} = 56 \frac{T_g \Delta c_p}{\Delta H_m} \qquad (7.132)$$

In Fig. 7.7 the two local fragilities are compared for 44 glass formers analyzed by Wang & Angell [2003]. Finally, we mention that other general, quantitative relationships can be constructed, as well, e.g., for the dependence of the fragility from the exponent of the stretched relaxation for viscous liquids [Xia & Wolynes, 2001].

We want to stress that all these quantitative predictions, however, rely on strong (sometimes somewhat arbitrary) and not verified assumptions, so

---

[14]Xia & Wolynes [2000] considered $D = 1/K$. We will keep the label D for the diffusion coefficient.

[15]To determine the number of beads Lubchenko & Wolynes [2003] compared the entropy of fusion per mole $\Delta H_m/T_m$ and the entropy of fusion in a Lennard-Jones system.

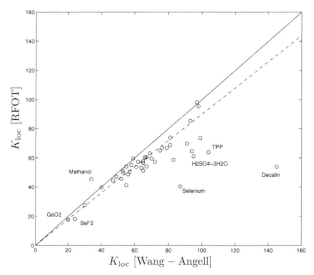

FIGURE 7.7

Local fragility as computed in the framework of the RFOT, Eq. (7.131), versus the local fragility obtained by the phenomenological law inferred from experimental data by Wang & Angell [2003], Eq. (7.132). Reprinted figure with permission from [Stevenson & Wolynes, 2005]. Copyright (2005) American Chemical Society.

that the strength of the whole approach eventually resides *a posteriori* in the promising comparison with experimental data.

The approximations adopted in the density functional approach to the RFOT theory for undercooled liquids can be summarized as follows.

1. Surface tension. The form of $\sigma(r)$ suffers from various inaccuracies. It has been derived in the assumption of large α and thin interfaces. Moreover, it has been computed at $T = T_K$, applying the theory of wetting to the interface formation between the rising new local amorphous region and the uniform state in which it is embedded and not the interface between two amorphous states [Xia & Wolynes, 2000].

2. Beads. The number of mobile units is uncertain, the different estimates (chemical counting, data fit, comparison between "per molecule" and "per bead" observables) yield compatible but usually different results [Lubchenko & Wolynes, 2003]. It is a serious problem, above all, for polymers.

3. Configurational entropy. The available experimental configurational entropy is, actually, the excess entropy, that coincides with the first one only assuming that the vibrational modes of the crystal and the glass are the same, cf. the dedicated Sec. 1.4.3.

4. Barrier softening. As T is not far below T_d, the interface tension goes to zero (indeed, above T_d, by definition, the motion is collision dominated and the liquid is uniform and not very viscous). The barriers, therefore, soften, the interfaces between mosaic tiles become thicker and less clearcut and the surface tension vanishes approaching T_d from below. The low viscosity case has been analyzed by Lubchenko & Wolynes [2003], but a theory connecting the low temperature mosaic scenario with the high temperature uniform liquid state is still lacking and the limit of validity of the mosaic description is far from being established as assessed, e.g., recently by Cavagna *et al.* [2007] in a soft sphere binary mixture analyzed below T_d.[16]

This situation is somewhat similar to what has been happening for about 40 years after the AG relation, that appears to be right and widely verified, though constructed on physically counterintuitive assumptions. In the present case, the assumptions behind the quantitative formulation of the RFOT theory are, actually, feasible, but, nevertheless, not justified *a priori* in a rigorous way. Moreover, the theory is based on the analogy with crystal nucleation, but the driving is related to a conceptually mysterious entropic term.

In order to be sure that the mosaic one is the proper description of the viscous liquid (as, presumably, of the amorphous solid), the microscopic hypothesis should be tested, or, otherwise, nothing can prevent us from thinking that the same predictions [e.g., besides the AG relation and the VF law, Eqs. (7.128)-(7.131)] might be built by alternative arguments. Moreover, an explanation of the generation of the mosaic state alternative to the entropic driven nucleation, would clarify the issue to the noninitiated. Two almost, but not completely, alternative ways of improving the comprehension of the physics hidden beyond the label of "entropic driving force" have been recently developed. We briefly report in the following the main lines of the reformulations of the RFOT theory according to the approach of Bouchaud & Biroli [2004] and the one of Lubchenko & Wolynes [2004].

7.5.4 Beyond entropic driving I: droplet partition function

The total configurational entropy of a system composed by different *independent* relaxing dominions of typical linear size ξ is

$$S_c \simeq \mathcal{N}(\xi) s^\star(\xi) \simeq \frac{V}{\xi^d} \cdot \xi^d \tilde{s}_c = V \tilde{s}_c \qquad (7.133)$$

where $s_c^\star(\xi)$ is the configurational entropy per tile of length ξ, and $\tilde{s}_c$ is the configurational entropy per unit volume. The total entropy of the "mosaic state" does not depend on the tile size and is rather puzzling to understand the convenience of having a heterogeneous mosaic instead of a homogeneous

---

[16]The same SSBM considered in Appendix 6.A.1 and in Secs. 2.8 and 7.3.

state, for which no surface tension would be paid. Moreover, even assuming the existence of a mosaic, it is not clear what physical condition would fix the length over which cooperative processes would occur. What is the precise nature of the "entropic force"?

We take the usual bubble of radius ξ. If $\tilde{s}_c \, \xi^d \gg O(1)$, the molecules inside the region $\mathcal{C}(\xi)$ can arrange into N_C configurations, each related to a metastable state of free energy F:

$$N_C(F) \sim \exp\left[\xi^d \frac{\tilde{s}_c(F,T)}{k_B}\right] \qquad (7.134)$$

where $\tilde{s}_c(F,T)$ is the droplet configurational entropy per unit volume counting all the subsets of configurations of $\mathcal{C}$ corresponding to local minima of the FEL at free energy F.

We start from a space-homogeneous metastable state "I" and we look what happens when a region of particles cooperatively rearranges itself in a configuration belonging to a different state, "b." On the boundary of $\mathcal{C}(\xi)$ the mismatch between I and b will cause an increase of the total free energy, thus the boundary acts as a random field on any configuration belonging to $b \neq I$ inside $\mathcal{C}(\xi)$. We denote by $f_{b/I}\xi^d$ the activation free energy density of the b state of $\mathcal{C}$, embedded in the macroscopically extended state I. We recall, furthermore, cf. Eq. (7.107), that $\Upsilon_{b/I}\xi^\theta$ is the surface free energy gain generated by the mismatch of b with the fixed boundary conditions belonging to I. If ξ is not too small, the surface coefficients $\Upsilon_{b/I}$ are *typically* equal to each other, implying $f_{b/I} = f_b$, independent of the uniform state.

According to Bouchaud & Biroli [2004], the partition function of the dominion $\mathcal{C}(\xi)$ embedded in the metastable state I is, then, written as

$$Z(\xi,T) \sim \sum_{b \neq I} e^{-\beta f_b \xi^d} + e^{-\beta f_I \xi^d + \beta \Upsilon \xi^\theta} \qquad (7.135)$$

$$\sim \int_0^\infty df N_C(f) \exp\left\{-\beta f \xi^d\right\} + \exp\left\{-\beta \xi^d f_I + \beta \Upsilon \xi^\theta\right\}$$

$$\sim \int_0^\infty df \exp\left\{-\beta \xi^d \left[f - T s_c(f,T)\right]\right\} + \exp\left\{-\beta \xi^d f_I + \beta \Upsilon \xi^\theta\right\}$$

where we have used Eq. (7.134) and we have subtracted from the f_I contribution the surface term $\Upsilon \xi^\theta$ (no mismatch). We have dropped the ground state free energy term yielding an irrelevant proportionality factor.

How stable is the homogeneous state I against fragmentation? We first consider the case in which it is a state with typical free energy excess, i.e., $f_I = f^\star$, where $f^\star$ is computed by the saddle point approximation of the integral of Eq. (7.135):

$$\left.\frac{\partial s_c(f,T)}{\partial f}\right|_{f^\star} = \frac{1}{T} \qquad (7.136)$$

The metastable states at $f^\star$ are the most numerous states and, hence, those at which the liquid glass former is most probably stuck when cooled down from high temperature. They are also called "threshold" states, for this reason, cf. Eq. (7.53) in Sec. 7.2.

The partition function is, in this case, approximated as

$$Z(\xi, T) \sim e^{-\beta f^{\star} \xi^d} \left[e^{s_c(f^{\star}, T) \xi^d / k_B} + e^{\beta \Upsilon \xi^{\theta}} \right] \qquad (7.137)$$

We recall that $\theta \leq d - 1$, and, if wetting is assumed, $\theta = d/2$. If ξ is small enough, the uniform state remains as it is (the second term dominates). This is the same situation as in mean-field. If, otherwise, ξ is large enough, the matched state is destroyed [the first term, the superposition of all metastable states, dominates, cf. Eq. (7.115)]. The crossover length $\xi^{\star}$ scales with temperature as $\xi^{\star} \sim (T s_c)^{-\frac{1}{d-\theta}} \sim (T - T_K)^{-\frac{1}{d-\theta}}$, in agreement with Eqs. (7.108), (7.111).[17]

We can reasonably assume that the free energy barrier scales as $\Delta F^{\ddagger} \sim (\xi^{\star})^{\psi}$, without imposing any relation with the scaling of the surface term in the activation free energy and without taking for granted the equality $\Delta F^{\ddagger} = F^{\ddagger}(r^{\ddagger} \sim \xi^{\star})$, employed in the previous section. We have, as a consequence, a generalized AG relation. In mean-field models, $s_c(f, T)$ goes to zero linearly in f, as $f \to 0$ (f is the excess entropy above the ground state), with $\partial^2 s_c / \partial f^2 < 0$. At low enough temperatures, this implies, for the saddle point values, that $T s_c(f^{\star}, T) \sim f^{\star} \sim T - T_K = \bar{t} T_K$, yielding the VF law, with a generalized exponent different from one:

$$\frac{\tau(\xi^{\star})}{\tau_0} \sim \exp\left(\frac{\Upsilon}{T - T_k} \right)^{\psi/(d-\theta)} \qquad (7.138)$$

In the previous sections, cf. Eqs. (7.109), (7.125), assuming the existence of a RFOT, the thermodynamic independence of the droplet subsystems, and the wetting of the interfaces of the tiles, we had, in $d = 3$, $\psi = \theta = 3/2$.

Regarding the determination of the exponents we have, once again, to notice that it cannot be other than sloppy, since

- the typical length for $\mathcal{C}(\xi^{\star})$ at T_g is smaller than $\mathcal{O}(10)$ in units of the average inter-particle distance a

- the surface term Υ also depends on temperature, and, more precisely, goes to zero near T_d (the mentioned barrier softening effect)

- a fit by the VF law is relatively insensitive to the precise value of the exponent

Anyway, the latter formulation allows for the explanation of the phenomenon of diffusion-viscosity decoupling detected in the viscous liquid, see, e.g., [Chang

---

[17]In this approach there is no difference between critical ($\xi^{\ddagger}$) and typical ($\xi^{\star}$) length-scales. This is not a big issue since, already in the approach of Sec. 7.5.3, they scale in the same way.

et al., 1994; Swallon *et al.*, 2003], for which diffusion and viscosity are no more inversely proportional to each other as formulated by the Stokes-Einstein relation, Eq. (1.19), cf. Sec. 1.6.

The diffusion coefficient D depends on the shortest timescale at which relaxation out of a state can occur, therefore it is

$$D \sim \frac{(\xi^\star)^2}{\tau(\xi^\star)} \tag{7.139}$$

The viscosity, instead, is connected to the *average* relaxation time, where the average is performed over all activated processes on all timescales above $\tau(\xi^\star)$. The processes activated on timescales longer than $\tau(\xi^\star)$ are linked to free energies lower than the typical ones. In order to probe free energies lower than the threshold one, Bouchaud & Biroli [2004] consider the fragmentation of the state I, whose free energy is $f_I = uf^\star$, with $u < 1$. Taking the saddle point of Eq. (7.135) one has, now,

$$Z(\xi, T) \sim \exp\left[-\beta f^\star \xi^d + s_c(f^\star, T)\xi^d/k_B\right] + \exp\left[-\beta u f^\star \xi^d + \beta \Upsilon \xi^\theta\right] \tag{7.140}$$

The correlation length of a region cooperatively rearranging itself inside a metastable state of free energy $uf^\star$ must now be at least

$$\xi_u \sim \left(\frac{\Upsilon}{Ts_c(f^\star, T) - f^\star + uf^\star}\right)^{\frac{1}{d-\theta}} \tag{7.141}$$

As said above, for $T \gtrsim T_K$, one has $Ts_c(f^\star, T) \sim T - T_K = \bar{t}\, T_K$, implying $Ts_c(f^\star, T) - f^\star \sim (\bar{t})^2$. This leads to

$$\xi_u \sim \left(\frac{\Upsilon}{t^2 + ut}\right)^{\frac{1}{d-\theta}} \sim \xi^\star \left(\frac{1}{t+u}\right)^{\frac{1}{d-\theta}} \tag{7.142}$$

Going towards the ground state, the excess free energy goes to zero ($u \to 0$) and the scaling with temperature of the correlation length of the CRR necessary to destroy the uniform state increases as

$$\xi_0 \approx (\xi^\star)^2 \tag{7.143}$$

As a consequence, the escape process from the lowest states has a characteristic time exponentially much longer than the timescale of activated processes out of a threshold state:

$$\frac{\tau(\xi_0)}{\tau_0} = \exp\left\{\left(\frac{1}{T - T_K}\right)^{\frac{2\psi}{d-\theta}}\right\} \tag{7.144}$$

Since the range of relaxation times is so broad, even though most of the processes occur on the minimum timescale available, the average relaxation time - and the viscosity - will be dominated by the longest processes. In this way this approach accounts for the experimental observation that the viscosity and the diffusion coefficient decouple.[18] We stress that the VF phenomenological law reproduces the temperature behavior of the viscosity and, therefore, of the average relaxation time, Eq. (7.144), rather than the shortest relaxation time, Eq. (7.138). If a strict estimate of the VF exponent would be possible in viscosity measurements, this should be compared to $2\psi/(d-\theta)$ and not to $\psi/(d-\theta)$.

After these last considerations we are able to clarify that, in the mosaic scenario, the relaxation is not due to any nucleation and that there is no need to introduce any weird entropic driving force. What happens is that domains of length-scale $\xi^\star$ remain basically unchanged for about $\tau(\xi^\star)$ and then, by fluctuation, rearrange themselves in a cooperative way. The system does not modify itself to lower the free energy (no nucleation then) but only by random fluctuations. Most of the times the fluctuation is between energetically (and statistically) equivalent states that are at the threshold level and have a scaling correlation length $\xi^\star$. Indeed, when a CRR appears of smaller linear dimension, the processes are faster but they lead nowhere, whereas, when a CRR has $\xi > \xi^\star$, it leads to a broader activated process but slower than those initiated by CRRs of linear dimension $\xi^\star$. Larger domains - even though rearranging more - will take much longer times to evolve, therefore, "large enough" domains are the fastest to relax and let the molecules diffuse to find their thermodynamically most convenient configurations. The activated processes involving the creation/destruction of threshold states (of free energy $f^\star$) carry out the relaxation (aging) dynamics.

7.5.5 Beyond entropic driving II: library of local states

To reformulate the mosaic scenario, avoiding explicitly basing it on the concept of entropic driving, Lubchenko & Wolynes [2004, 2007] follow yet another way.

Let I be one of the very many states (in the mean-field sense), or basins (in the PEL sense of Chapter 6), in which the uniform liquid can find itself, with free energy $F_I^{\text{lib}} = u_I - Ts_{\text{vib}}$. The free energies of all basins available to the system form a global library. If the system is large enough, the spectrum of the F^{lib} distribution will be more and more dense as the size increases.

The further step is to construct a local library. One takes a system in a basin of free energy F_I^{lib} from the global library and cuts out a region $\mathcal{C}(\mathbf{r}, n)$ centered around the point $\mathbf{r}$ in space and consisting in $n \ll N$ mobile units. Then, one blocks all external beads (practically imposing fixed boundary con-

---

[18]Moreover, since the width of the distribution of the logarithm of the relaxation times increases as temperature is lowered, $\log \tau(\xi_0) \sim \bar{t}^\gamma \log \tau(\xi^\star)$, the relaxation functions of thermodynamic observables - generically depending on a superposition of different processes - become more and more stretched, as actually happens in real glass formers [Kohlrausch, 1847].

ditions for the motion inside $\mathcal{C}$). Eventually, new minima are looked for, changing the configuration of the n beads inside $\mathcal{C}$. The set of the new structures possibly found corresponds to a set of free energies similar to a subset of the global library.

Since $n \ll N$, structures different from the initial ones will correspond to higher lying local minima in free energy (in the average). This takes place because of the constraint imposed by the fixed boundary conditions and also because, according to the distribution of the free energy, the (far) larger number of alternative states is available at higher values.

If n increases, the free energy increases in the average, but also the density of its spectrum and the spread of its distribution. This implies that, sooner or later, for some n^*, the lowest local level available will be reachable from the initial configuration by thermal fluctuation and a transition to a different basin becomes feasible. The time needed will be long, since the mismatch at the interface will generate a barrier. Too small regions will not be able to overcome the barrier. As we have seen previously, a critical size exists, above which the system is able to expand in the packing configurations belonging to a new basin, cf. Eqs. (7.108) or (7.127).

To estimate the critical size, the barrier height and the relaxation time, or, at least, their scaling with temperature, Lubchenko & Wolynes [2004] introduce a "bulk" free energy $\Phi(\mathbf{r}, n)$ for the isolated $\mathcal{C}$ region. The bonds with the interface are not included in the definition of Φ.

The difference between the free energy of a new state b and the initial, uniform, state I is

$$F_b^{\text{lib}} - F_I^{\text{lib}} = \Phi_b(\mathbf{r}, n) - \Phi_I(\mathbf{r}, n) + \Gamma_{b/I} \qquad (7.145)$$

where, quite likely, $F_b^{\text{lib}} > F_I^{\text{lib}}$ so that $\Gamma > 0$. Since Γ is due to the mismatch of the interactions at the interface, it will grow, at most, as the surface of $\mathcal{C}$ itself. Generically, it will scale as n^x with the size of the droplet. It is now possible to reproduce previous results in terms of free energy differences, without evoking entropically driven nucleation.

If $\omega(\Phi)$ is the number of minimal packings of bulk free energy Φ feasible in the region $\mathcal{C}$, ignoring the rest of the system, the rate of flow to a localized state of size n can be written as

$$k(n) = \frac{1}{\tau_0} \int dF \; \omega(\Phi) e^{-\beta(F - F_I^{\text{lib}})} \simeq \frac{1}{\tau_0} e^{S_c(\bar{\Phi})/k_B} e^{-\beta(\bar{F} - F_I^{\text{lib}})} \qquad (7.146)$$

where $S_c(\Phi)$ is the configurational entropy of the isolated droplet. It depends on the average bulk free energy $\bar{\Phi}$, that depends on T and n: $S_c(T, n)$. The variable F is the non-mean-field generalization of the free energy of pure states (cf. Sec. 7.2).

We meet again the activation energy $F^{\ddagger} = \bar{F} - F_I^{\text{lib}} - T S_c(T, n)$, that can now be rewritten as

$$F^{\ddagger} = \bar{\Phi} - \Phi_I + \Gamma - T S_c(T, n) = F_{\text{bulk}} - \Phi_I + \Gamma \qquad (7.147)$$

Both $F_{\text{bulk}} \equiv \bar{\Phi} - TS_{\text{c}}(T,n) = n\, f_{\text{bulk}}$ and $\Phi_I = n\, \phi_I$ are extensive in the droplet size. The activation free energy, then, scales like

$$F^{\ddagger} = n(f_{\text{bulk}} - \phi_I) + \gamma_0 n^x + \delta F \qquad (7.148)$$

where the fluctuation δF scales as $\sqrt{n}$ and is neglected.

The minimum of $k(n)$ yields the size at which the activation barrier is crossed

$$n^{\ddagger} \simeq \left(\frac{\phi_I - f_{\text{bulk}}}{x\, \gamma_0} \right)^{1/(x-1)} \qquad (7.149)$$

and, eventually, the barrier scales as

$$\Delta F^{\ddagger} \simeq \gamma_0 (1-x) \left(\frac{\phi_I - f_{\text{bulk}}}{x\gamma_0} \right)^{x/(x-1)} \qquad (7.150)$$

The wetting argument put forward in Sec. 7.5.3 would lead to an exponent $x = 1/2$ (instead of $2/3$ for a purely surface scaling of the mismatch term). We notice that, actually, the neglected fluctuations δF also were supposed to scale as $\sqrt{n}$, but they are neglected in the derivation of the critical size and the barrier. We notice also that the number n of beads involved turns out to be of the order of 10^2 (according to the estimates presented and to the experiments mentioned above, see, e.g., the review of Ediger [2000]) and, therefore, also in this approach, all scalings are somewhat arbitrary.

To reconnect with the entropic driving, one has to consider that f_{bulk} is the local equivalent of the mean-field total free energy computed summing over all metastable states, whereas ϕ_I is the state free energy f, therefore yielding [cf. Eq. (7.54)]

$$f_{\text{bulk}} - \phi_I = -Ts_{\text{c}} \qquad (7.151)$$

This equation, inserted in Eq. (7.150), leads to a generalized AG relation for $\tau_{\text{eq}} \simeq 1/k(n^{\star})$. If $x = 1/2$, the original AG relation, Eq. (1.14), is obtained.

Bibliography

Adam, G., & Gibbs, J.H. 1965. *On the temperature dependence of cooperative relaxation properties in glass-forming liquids.* J. Chem. Phys., **43**, 139.

Allahverdyan, A., & Nieuwenhuizen, Th.M. 2000. *Extracting work from a single thermal bath in the quantum regime.* Phys. Rev. Lett., **85**, 1799.

Allahverdyan, A., & Nieuwenhuizen, Th.M. 2002a. *Statistical thermodynamics of quantum Brownian motion: Construction of perpetuum mobile of the second kind.* Phys. Rev. E, **66**, 036102.

Allahverdyan, A.E., & Nieuwenhuizen, Th.M. 2002b. *Entropy production, energy dissipation and violation of Onsager relations in the steady adiabtic state.* Phys. Rev. E, **62**, 845.

Allahverdyan, A.E., & Nieuwenhuizen, T.M. 2005. *Fluctuations of work from quantum subensembles: The case against quantum work-fluctuation theorems.* Phys. Rev. E, **71**, 066102.

Angelani, L., Di Leonardo, R., Ruocco, G., Scala, A., & Sciortino, F. 2000. *Saddles in the energy landscape probed by supercooled liquids.* Phys. Rev. Lett., **85**, 5356–5359.

Angelani, L., Di Leonardo, R., Parisi, G., & Ruocco, G. 2001. *Topological description of the aging Dynamics in simple glasses.* Phys. Rev. Lett., **87**, 055502.

Angell, C.A. 1985. *Spectroscopy simulation and scattering, and the medium range order problem in glass.* J. Non-Cryst. Solids, **73**, 1–17.

Angell, C.A. 1991. *Relaxation in liquids, polymers and plastic crystals - strong/fragile patterns and problems.* J. Non-Cryst. Solids, **131-133**, 13–31.

Angell, C.A. 1995. *Formation of glasses from liquids and biopolymers.* Science, **267**, 1924–1935.

Angell, C.A. 1997. *Entropy and fragility in supercooling liquids.* J. Res. NIST, **102**, 171.

Angell, C.A., & Smith, D.L. 1982. *Test of the entropy basis of the Vogel-Tammann-Fulcher equation. Dielectric relaxation of polyalcohols near T_g.* J. Phys. Chem., **86**, 3845–3852.

Angell, C.A., & Torell, L.M. 1983. *Short time relaxation processes in liquids: Comparison of experimental and computer simulation glass transitions on picosecond time scales.* J. Chem. Phys., **78**, 937.

Angell, C.A., Cheeseman, P.A., Clarke, J.H.R., & Woodcock, L.V. 1977. *Molecular dynamics modelling of amorphous solid structures. Page 191 of:* Gaskell, P.H. (ed), *The structure of non-crystalline materials.* Taylor and Francis (London).

Angell, C.A., Ngai, K.L., McKenna, G.B., McMillian, P.F., & Martin, S.W. 2000. *Relaxation in glassforming liquids and amorphous solids.* J. App. Phys., **88**, 3113.

Aquino, G., Leuzzi, L., & Nieuweunhuizen, T.M. 2006a. *Kovacs effect in a fragile glass model.* Phys. Rev. B, **73**, 094205.

Aquino, G., Leuzzi, L., & Nieuweunhuizen, T.M. 2006b. *Kovacs effect in solvable model glasses.* J. Phys.: Conf. Ser., **40**, 50.

Arenzon, J.J., & Sellitto, M. 2004. *Kovacs effect in facilitated spin models of strong and fragile glasses.* Eur. Phys. J. B, **42**, 543.

Arora, D., Bhatia, D.P., & Prasad, M.A. 1999. *Exact solutions of some urn models of relaxation in glassy dynamics.* Phys. Rev. E, **60**, 145–148.

Baldassarri, A., Barrat, A., D'Anna, G., Loreto, V., Mayor, P., & Puglisi, A. 2005. *What is the temperature of a granular medium?* J. Phys.: Cond. Matt., **17**, S2405–S2428.

Barrat, A., Burioni, R., & Mézard, M. 1996. *Ageing classification in glassy dynamics.* J. Phys. A: Math. Gen., **29**, 1311–1330.

Barrat, A., Loreto, V., & Puglisi, A. 2004. *Temperature probes in binary granular gases.* Physica A, **334**, 513.

Barrat, J.-L., Roux, J.-N., & Hansen, J.-P. 1990. *Diffusion, viscosity and structural slowing down in soft sphere alloys near the kinetic glass transition.* Chem. Phys., **149**, 197.

Barrat, J.L., & Kob, W. 1999. *Fluctuation-dissipation ratio in an aging Lennard-Jones glass.* Europhys. Lett., **46**, 637–642.

Bässler, H. 1987. *Viscous flow in supercooled liquids analyzed in terms of transport theory for random media with energetic disorder.* Phys. Rev. Lett., **58**, 767.

Bellon, L., & Ciliberto, S. 2002. *Experimental study of the fluctuation-dissipation relation during an aging process.* Physica D, **168**, 325–335.

Bellon, L., Ciliberto, S., & Laroche, C. 2001. *Violation of the fluctuation-dissipation relation during the formation of a colloidal glass.* Europhys. Lett., **53**, 511.

Bellon, L., Ciliberto, S., & Laroche, C. 2002. *Advanced memory effects in the aging of a polymer glass.* Eur. Phys. J. B, **25**, 223.

Benaïm, M., Schreiber, S.J., & Tarrès, P. 2004. *Generalized urn models of evolutionary processes.* Ann. Appl. Probab., **14**, 1445.

Bengtzelius, U., Götze, W., & Sjölander, A. 1984. *Dynamics of supercooled liquids and the glass transition.* J. Phys. C, **17**, 5915.

Berendsen, H.J.C., Grigera, J.R., & Straatsma, T.P. 1987. *The missing term in effective pair potentials.* J. Phys. Chem., **91**, 6269–6271.

Bernu, B., Hiwatari, Y., & Hansen, J.-P. 1985. *A molecular dynamics study of the glass transition in binary mixtures of soft spheres.* J. Phys. C, **18**, L371–L376.

Bernu, B., Hansen, J.-P., Hiwatari, Y., & Pastore, G. 1987. *Soft sphere model for the glass transition in binary alloys: Pair structure and self-diffusion.* Phys. Rev. A, **36**, 4891.

Berthier, L., & Barrat, J.-L. 2002. *Non-equilibrium dynamics and fluctuation-dissipation relation in a sheared fluid.* J. Chem. Phys., **116**, 6228.

Berthier, L., & Bouchaud, J-P. 2002. *Geometrical aspects of aging and rejuvenation in the Ising spin glass: A numerical study.* Phys. Rev. B, **66**, 054404.

Bertin, E.M., Bouchaud, J-P., Drouffe, J.M., & Godréche, C. 2003. *The Kovacs effect in model glasses.* J. Phys. A: Math. Gen., **36**, 10701.

Bialas, P., Burda, Z., & Johnston, D. 1997. *Condensation in the backgammon model.* Nucl. Phys. B, **493**, 505.

Binder, K., & Kob, W. 2005. *Glassy Materials and Disordered Solids.* World Scientific

(Singapore).

Biroli, G., Bouchaud, J.-P., Miyazaki, K., & Reichman, D.R. 2006. *Inhomogeneous mode-coupling theory and growing dynamic length in supercooled liquids.* Phys. Rev. Lett., **97**, 195701.

Blair, D. 1973. *A History of Glass in Japan.* Kodansha International Ltd. & The Corning Museum of Glass (Tokyo).

Bonilla, L.L., Padilla, F.G., Parisi, G., & Ritort, F. 1996a. *Analytical solution of the Monte Carlo dynamics of a simple spin-glass model.* Europhys. Lett., **34**, 159.

Bonilla, L.L., Padilla, F.G., Parisi, G., & Ritort, F. 1996b. *Closure of the Monte Carlo dynamical equations in the spherical Sherrington-Kirkpatrick model.* Phys. Rev. B, **54**, 4170.

Bonilla, L.L., Padilla, F.G., & Ritort, F. 1998. *Aging in the linear harmonic oscillator.* Physica A, **250**, 315.

Bouchaud, J-P., & Biroli, G. 2004. *On the Adam-Gibbs-Kirkpatrick-Thirumalai-Wolynes scenario for the viscosity increase in glasses.* J. Chem. Phys. , **121**, 7347.

Bouchaud, J.P. 1992. *Weak ergodicity breaking and aging in disordered systems.* J. Phys. I (France), **2**, 1705.

Bouchaud, J.P. 1994. *Towards an experimental determination of the number of metastable states in spin-glasses?* J. Phys. I (France), **4**, 139.

Bouchaud, J.P., & Dean, D.S. 1995. *Aging on Parisi's tree.* J. Phys. I (France), **5**, 265.

Bouchaud, J.P., Cugliandolo, L., Kurchan, J., & Mézard, M. 1998. *Out of equilibrium dynamics in spin-glasses and other glassy systems. Page 161 of:* Young, A.P. (ed), *Spin Glasses and Random Fields.* World Scientific (Singapore).

Bouchaud, J.P. Cugliandolo, L.F., Kurchan, J., & Mézard, M. 1996. *Mode-coupling approximations, glass theory and disordered systems.* Physica A, **226**, 243–273.

Boucheron, S., & Gardy, D. 1997. *An urn model from learning theory.* Random Structures and Algorithms, **10**, 43–67.

Broderix, K., Bhattacharya, K.K., Cavagna, A., Zippelius, A., & Giardina, I. 2000. *Energy Landscape of a Lennard-Jones Liquid: Statistics of Stationary Points.* Phys. Rev. Lett., **85**, 5360.

Büchner, S., & Heuer, A. 1999. *Potential energy landscape of a model glass former: Thermodynamics, anharmonicities, and finite size effects.* Phys. Rev. E, **60**, 6507–6518.

Buhot, A. 2003. *Kovacs effect and fluctuation-dissipation relations in 1D kinetically constrained models.* J. Phys. A: Math. Gen., **36**, 12367.

Buhot, A., & Garrahan, J.P. 2002. *Fluctuation-dissipation relations in the activated regime of simple strong-glass models.* Phys. Rev. Lett., **88**, 225702.

Buisson, L., Bellon, L., & Ciliberto, S. 2003a. *Intermittency in aging.* J. Phys.: Cond. Matt., **15**, S1163.

Buisson, L., Ciliberto, S., & Garcimartín, A. 2003b. *Intermittent origin of the large violations of the fluctuation-dissipation relations in an aging polymer glass.* Europhys. Lett., **63**, 603–609.

Caiazzo, A., Coniglio, A., & Nicodemi, M. 2004. *Glass-glass transition and new dynamical singularity points in an analytically solvable p-spin glasslike model.* Phys. Rev. Lett., **93**, 215701.

Callen, H.B., & Welton, T.A. 1951. *Irreversibility and generalized noise.* Phys. Rev., **83**, 34.

Cape, J.N., & Woodcock, L.W. 1980. *Glass transition in a soft-sphere model.* J. Chem. Phys., **72**, 976.

Carnot, Sadi. 1824. *Réflections sur la puissance motrice du feu et sur les machines propres à développer cette puissance.* Bachelier (Paris).

Cavagna, A., & Castellani, T. 2005. *Spin-glass theory for pedestrians.* J. Stat. Mech., P05012.

Cavagna, A., Giardina, I., & Parisi, G. 1998. *Stationary points of the Thouless-Anderson-Palmer free energy.* Phys. Rev. B, **57**, 11251.

Cavagna, A., Giardina, I., & Parisi, G. 1999. *Analytic computation of the instantaneous normal modes spectrum in low-density liquids.* Phys. Rev. Lett., **83**, 108.

Cavagna, A., Grigera, T.S., & Verrocchio, P. 2007. *Mosaic multi-state scenario vs. one-state description of supercooled liquids.* Phys. Rev. Lett., **98**, 187801.

Chang, I., Fujara, F., Geil, B., Heurberger, G., Mangel, T., & Sillescu, H. 1994. *Translational and rotational molecular motion in supercooled liquids studied by NMR and forced Rayleigh scattering.* J. Non-Cryst. Solids, **172-174**, 248–255.

Chayes, L., Emery, V.J., Kivelson, S.A., Nussinov, Z., & Tarjus, G. 1996. *Avoided critical behavior in a uniformly frustrated system.* Physica A, **225**, 129–153.

Chen, S.-H., Chen, W.R., & Mallamace, F. 2003. *The glass-to-glass transition and its end point in a copolymer micellar system.* Science, **300**, 619.

Coluzzi, B., & Parisi, G. 1998. *On the approach to the equilibrium and the equilibrium properties of a glass-forming model.* J. Phys. A: Math. Gen., **31**, 4349.

Coluzzi, B., Mézard, M., Parisi, G., & Verrocchio, P. 1999. *Thermodynamics of binary mixture glasses.* J. Chem. Phys., **111**, 9039.

Coluzzi, B., Parisi, G., & Verrocchio, P. 2000a. *Lennard-Jones binary mixture: A thermodynamical approach to glass transition.* J. Chem. Phys., **112**, 2933–2944.

Coluzzi, B., Parisi, G., & Verrocchio, P. 2000b. *Thermodynamical liquid-glass transition in a lennard-jones binary mixture.* Phys. Rev. Lett., **84**, 306.

Corberi, F., Lippiello, E., & Zannetti, M. 2001a. *Interface fluctuations, bulk fluctuations, and dimensionality in the off-equilibrium response of coarsening systems.* Phys. Rev. E, **63**, 061506.

Corberi, F., Lippiello, E., & Zannetti, M. 2001b. *On the connection between off-equilibrium response and statics in non disordered coarsening systems.* Eur. Phys. J. B, **24**, 359–376.

Corberi, F., Lippiello, E., & Zannetti, M. 2002a. *Slow relaxation in the large-N model for phase ordering.* Phys. Rev. E, **65**, 046136.

Corberi, F., Castellano, C., Lippiello, E., & Zannetti, M. 2002b. *Universality of the off-equilibrium response function in the kinetic Ising chain.* Phys. Rev. E, **65**, 066114.

Corberi, F., Lippiello, E., & Zannetti, M. 2003a. *Comment on "Aging, Phase Ordering, and Conformal Invariance."* Phys. Rev. Lett., **90**, 099601.

Corberi, F., Lippiello, E., & Zannetti, M. 2003b. *Scaling of the linear response function from zero-field-cooled and thermoremanent magnetization in phase-ordering kinetics.* Phys. Rev. E, **68**, 046131.

Corberi, F., Castellano, C., Lippiello, E., & Zannetti, M. 2004. *Generic features of the fluctuation dissipation relation in coarsening systems.* Phys. Rev. E, **70**, 017103.

Crisanti, A., & Leuzzi, L. 2004. *Spherical* 2 + *p spin-glass model: An exactly solvable model for glass to spin-glass transition.* Phys. Rev. Lett., **93**, 217203.

Crisanti, A., & Leuzzi, L. 2005. *Stable solution of the simplest spin model for inverse freezing.* Phys, Rev. Lett., **95**, 087201.

Crisanti, A., & Leuzzi, L. 2006. *Spherical* 2 + *p spin-glass model: An analytically solvable model with a glass-to-glass transition.* Phys. Rev. B, **73**, 014412.

Crisanti, A., & Ritort, F. 2000a. *Activated processes and inherent structure dynamics of finite-size mean-field models for glasses.* Europhys. Lett., **52**, 640.

Crisanti, A., & Ritort, F. 2000b. *Are mean-field spin-glass models relevant for the structural glass transition?* Physica A, **280**, 155.

Crisanti, A., & Ritort, F. 2000c. *Potential energy landscape of finite-size mean-field models for glasses.* Europhys. Lett., **51**, 147.

Crisanti, A., & Ritort, F. 2003. *Topical Review: Violation of the fluctuation dissipation theorem in glassy systems: basic notions and the numerical evidence.* J. Phys. A: Math. Gen., **36**, 181.

Crisanti, A., & Ritort, F. 2004. *Intermittency of glassy relaxation and the emergence of a non-equilibirum spontaneous measure in the aging regime.* Europhys. Lett., **66**, 253.

Crisanti, A., & Sommers, H.J. 1992. *The spherical p-spin interaction spin-glass model - the statics.* Z. Phys. B, **87**, 341.

Crisanti, A., & Sommers, H.J. 1995. *Thouless-Anderson-Palmer approach to the spherical p-spin spin glass model.* J. Phys. I (France), **5**, 805–813.

Crisanti, A., Horner, H., & Sommers, H.J. 1993. *The spherical p-spin interaction spin-glass model - the dynamics.* Z. Phys. B, **92**, 257.

Crisanti, A., Ritort, F., Rocco, A., & Sellitto, M. 2000. *Inherent structures and nonequilibrium dynamics of one-dimensional constrained kinetic models: A comparison study.* J. Chem. Phys., **113**, 10615.

Crisanti, A., Leuzzi, L., & Rizzo, T. 2003. *The Complexity of the Spherical p-spin spin glass model, revisited.* Eur. Phys. J. B, **36**, 129–136.

Crisanti, A., Leuzzi, L., & Rizzo, T. 2005. *Complexity in mean-field spin glass models: The Ising p-spin.* Phys. Rev. B, **71**, 094202.

Cugliandolo, L., Kurchan, J., & Peliti, L. 1997. *Energy flow, partial equilibration, and effective temperatures in systems with slow dynamics.* Phys. Rev. E, **55**, 3898.

Cugliandolo, L.F. 2003. *Dynamics of glassy systems. In:* Barrat, J.-L., Feigelmann, M., Kurchan, J., & Dalibard, J. (eds), *Slow Relaxation and Nonequilibrium Dynamics in Condensed Matter - Les Houches Session LXXVII.* EDP Sciences, Springer-Verlag (Berlin).

Cugliandolo, L.F., & Kurchan, J. 1993. *Analytical solution of the off-equilibrium dynamics of a long-range spin-glass model.* Phys. Rev. Lett., **71**, 173.

Cugliandolo, L.F., & Kurchan, J. 1994. *On the out-of-equilibrium relaxation of the Sherrington-Kirkpatrick model.* J. Phys. A: Math. Gen., **27**, 5749.

Cugliandolo, L.F., & Kurchan, J. 1995. *Weak-ergodicity breaking in mean-field spin-glass models.* Phil. Mag., **71**, 501.

Cugliandolo, L.F., & Kurchan, J. 1999. *Thermal properties of slow dynamics.* Physica A, **263**, 242–251.

Cugliandolo, L.F., Grempel, D.R., Kurchan, J., & Vincent, E. 1999. *A search for fluctuation-dissipation theorem violations in spin-glasses from susceptibility data*. Europhys. Lett., **48**, 699–705.

Cugliandolo, L.F., Lozano, G., & Lozza, H. 2004. *Memory effects in classical and quantum mean-field disordered models*. Eur. Phys. J. B, **41**, 87.

Cummins, H.Z. 1999. *The liquid-glass transition: A mode-coupling perspective*. J. Phys.: Cond. Matt., **11**, A95.

Daley, D.J., & Gani, J. 2001. *Epidemic Modeling: An Introduction*. Cambridge University Press (Cambridge, U.K.).

D'Anna, G., Mayor, P., Barrat, A., Loreto, V., & Nori, F. 2003. *Observing Brownian motion in vibration-fluidized granular matter*. Nature, **424**, 909.

Das, S.P. 2004. *Mode-coupling theory and the glass transition in supercooled liquids*. Rev. Mod. Phys., **76**, 785.

Das, S.P., & Mazenko, G.F. 1986. *Fluctuating nonlinear hydrodynamics and the liquid-glass transition*. Phys. Rev. A, **34**, 2265–2282.

Das, S.P., Mazenko, G.F., Ramaswamy, S., & Toner, J.J. 1985. *Hydrodynamic theory of the glass transition*. Phys. Rev. Lett., **54**, 118–121.

Davies, R.O, & Jones, G.O. 1953. *Thermodynamic and kinetic properties of glasses*. Adv. Phys., **2**, 370–410.

Dawson, K., Foffi, G., Fuchs, M., Götze, W., Sciortino, F., Sperl, M., Tartaglia, P., Voigtmann, Th., & Zaccarelli, E. 2000. *Higher-order glass-transition singularities in colloidal systems with attractive interactions*. Phys. Rev. E, **63**, 011401.

De Dominicis, C., & Giardina, I. 2006. *Random Fields and Spin Glasses*. Cambridge University Press (Cambridge, U.K.).

de Gennes, P.G. 1979. *Scaling Concepts in Polymer Physics*. Cornell University Press (New York).

de Groot, S.R., & Mazur, P. 1962. *Non-Equilibrium Thermodynamics*. North-Holland (Amsterdam, The Netherlands).

de Jong, B.H.W.S. 2002. *Glass*. Page 365 of *Ullmann's Encyclopedia of Industrial Chemistry*. Wiley-VCH (Weinheim, Germany).

Deb, S.K., Wilding, M., Somayazulu, M., & McMillan, P.F. 2001. *Pressure-induced amorphization and an amorphous-amorphous transition in densified porous silicon*. Nature, **414**, 528.

Debendetti, P.G., & Stillinger, F.H. 2001. *Supercooled liquids and the Glass Transition*. Nature, **410**, 259–267.

Debenedetti, P.G., Stillinger, F.H., & Truskett, T.M. 1999. *The Equation of state of an energy landscape*. J. Phys. Chem., **103**, 7390–7397.

Derrida, B. 1980. *Random-energy model: Limit of a family of disordered models*. Phys. Rev. Lett., **45**, 79.

Derrida, B. 1981. *Random-energy model: An exactly solvable model of disordered systems*. Phys. Rev. B, **24**, 2613.

di Leonardo, R., Angelani, L., Parisi, G., & Ruocco, G. 2000. *Off-Equilibrium effective temperature in monatomic Lennard-Jones glass*. Phys. Rev. Lett., **84**, 6054.

Di Marzio, E.A. 1974. *Validity of the Ehrenfest relation for a system with more than one order parameter.* J. App. Phys., **45**, 4143.

Di Marzio, E.A. 1981. *Equilibrium theory of glasses.* Ann. New York Acad. Sci., **371**, 1.

Di Marzio, E.A., & Gibbs, J.H. 1958. *Chain stiffness and the lattice theory of polymer phases.* J. Chem. Phys., **28**, 807.

Doi, M., & Edwards, S.F. 1986. *The Theory of Polymer Dynamics.* Oxford University Press (Oxford, U.K.).

Donth, E. 2001. *The Glass Transition.* Springer (Berlin).

Drmota, M., Gardy, D., & Gittenberger, B. 2001. *A unified presentation of some urn models.* Algorithmica, **29**, 120–147.

Drouffe, J.M., Godréche, C., & Camia, F. 1998. *A simple stochastic model for the dynamics of condensation.* J. Phys. A: Math. Gen., **31**, L19.

Eckert, T., & Bartsch, E. 2002. *Re-entrant glass transition in a colloid-polymer mixture with depletion attractions.* Phys. Rev. Lett., **89**, 125701.

Ediger, M.D. 2000. *Spatially etherogeneous dynamics in supercooled liquids.* Annu. Rev. Phys. Chem., **51**, 99–128.

Ehrenfest, P. 1933. *Phasenumwandlungen im ueblichen und erweiterten Sinn, klassifiziert nach dem entsprechenden Singularitaeten des thermodynamischen Potentiales.* Proc. Roy. Acad. Amsterdam, **36**, 154.

Ehrenfest, P., & Ehrenfest, T. 1907. *Über zwei bekannte Einwände gegen das Boltzmannsche H-Theorem.* Phys. Zeitschrift, **8**, 311.

Emch, G.G., & Liu, C. 2002. *The Logic of Thermostatistical Physics.* Springer (Berlin).

Exartier, R., & Peliti, L. 2000. *Measuring effective temperatures in out-of-equilibrium driven systems.* Eur. Phys. J. B, **16**, 119.

Falcioni, M., & Vulpiani, A. 1995. *The relevance of chaos for the linear response theory.* Physica A, **215**, 481–494.

Falkovich, G., Gawedzki, K., & Vergassola, M. 2001. *Particles and fields in fluid turbulence.* Rev. Mod. Phys., **73**, 913.

Feller, W. 1993. *An Introduction to Probability Theory and Its Applications.* Vol. **1**. Wiley (New York).

Ferry, J.D. 1961. *Viscoelastic Properties of Polymers.* John Wiley & Sons, Inc. (New York).

Fielding, S., & Sollich, P. 2002. *Observable dependence of fluctuation-dissipation relations and effective temperatures.* Phys. Rev. Lett., **88**, 050603.

Forgacs, G., Lipowsky, R., & Nieuwenhuizen, Th.M. 1991. *The behavior of interfaces in ordered and in disordered systems. Page 135 of:* Domb, C., & Lebowitz, J. (eds), *Phase Transitions and Critical Phenomena 14.* Academic (New York).

Franz, S. 2005. *First Steps of a Nucleation Theory in Disordered Systems.* J. Stat. Mech., **5**, P04001.

Franz, S. 2006. *Metastable states, relaxation times and free-energy barriers in finite-dimensional glassy systems.* Europhys. Lett., **73**, 492.

Franz, S., & Parisi, G. 1995. *Recipes for metastable states in spin-glasses.* J. Phys. I (France), **5**, 1401–1415.

Franz, S., & Parisi, G. 1997. *Phase diagram of coupled glassy systems: A mean-field study.* Phys. Rev. Lett., **79**, 2486.

Franz, S., & Ritort, F. 1995. *Dynamical solution of a model without energy barriers.* Europhys. Lett., **31**, 507.

Franz, S., & Ritort, F. 1996. *Glassy mean-field dynamics of the backgammon model.* J. Stat. Phys., **85**, 131.

Franz, S., & Ritort, F. 1997. *Relaxation processes and entropic traps in the backgammon model.* J. Phys. A: Math. Gen., **30**, L359.

Franz, S., & Virasoro, M.A. 2000. *Quasi-equilibrium interpretation of ageing dynamics.* J. Phys. A: Math. Gen., **33**, 891.

Fredrickson, G.H., & Andersen, H.C. 1984. *Kinetic Ising model of the glass transition.* Phys. Rev. Lett., **53**, 1244.

Fulcher, G.S. 1925. *Analysis of recent measurements of the viscosity of glasses.* J. Am. Ceram. Soc., **8**, 339.

Gani, J. 2004. *Random-allocation and urn models.* J. Appl. Probab., **41**, 313.

Gardner, E. 1985. *Spin glasses with p-spin interactions.* Nucl. Phys. B, **257**, 747.

Gardy, D., & Louchard, G. 1995. *Dynamic analysis of some relational databases parameters.* Theoretical Computer Science, **144**, 125–159.

Garriga, A., & Ritort, F. 2001a. *Heat transfer and Fourier's law in off-equilibirum systems.* Eur. Phys. J. B, **21**, 115.

Garriga, A., & Ritort, F. 2001b. *Validity of the zero-thermodynamic law in off-equilibrium coupled harmonic oscillators.* Eur. Phys. J. B, **20**, 105.

Garriga, A., & Ritort, F. 2005. *Mode-dependent nonequilibrium temperature in aging systems.* Phys. Rev. E, **72**, 031505.

Gibbs, J.H. 1956. *Nature of the glass transition in polymers.* J. Chem. Phys., **25**, 185.

Gibbs, J.H., & Di Marzio, E.A. 1958. *Nature of the glass transition and the glassy state.* J. Chem. Phys., **28**, 373.

Giovambattista, N., Stanley, H.E., & Sciortino, F. 2003. *Potential-energy landscape study of the amorphous-amorphous transformation in H_2O.* Phys. Rev. Lett., **91**, 115504.

Glauber, R.J. 1963. *Time-dependent statistics of the Ising model.* J. Math. Phys., **4**, 294.

Godrèche, C., & Luck, J.M. 1996. *Long-time regime and scaling of correlations in a simple model with glassy behavior.* J. Phys. A: Math. Gen., **29**, 1915.

Godrèche, C., & Luck, J.M. 1997. *Non-equilibrium dynamics of a simple stochastic model.* J. Phys. A: Math. Gen., **30**, 6245.

Godrèche, C., & Luck, J.M. 1999. *Correlation and response in the backgammon model: The Ehrenfest legacy.* J. Phys. A: Math. Gen., **32**, 6033.

Godrèche, C., & Luck, J.M. 2000. *Response of non-equilibrium systems at criticality: Exact results for the Glauber-Ising chain.* J. Phys. A: Math. Gen., **33**, 1151.

Godrèche, C., & Luck, J.M. 2001. *Nonequilibrium dynamics of the zeta urn models.* Eur. Phys. J. B, **23**, 473.

Godrèche, C., & Luck, J.M. 2002. *Nonequilibrium dynamics of urn models.* J. Phys: Cond. Matt., **14**, 1601.

Godrèche, C., & Luck, J.M. 2005. *Dynamics of the condensate in zero-range processes.* J. Phys. A: Math. Gen., **38**, 7215.

Godrèche, C. Bouchaud, J.P., & Mézard, M. 1995. *Entropy barriers and slow relaxation in some random walk models.* J. Phys. A: Math. Gen., **28**, L603.

Goldstein, M. 1969. *Viscous liquid and the glass transition: a potential energy barrier picture.* J. Chem. Phys., **51**, 3728–3739.

Goldstein, M. 1973. *Viscous liquids and the glass transition. IV. Thermodynamic equations and the transition.* J. Chem. Phys., **77**, 667.

Goldstein, M. 1976. *Viscous liquids and the glass transition. V. Sources of the excess heat of the liquid.* J. Chem. Phys., **64**, 4767.

Götze, W. 1991. *Liquids, freezing and glass transition. In:* Hansen, J.P., Levesque, D., & Zinn-Justin, J. (eds), *Les Houches Session 1989.* North Holland (Amsterdam).

Götze, W. 1999. *Recent tests of the mode-coupling theory for glassy dynamics.* J. Phys.: Cond. Matt., **11**, A1.

Götze, W.G. 1984. *Some aspects of phase transitions described by the self consistent current relaxation theory.* Z. Phys. B, **56**, 139–154.

Götze, W.G. 1985. *Properties of the glass instability treated within a mode coupling theory.* Z. Phys. B, **60**, 195–203.

Götze, W.G., & Sjögren, L. 1992. *Relaxation processes in supercooled liquids.* Rep. Prog. Phys., **55**, 241–376.

Greer, A.L. 2000. *Condensed matter: Too hot to melt.* Nature, **404**, 134.

Griffiths, R.B. 1964. *Nonanalytic behavior above the critical point in a random Ising ferromagnet.* Phys. Rev. Lett., **23**, 17.

Grigera, T.S., & Israeloff, N.E. 1999. *Observation of fluctuation-dissipation-theorem violations in a structural glass.* Phys. Rev. Lett., **83**, 5038.

Grigera, T.S., Martin-Mayor, V., Parisi, G., & Verrocchio, P. 2004. *Asymptotic aging in structural glasses.* Phys. Rev. B, **70**, 014202.

Gross, D.J., & Mézard, M. 1984. *The simplest spin glass.* Nucl. Phys. B, **240**, 431.

Gross, D.J., Kanter, I., & Sompolinsky, H. 1985. *Mean-field theory of the Potts glass.* Phys. Rev. Lett., **55**, 304.

Gupta, P.K. 1980. *Fictive pressure effects in structural relaxation.* J. Non-Cryst. Solids, **102**, 231.

Gupta, P.K., & Moynihan, C.T. 1976. *Prigogine-Defay ratio for systems with more than one order parameter.* J. Chem. Phys., **65**, 4136.

Hall, R.W., & Wolynes, P.G. 2003. *Microscopic theory of network glasses.* Phys. Rev. Lett., **90**, 085505.

Hansen, J.-P., & Yip, S. 1995. *Molecular dynamics investigations of slow relaxations in supercooled liquids.* Transp. Theory Stat. Phys., **24**, 1149.

Hansen, J.P., & McDonald, I.R. 2006. *Theory of Simple Liquids.* Academic Press (London).

Haus, J.W., & Kehr, K.W. 1987. *Diffusion in regular and disordered lattices.* Phys. Rep., **150**, 263–406.

Haus, J.W., Kehr, K.W., & Lyklema, J.W. 1982. *Diffusion in a disordered medium.* Phys. Rev. B, **25**, 2905.

Herisson, D., & Ocio, M. 2002. *Fluctuation-dissipation ratio of a spin glass in the aging regime.* Phys. Rev. Lett., **88**, 257202.

Hernandez-Suarez, C.M., & Castillo-Chavez, C. 2000. *Urn models and vaccine efficacy estimation.* Stat. Med., **19**, 827.

Hess, F.G. 1954. Am. Math. Mon., **61**, 323.

Heuer, A., & Büchner, S. 2000. *Why is the density of inherent structures of a Lennard-Jones-type system Gaussian?* J. Phys.: Cond. Matt., **12**, 6535.

Hodge, I.M. 1994. *Enthalpy relaxation and recovery in amorphous materials.* J. Non-Cryst. Solids, **169**, 211.

Hodge, I.M., & O'Reilly, J.M. 1999. *Nonlinear kinetic and thermodynamic properties of monomeric organic glasses.* J. Phys. Chem. B, **103**, 4171.

Hohenberg, P.C., & Shramian, B.I. 1989. *Chaotic behavior of an extended system.* Physica D, **37**, 109–115.

Hoppe, F.M. 1987. *The sampling theory of neutral alleles and an urn model in population genetics.* J. Math. Biol., **25**, 123.

Horner, H. 1992a. *Dynamics of learning and generalization in a binary perceptron model.* Z. Phys. B, **87**, 371.

Horner, H. 1992b. *Dynamics of learning for the binary perceptron problem.* Z. Phys. B, **86**, 291.

Jabbari-Farouji, S., Mizuno, D., Atakhorrami, M., MacKintosh, F.C., Schmidt, C.F., Eiser, E., Wegdam, G.H., & Bonn, D. 2007. *Fluctuation-dissipation theorem in an aging colloidal glass* . Phys. Rev. Lett. , **98**, 108302.

Jack, R.L., Berthier, L., & Garrahan, J.P. 2006. *Fluctuation-Dissipation relations in plaquette spin systems with multi-stage relaxation.* J. Stat. Mech., P12005.

Jäckle, J. 1989. *On the applicability of the mode-coupling theory of glass transitions to good glass formers.* J. Phys.: Cond. Matt., **1**, 267.

Johari, G.P. 2000. *A resolution for the enigma of a liquid's configurational entropy-molecular kinetics relation.* J. Chem. Phys., **112**, 8958.

Johari, G.P., & Goldstein, M. 1971. *Viscous liquids and the glass transition. III. secondary relaxations in aliphatic alcohols and other nonrigid molecules.* J. Chem. Phys., **55**, 4245.

Josserand, C., Tkachenko, A., Mueth, D.M., & Jaeger, H.M. 2000. *Memory Effects in Granular Materials* . Phys. Rev. Lett., **85**, 3632.

Kac, M. 1947. *Random Walk and the theory of Brownian motion.* Am. Math. Mon., **54**, 369.

Kac, M. 1959. *Probability and Related Topics in Physical Sciences.* Interscience (New York).

Kac, M., & Logan, J. 1987. *Fluctuations. Page 1 of:* Montroll, E.W., & Lebowitz, J.L. (eds), *Fluctuation Phenomena.* North-Holland (Amsterdam).

Kauzmann, W. 1948. *The nature of the glassy state and the behavior of liquids at low temperatures.* Chem. Rev., **43**, 219.

Keesom, W.H. 1933. *On the jump in the expansion coefficient of liquid helium in passing the lambda-point.* Proc. Roy. Acad. Amsterdam, **36**, 147.

Keyes, T. 2000. *Entropy, dynamics, and instantaneous normal modes in a random energy model.* Phys. Rev. E, **62**, 7905–7908.

Keyes, T., & Chowdhary, J. 2004. *Potential-energy-landscape-based extended van der Waals equation.* Phys. Rev. E, **69**, 041104.

Kim, B.J., Jeon, G.S., & Choi, M.Y. 1996. *Comment on "Glassiness in a Model without Energy Barriers."* Phys. Rev. Lett., **76**, 4648.

Kirkpatrick, T. R., & Wolynes, P. G. 1987a. *Connections between some kinetic and equilibrium theories of the glass transition.* Phys. Rev. A, **35**, 3072.

Kirkpatrick, T.R., & Thirumalai, D. 1987a. *Dynamics of the structural glass transition and the p-spin interaction spin-glass model.* Phys. Rev. Lett., **58**, 2091.

Kirkpatrick, T.R., & Thirumalai, D. 1987b. *p-spin-interaction spin-glass models: Connections with the structural glass problem.* Phys. Rev. B, **36**, 5388.

Kirkpatrick, T.R., & Wolynes, P.G. 1987b. *Stable and metastable states in mean-field Potts and structural glasses.* Phys. Rev. B, **36**, 8552.

Kirkpatrick, T.R., Thirumalai, D., & Wolynes, P.G. 1989. *Scaling concepts for the dynamics of viscous liquids near an ideal glassy state.* Phys. Rev. A, **40**, 1045.

Kivelson, D., Kivelson, S.A., Zhao, X.L., Nussinov, Z., & Tarjus, G. 1995. *A thermodynamic theory of supercooled liquids.* Physica A, **219**, 27–38.

Kivelson, D., Tarjus, G., Zhao, X., & Kivelson, S.A. 1996. *Fitting of viscosity: Distinguishing the temperature dependences predicted by various models of supercooled liquids.* Phys. Rev. E, **53**, 751.

Kivelson, D., Tarjus, G., & Kivelson, S.A. 1997. *A viewpoint, model and theory for supercooled liquids.* Prog. Teor. Phys. Supp., **126**, 289–299.

Kob, W., & Andersen, H.C. 1993. *Relaxation dynamics in a lattice gas: A test of the mode-coupling theory of the ideal glass transition.* Phys. Rev. E, **47**, 3281.

Kob, W., & Andersen, H.C. 1994. *Scaling behavior in the α-relaxation regime of a supercooled Lennard-Jones mixture.* Phys. Rev. Lett., **73**, 1376.

Kob, W., & Andersen, H.C. 1995a. *Testing mode-coupling theory for a supercooled binary Lennard-Jones mixture. I: The van Hove correlation function.* Phys. Rev. E, **51**, 4626.

Kob, W., & Andersen, H.C. 1995b. *Testing mode-coupling theory for a supercooled binary Lennard-Jones mixture. II. Intermediate scattering function and dynamic susceptibility.* Phys. Rev. E, **52**, 4134.

Kob, W., Sciortino, S., & Tartaglia, P. 2000. *Aging as dynamics in configuration space.* Europhys. Lett., **49**, 590.

Kohlrausch, R. 1847. *Nachtrag ber die Elastische Nachwirkung beim Cocon- und Glasfaden, und die Hygroskopische Eigenschaft des Ersteren.* Ann. Phys. (Leipzig), **12**, 392.

Kovacs, A. J. 1963. *Glass transition in amorphous polymers: A phenomenological study.* Adv. Polym. Sci., **3**, 394.

Kroon, M., Wegdam, G.H., & Sprik, R. 1996. *Dynamic light scattering studies on the sol-gel transition of a suspension of anisotropic colloidal particles.* Phys. Rev. E, **54**, 6541.

Krzakala, F. 2005. *Glassy Properties of the Kawasaki Dynamics of Two-Dimensional Ferromagnets.* Phys. Rev. Lett., **94**, 077204.

Kubo, R. 1957. *Statistical-mechanical theory of irreversible processes. I. general theory and simple applications to magnetic and conduction problems.* J. Phys. Soc. Jpn., **12**, 570.

Kubo, R. 1985. *Statistical Physics II: Non Equilibrium Statistical Physics.* Springer-Verlag (Berlin).

Kurchan, J. 2005. *In and out of equilibrium.* Nature, **433**, 222.

Kurchan, J., Parisi, G., & Virasoro, M.A. 1993. *Barriers and metastable states as saddle points in the replica approach.* J. Phys. I (France), **3**, 1819–1838.

La Nave, E., Mossa, S., & Sciortino, F. 2002. *Potential energy landscape equation of state.* Phys. Rev. Lett., **88**, 225701.

La Nave, E., Sciortino, F., Tartaglia, P., De Michele, C., & Mossa, S. 2003a. *Numerical evaluation of the statistical properties of a potential energy landscape.* J. Phys.: Cond. Matt., **15**, S1085.

La Nave, E., Sciortino, F., Tartaglia, P., Shell, M.S., & Debenedetti, P.G. 2003b. *Test of nonequilibrium thermodynamics in glassy systems: the Soft-Sphere Case.* Phys. Rev. E, **68**, 032103.

Landau, L.D., & Lifshitz, E.M. 1980. *Course of Theoretical Physics. Vol. 5 Statistical Physics.* Pergamon Press (Oxford, U.K.).

Lefloch, F., Hammann, J., Ocio, M., & Vincent, E. 1992. *Can aging phenomena discriminate between the droplet model and a hierarchical description in spin glasses?* Europhys. Lett., **18**, 647.

Leonard, S., Mayer, P., Sollich, P., Berthier, L., & Garrahan, J.P. 2007. *Non-equilibrium dynamics of spin facilitated glass models. In:* Lefèvre, A., & Biroli, G. (eds), *JSTAT Focus Issue: Dynamics of Nonequilibrium Systems.* IOP (Bristol) and SISSA (Trieste).

Leutheusser, E. 1984. *Dynamical model of the liquid-glass transition.* Phys. Rev. A, **29**, 2765.

Leuzzi, L. 2002. Thermodynamics of Glassy Systems: Glasses, Spin-Glasses and Optimization. Universiteit van Amsterdam (Amsterdam).

Leuzzi, L. 2007. *Spin-glass model for inverse freezing.* Phil. Mag. B, **87**, 543–551.

Leuzzi, L., & Nieuwenhuizen, Th.M. 2001a. *Effective temperatures in an exactly solvable model for a fragile glass.* Phys. Rev. E, **64**, 011508.

Leuzzi, L., & Nieuwenhuizen, Th.M. 2001b. *Inherent structures in models for fragile and strong glass.* Phys. Rev. E, **64**, 066125.

Leuzzi, L., & Nieuwenhuizen, Th.M. 2002. *Exactly solvable model glass with facilitated dynamics.* J. Phys.: Cond. Matt., **14**, 1637.

Leuzzi, L., & Ritort, F. 2002. *Disordered backgammon model.* Phys. Rev. E, **65**, 056125.

Lewis, L.J., & Wahnström, G. 1993. *Relaxation of a molecular glass at intermediate times.* Solid State Commun., **86**, 295.

Lewis, L.J., & Wahnström, G. 1994a. *Molecular-dynamics study of supercooled orthoterphenyl.* Phys. Rev. E, **50**, 3865.

Lewis, L.J., & Wahnström, G. 1994b. *Rotational dynamics in ortho-terphenyl: A microscopic view.* J. Non-Cryst. Solids, **172-174**, 69–76.

Lifshitz, I.M. 1964. *The energy spectrum of disordered systems.* Adv. in Phys., **13**, 483.

Lipowsky, A. 1997. *Absorption time in certain urn models.* J. Phys. A: Math. Gen., **30**, L91.

Lippiello, E., & Zannetti, M. 2000. *Fluctuation dissipation ratio in the one-dimensional kinetic Ising model.* Phys. Rev. E, **61**, 3369–3374.

Lubchenko, V., & Wolynes, G.P. 2007. *Theory of structural glasses and supercooled liquids.*

Ann. Rev. Phys. Chem., **58**, 235–266.

Lubchenko, V., & Wolynes, P.G. 2003. *Barrier softening near the onset of nonactivated transport in supercooled liquids: Implications for establishing detailed connection between thermodynamic and kinetic anomalies in supercooled liquids.* J. Chem. Phys., **119**, 9088.

Lubchenko, V., & Wolynes, P.G. 2004. *Theory of aging in structural glasses.* J. Chem. Phys., **121**, 2852.

Macfarlane, A., & Martin, G. 2002. *The Glass Bathyscaphe. How Glass Changed the World.* Profile Books, Ltd (London).

Maggi, C. 2006. *Relazione di fluttuazione-dissipazione generalizzata in un sistema termodinamico fuori dall'equilibrio.* Master thesis, Department of Physics, University of Rome *Sapienza.*

Marinari, E., Parisi, G., & Ritort, F. 1994a. *Replica field theory for deterministic models: I. Binary sequences with low autocorrelation.* J. Phys. A: Math. and Gen., **27**, 7615.

Marinari, E., Parisi, G., & Ritort, F. 1994b. *Replica field theory for deterministic models. II. A non-random spin glass with glassy behaviour.* J. Phys. A: Math. and Gen., **27**, 7647.

Marinari, E., Parisi, G., Ricci-Tersenghi, F., & Ruiz-Lorenzo, J.J. 1998. *Violation of the fluctuation-dissipation theorem in finite-dimensional spin glasses.* J. Phys. A: Math. Gen., **31**, 2611–2620.

Martinez, L.-M., & Angell, C.A. 2001. *A thermodynamic connection to the fragility of glass-forming liquids.* Nature, **410**, 663.

Matsuoka, S. 1992. *Relaxation Phenomena in Polymers.* Hanser (Munich).

Mayer, P., & Sollich, P. 2005. *Observable dependent quasiequilibrium in slow dynamics.* Phys. Rev. E, **71**, 046113.

Mayer, P., Berthier, L., Garrahan, J.P., & Sollich, P. 2003. *Fluctuation-dissipation relations in the nonequilibrium critical dynamics of Ising models.* Phys. Rev. E, **68**, 016116.

Mayer, P., Berthier, L., Garrahan, J.P., & Sollich, P. 2004. *Reply to* Comment on *"Fluctuation-dissipation relations in the nonequilibrium critical dynamics of Ising models."* Phys. Rev. E, **70**, 018102.

Mayer, P., Léonard, S., Berthier, L., Garrahan, J.P., & Sollich, P. 2006. *Activated aging dynamics and negative fluctuation-dissipation ratios.* Phys. Rev. Lett., **96**, 030602.

McGrath, R., & Frost, A.C. 1937. *Glass in Architecture and Decoration.* Architectural Press (London).

McKenna, G.B. 1989. *Glass formation and glassy behavior. Pages 311–362 of:* Booth, C., & Price, C. (eds), *Comprehensive Polymer Science: The Synthesis, Characterization, Reactions & Applications of Polymers. Vol. 2. Polymer Properties.* Pergamon (Oxford, U.K.).

Mézard, M. 1999. *How to compute the thermodynamics of a glass using a cloned liquid.* Physica A, **265**, 352–369.

Mézard, M., & Parisi, G. 1996. *A tentative replica study of the glass transition.* J. Phys. A: Math. and Gen., **29**, 6515–6524.

Mézard, M., & Parisi, G. 1999a. *A first-principle computation of the thermodynamics of glasses.* J. Chem. Phys., **111**, 1076.

Mézard, M., & Parisi, G. 1999b. *Thermodynamics of glasses: A first principles computation.* Phys. Rev. Lett., **82**, 747.

Mézard, M., & Parisi, G. 2000. *Statistical physics of structural glasses.* J. Phys.: Cond. Matt., **12**, 6655.

Mézard, M., & Parisi, G. 2001. *The Bethe lattice spin glass revisited.* Eur. Phys. J. B, **20**, 217.

Mézard, M., Parisi, G., & Virasoro, M. 1987. *Spin Glass Theory and Beyond.* World Scientific (Singapore).

Mishima, O., & Stanley, H.E. 1998. *Decompression-induced melting of ice IV and the liquidliquid transition in water.* Nature, **392**, 164.

Mishima, O., Calvert, L.D., & Whalley, E. 1985. *An apparently first-order transition between two amorphous phases of ice induced by pressure.* Nature, **314**, 76.

Monasson, R. 1995. *Structural glass transition and the entropy of the metastable states.* Phys. Rev. Lett., **75**, 2847–2850.

Montanari, A., Müller, M., & Mézard, M. 2004. *Phase diagram of random heteropolymers.* Phys. Rev. Lett., **92**, 185509.

Monthus, C., & Bouchaud, J.P. 1996. *Models of traps and glass phenomenology.* J. Phys. A: Math. Gen., **29**, 3847.

Mossa, S., & Sciortino, F. 2004. *Crossover (or Kovacs) effect in an aging molecular liquid.* Phys. Rev. Lett., **92**, 045504.

Mossa, S., La Nave, E., Stanley, H.E., Donati, C., Sciortino, F., & Tartaglia, P. 2002. *Dynamics and configurational entropy in the Lewis-Wahnström model for supercooled orthoterphenyl.* Phys. Rev. E, **65**, 41205.

Moynihan, C.T., & Angell, C.A. 2000. *Bond lattice or excitation model analysis of the configurational entropy of molecular liquids.* J. Non-Cryst. Solids, **274**, 131.

Moynihan, C.T., & Lesikar, A.V. 1981. *Structure and mobility in molecular and atomic glasses.* Ann. NY Acad. Sci., **371**, 152.

Moynihan, C.T., Easteal, A.J., Debolt, M.A., & Tucker, J.C. 1976. *Dependence of the fictive temperature of glass on cooling rate.* J. Am. Ceram. Soc., **59**, 16.

Müller, M., Mézard, M., & Montanari, A. 2004. *Glassy phases in random heteropolymers with correlated sequences.* J. Chem. Phys., **92**, 11233.

Mydosh, J.A. 1993. *Spin Glasses: An Experimental Introduction.* Taylor and Francis (London).

Mysen, B.O., & Richet, P. 2005. *Silicate Glasses And Melts.* Elsevier Science & Technology (Oxford, U.K.).

Narayanaswamy, O.S. 1971. *A model of structural relaxation in glass.* J. Am. Ceram. Soc., **54**, 491–498.

Nicodemi, M. 1999. *Dynamical response functions in models of vibrated granular media.* Phys. Rev. Lett., **82**, 3734–3737.

Nieuwenhuizen, Th.M. 1989a. *Griffiths singularities in two-dimensional Ising models: Relation Lifshitz band tails.* Phys. Rev. Lett., **63** , 1760.

Nieuwenhuizen, Th.M. 1989b. *Trapping and Lifshitz tails in random media, self-attracting polymers and the number of distinct sites visited: A renormalized instanton approach in three dimensions.* Phys. Rev. Lett., **62**, 357.

Nieuwenhuizen, Th.M. 1995. *To maximize or not to maximize the free energy of glassy systems.* Phys. Rev. Lett., **74**, 3463.

Nieuwenhuizen, Th.M. 1997a. *Ehrenfest relations at the glass transition: Solution to an old paradox.* Phys. Rev. Lett., **79**, 1317.

Nieuwenhuizen, Th.M. 1997b. *Solvable glassy system: Static versus dynamical transition.* Phys. Rev. Lett., **78**, 3491.

Nieuwenhuizen, Th.M. 1998a. *Thermodynamic description of a dynamical glassy transition.* J. Phys. A: Math. Gen., **31**, L201.

Nieuwenhuizen, Th.M. 1998b. *Thermodynamics of black holes, an analogy with glasses.* Phys. Rev. Lett., **81**, 2201.

Nieuwenhuizen, Th.M. 1998c. *Thermodynamics of the glassy state: effective temperature as an additional system parameter.* Phys. Rev. Lett., **80**, 5580.

Nieuwenhuizen, Th.M. 1999. *Solvable model for the standard folklore of the glassy state.* Unpublished, arXiv:cond–mat/9911052.

Nieuwenhuizen, Th.M. 2000. *Thermodynamic picture of the glassy state gained from exactly solvable models.* Phys. Rev. E, **61**, 267.

Nieuwenhuizen, Th.M., & Ernst, M.H. 1985. *Excess noise in a hopping model for a resistor with quenched disorder.* J. Stat. Phys., **41**, 773.

Nieuwenhuizen, Th.M., & Luck, J.M. 1989. *Singularities in spectra of disordered systems: An instanton for arbitrary dimension and randomness.* Europhys. Lett., **9**, 407.

Nyquist, H. 1928. *Thermal agitation of electric charge in conductors.* Phys. Rev., **32**, 110.

O'Reilly, J.M. 1962. *The effect of pressure on glass temperature and dielectric relaxation time of polyvinyl acetate.* J. Polym. Sci., **57**, 429.

O'Reilly, J.M. 1977. *Conformational specific heat of polymers.* J. Appl. Phys., **48**, 4043–4048.

Oxtoby, D. 1991. *Crystallization of liquids: A density functional approach. Page 145 of:* Hansma, J.P., Levesque, D., & Zinn-Justin, J. (eds), *Liquids, Freezing and the Glass Transition.* North-Holland (Amsterdam).

Parisi, G. 1980. *A sequence of approximated solutiona to the S-K model for spin glasses.* J. Phys. A: Math. Gen., **13**, L115.

Parisi, G. 1995. *Gauge theories, spin Glasses and real glasses. In:* Lindström, U. (ed), *The Oscar Klein Centenary.* World Scientific (Singapore).

Parisi, G. 1997a. *Numerical indications for the existence of a thermodynamic transition in binary glasses.* J. Phys. A: Math. and Gen., **30**, 8523.

Parisi, G. 1997b. *Off-equilibrium fluctuation-dissipation relation in fragile glasses.* Phys. Rev. Lett., **79**, 3660.

Parisi, G. 1997c. *Short-time aging in binary glasses.* J. Phys. A: Math. and Gen., **30**, L765.

Parisi, G. 2003. *Glasses, replicas and all that. Page 271 of:* Barrat, J.-L., Feigelmann, M., Kurchan, J., & Dalibard, J. (eds), *Slow Relaxation and Nonequilibrium Dynamics in Condensed Matter.* EDP Sciences, Springer-Verlag (Berlin).

Parisi, G., & Potters, M. 1995. *Mean-field equations for spin models with orthogonal interaction matrices.* J. Phys. A: Math. and Gen., **28**, 5267.

Pham, K.N. 2002. *Multiple glassy states in a simple model system.* Science, **296**, 104.

Pleiming, M. 2004. *Comment on* Fluctuation-dissipation relations in the nonequilibrium critical dynamics of Ising models. Phys. Rev. E, **70**, 018101.

Pliny. 1972. Natural History, books 36-37, *translated by Eichholz, D.E.* Loeb Classical Library (Cambridge, MA).

Poole, P.H., Sciortino, F., Essmann, U., & Stanley, H.E. 1992. *Phase behaviour of metastable water.* Nature, **360**, 324.

Potts, R.B. 1952. *Some generalized order - disorder transformations.* Proc. Camb. Phil. Soc., **48**, 106.

Prados, A. Brey, A.J., & Sanchez-Rey, B. 1997. *Glassy behavior in a simple model with entropy barriers.* Phys. Rev. B, **55**, 6343.

Prigogine, I., & Defay, R. 1954. *Chemical Thermodynamics.* Longmans, Green and Co. (New York).

Puglisi, A., Baldassarri, A., & Loreto, V. 2002. *Fluctuation-dissipation relations in driven granular gases.* Phys. Rev. E, **66**, 061305.

Qin, Q., & McKenna, G.B. 2006. *Correlation between dynamic fragility and glass transition temperature for different classes of glass forming liquids.* J. Non-Cryst. Solids, **352**, 2977.

Ramakrishnan, T.V., & Yussouff, M. 1979. *First-principles order-parameter theory of freezing.* Phys. Rev. B, **19**, 2775–2794.

Rao, K.J. 2002. *Structural Chemistry of Glasses.* Elsevier Science (Amsterdam, The Netherlands).

Rastogi, S., Höhne, G.W.H., & Keller, A. 1999. *Unusual pressure-induced phase behavior in crystalline poly(4-methylpentene-1): Calorimetric and spectroscopic results and further implications.* Macromolecules, **32**, 8897.

Rehage, G., & Oels, H.J. 1977. *Pressure-volume-temperature measurements on atactic polystyrene. A thermodynamic view.* Macromolecules, **10**, 1036.

Ricci-Tersenghi, F. 2003. *Measuring the fluctuation-dissipation ratio in glassy systems with no perturbing field.* Phys. Rev. E, **68**, 065104(R).

Richert, R., & Angell, C.A. 1998. *Dynamics of glass forming liquids. V. On the link between molecular dynamics and configurational entropy.* J. Chem. Phys., **108**, 9016–9026.

Rinn, B., Maas, P., & Bouchaud, J.P. 2000. *Multiple scaling regimes in simple aging models.* Phys. Rev. Lett., **84**, 5403.

Ritort, F. 1995. *Glassiness in model without energy barriers.* Phys. Rev. Lett., **75**, 1190.

Ritort, F. 2004. *Spontaneous relaxation in generalized oscillator models with glassy dynamics.* J. Phys. Chem. B, **108**, 6893.

Ritort, F., & Sollich, P. 2003. *Glassy dynamics of kinetically constrained models.* Advances in Physics, **52**, 219.

Roberts, C.J., Debenedetti, P.G., & Stillinger, F.H. 1999. *Equation of state of the energy landscape of SPC/E water.* J. Phys. Chem. B, **103**, 10258.

Ruocco, G., Sciortino, F., Zamponi, F., De Michele, C., & Scopigno, T. 2004. *Landscapes and fragilities.* J. Chem. Phys., **120**, 10666.

Sastry, S. 2000. *Liquid limits: Glass transition and liquid-gas spinodal boundaries of metastable liquids.* Phys. Rev. Lett., **85**, 590.

Sastry, S. 2001. *The relationship between fragility, configurational entropy and the potential energy landscape of glass-forming liquids.* Nature, **409**, 164–167.

Sastry, S., Debenedetti, P.G., & Stillinger, F.H. 1997. *Statistical geometry of particle packings II. "Weak spots" in liquids.* Phys. Rev. E, **56**, 5533.

Sastry, S., Debendetti, P.G., & Stillinger, F.H. 1998. *Signatures of distinct regimes in the energy landscape of a glass-forming liquid.* Nature, **393**, 554–557.

Scala, A., Starr, F.W., La Nave, E., Sciortino, F., & Stanley, H.E. 2000. *Configurational entropy and diffusivity of supercooled water.* Nature, **406**, 166.

Scherer, G.W. 1986. *Relaxation in Glasses and Composites.* Wiley (New York).

Schmitz, R. 1988. *Fluctuations in nonequilibrium fluids.* Physics Reports, **171**, 1.

Schmitz, R., Dufty, J. W., & De, P. 1993. *Absence of a sharp glass transition in mode coupling theory.* Phys. Rev. Lett., **71**, 2066–2069.

Schreiber, S.J. 2001. *Urn models, replicator processes, and random genetic drift.* SIAM J. Appl. Math., **61**, 2148.

Schulz, M. 1998. *Energy landscape, minimum points, and non-Arrhenius behavior of supercooled liquids.* Phys. Rev. B, **57**, 11319.

Schupper, N., & Shnerb, N.M. 2004. *Spin model for inverse melting and inverse glass transition.* Phys. Rev. Lett., **93**, 037202.

Sciortino, F. 2002. *One liquid, two glasses.* Nature Materials, **1**, 145.

Sciortino, F. 2005. *Potential energy landscape description of supercooled liquids and glasses.* J. Stat. Mech., P05015.

Sciortino, F., & Tartaglia, P. 2001. *Extension of the fluctuation-dissipation theorem to the physical aging of a model glass-forming liquid.* Phys. Rev. Lett., **86**, 107–110.

Sciortino, F., Kob, W., & Tartaglia, P. 1999. *Inherent structure entropy of supercooled liquids.* Phys. Rev. Lett., **83**, 3214–3217.

Sciortino, F., La Nave, E., & Tartaglia, P. 2003. *Physics of the liquid-liquid critical point.* Phys. Rev. Lett., **91**, 155701.

Scully, M.O. 2001. *Extracting work from a single thermal bath via quantum negentropy.* Phys. Rev. Lett., **87**, 220601.

Sellitto, M. 2006. *Inverse freezing in mean-field models of fragile glasses.* Phys. Rev. B, **73**, 180202.

Sen, S.N., & Chaudhary, M. 1985. *Ancient Glass and India.* Indian National Science Academy (New Delhi).

Shell, M.S., Debenedetti, P.G., La Nave, E., & Sciortino. 2003. *Energy landscapes, ideal glasses, and their equation of state.* J. Chem. Phys., **118**, 8821.

Siegert, A.J.F. 1949. *On the Approach to Statistical Equilibrium.* Phys. Rev., **76**, 1708.

Simon, S.L., & McKenna, G.B. 1997. *Interpretation of the dynamic heat capacity observed in glass-forming liquids.* J. Chem. Phys., **107**, 8678.

Singh, Y., Stoessel, J.P., & Wolynes, P.G. 1985. *Hard-spheres gas and the density-functional theory of aperiodic crystals.* Phys. Rev. Lett., **54**, 1059.

Sjölander, A. 1965. *Theory of the neutron scattering. Pages 291–345 of:* Egelstaff, P.A. (ed), *Thermal Neutron Scattering.* Academic Press Inc. (New York).

Sollich, P., Fielding, S., & Mayer, P. 2002. *Fluctuation-dissipation relations and effective temperatures in simple non-mean field systems.* J. Phys.: Cond. Matt., **14**, 1683.

Sommers, H.J., & Dupont, W. 1984. *Distribution of frozen fields in the mean-field theory of spin glasses.* J. Phys. C, **17**, 5785.

Sompolinsky, H. 1981. *Time-dependent order parameters in spin-glasses.* Phys. Rev. Lett., **47**, 935.

Sompolinsky, H., & Zippelius, A. 1982. *Relaxational dynamics of the Edwards-Anderson model and the mean-field theory of spin-glasses.* Phys. Rev. B, **25**, 6860.

Song, C., Wang, P., & Makse, H.A. 2005. *Experimental measurements of an effective temperature for jammed granular materials.* PNAS, **102**, 2299.

Starr, F.W., Sastry, S., La Nave, E., Scala, A. Stanley, H.E., & Sciortino, F. 2001. *Thermodynamic and structural aspects of the potential energy surface of simulated water.* Phys. Rev. E, **63**, 41201.

Steinhardt, P.J., & Ostlund, S. 1987. *The Physics of Quasicrystals.* World Scientific (Singapore).

Stevenson, J.D., & Wolynes, P.G. 2005. *Thermodynamic-kinetic correlations in supercooled liquids: A critical survey of experimental data and predictions of the random first-order transition theory of glasses.* J. Phys. Chem. B, **109**, 15093–15097.

Stillinger, F.H. 1995. *A topographic view of supercooled liquids and glass formation.* Science, **267**, 1935–1939.

Stillinger, F.H., & Weber, T.A. 1982. *Hidden structure in liquids.* Phys. Rev. A, **25**, 978.

Stillinger, F.H., & Weber, T.A. 1983. *Dynamics of structural transitions in liquids.* Phys. Rev. A, **28**, 2408–2416.

Stillinger, F.H., & Weber, T.A. 1984. *Packing Structures and Transitions in Liquids and Solids.* Science, **225**, 983–989.

Stillinger, F.H., Debenedetti, P.G., & Truskett, T.M. 2001. *The Kauzmann paradox revisited.* J. Phys. Chem. B, **105**, 11809–11816.

Struik, L.C.E. 1978. Physical Aging in Amorphous Polymers and Other Materials. Elsevier (Houston).

Subbrarayappa, B.V. 1999. *Chemistry and Chemical Techniques in India.* Center for Studies in Civilization (New Delhi).

Swallon, S.F., Bonvallet, P.A., McMalhon, R.J., & Ediger, M.D. 2003. *Self-diffusion of tris-naphthylbenzene near the glass transition temperature.* Phys. Rev. Lett., **90**, 015901.

Takahasi, H. 1952. *Generalized theory of thermal fluctuations.* J. Phys. Soc. Jpn., **7**, 439.

Tammann, G. 1903. Kristallisieren und Schmelzen. Metzger und Wittig (Leipzig).

Tammann, G., & Hesse, G. 1926. *Die Abhängigkeit der Viscosität von der Temperatur bie unterkühlten Fluessigkeiten.* Z. Anorg. Allgem. Chem., **156**, 245.

Tanaka, H. 1996. *A self-consistent phase diagram for supercooled water.* Nature, **380**, 328.

Tarjus, G., & Kivelson, D. 1995. *Breakdown of the Stokes-Einstein relation in supercooled liquids.* J. Chem. Phys., **103**, 3071.

Tarjus, G., Kivelson, D., & Kivelson, S. 1997. *Frustration-limited domain theory of supercooled liquids and the glass transition. Page 67 of:* Fourkas, J. et al. (ed), *Supercooled Liquids: Advances and Novel Applications*, vol. 676. ACS Symposium Series (Washington, D.C.).

Tarjus, G., Kivelson, S.A., Nussinov, Z., & Viot, P. 2005. *The frustration-based approach of supercooled liquids and the glass transition: A review and critical assessment.* J. Phys.:

Cond. Matt., **17**, R1143.

Thirumalai, D., & Kirkpatrick, T.R. 1988. *Mean-field Potts glass model: Initial-condition effects on dynamics and properties of metastable states.* Phys. Rev. B, **38**, 4881.

Thouless, D.J., Anderson, P.W., & Palmer, R.G. 1977. *Solvable model of a spin glass.* Phil. Mag., **35**, 593.

Tool, AQ. 1946. *Relation between inelastic deformation and thermal expansion of glass in its annealing. range.* J. Amer. Ceram. Soc., **29**, 240.

Torell, L.M. 1982. *Brillouin scattering study of hypersonic relaxation in a $Ca(NO_3)_2$-KNO_3 mixture.* J. Chem. Phys., **76**, 3467.

Toulouse, G. 1977. *Theory of the frustration effect in spin glasses: I.* Commun. Phys., **2**, 115.

Tracht, U., Wilhelm, M., Heuer, A., Feng, H., Schmidt-Rohr, K., & Spiess, H.W. 1998. *Length scale of dynamic heterogeneities at the glass transition determined by multidimensional nuclear magnetic resonance.* Phys. Rev. Lett., **81**, 2727.

Tsiok, O.B., Brazhkin, V.V., Lyapin, A.G., & Khvostantsev, L.G. 1998. *Logarithmic kinetics of the amorphous-amorphous transformations in SiO_2 and GeO_2 glasses under high pressure.* Phys. Rev. Lett., **80**, 999.

Turnbull, D., & Cohen, M.H. 1961. *Free-volume model of the amorphous phase: glass transition.* J. Chem. Phys., **34**, 120.

Turnbull, D., & Cohen, M.H. 1970. *On the free-volume model of the liquid-glass transition.* J. Chem. Phys., **52**, 3038.

Utz, M., Debenedetti, P.G., & Stillinger, F.H. 2001. *Isotropic tensile strength of molecular glasses.* J. Chem. Phys., **114**, 10049.

van Kampen, N.G. 1981. *Stochastic Processes in Physics and Chemistry.* North-Holland (Amsterdam).

van Mourik, P.J., & Coolen, A.C.C. 2001. *Cluster derivation of Parisi's RSB solution for disordered systems.* J. Phys. A: Math. Gen., **34**, L111.

van Ruth, N.J.L., & Rastogi, S. 2004. *Nonlinear changes in specific volume. A route to resolve an entropy crisis.* Macromolecules, **37**, 8191.

Villain, J. 1985. *Equilibrium critical properties of random field systems: New conjectures.* J. Phys. (France), **46**, 1843.

Vincent, E., Hammann, J., Ocio, M., Bouchaud, J.-P., & Cugliandolo, L. 1997. *Slow dynamics and aging in spin glasses. Page 184 of:* Rubi, M. (ed), *Complex Behavior of Glassy Systems.* Lecture Notes in Physics. Springer-Verlag (Berlin).

Vogel, H. 1921. *Temperaturabhängigkeitsgesetz der Viskosität von Flüssigkeiten.* Physik. Z., **22**, 645.

Vollmayr, K., Kob, W., & Binder, K. 1996. *How do the properties of a glass depend on the cooling rate? A computer simulation study of a Lennard-Jones system.* J. Chem. Phys., **105**, 4714.

Wahnström, G., & Lewis, L.J. 1993. *Molecular dynamics simulation of a molecular glass at intermediate times.* Physica A, **201**, 150.

Wales, D.J. 2003. *Energy Landscapes.* Cambridge University Press (Cambridge, U.K.).

Wang, L.M., & Angell, C.A. 2003. *Reply to Comment on "Direct determination of the fragility indices of glassforming liquids by differential scanning calorimetry: Kinetic versus thermodynamic fragilities.*" J. Chem. Phys., **118**, 10353.

Weber, T.A., & Stillinger, F.H. 1985. *Local order and structural translations in amorphous metal-metalloid alloys.* Phys. Rev. B, **31**, 1954.

Weeks, E., Croker, J.C., Levitt, A.C., Schofield, A., & Weitz, D.A. 2000. *Three-dimensional direct imaging of structural relaxation near the colloidal glass transition.* Science, **287**, 627.

Wendorff, J.H., & Fischer, E.W. 1973. *Thermal density fluctuations in amorphous polymers as revealed by small angle X-ray diffraction.* Kolloid-Z. Z. Polym., **251**, 876.

Williams, M.L., Landel, R.F., & Ferry, J.D. 1955. *The temperature dependence of relaxation mechanisms in amorphous polymers and other glass-forming liquids.* J. Am. Chem. Soc., **77**, 3701–3707.

Wolfgardt, M., Baschnagel, J., Paul, W., & Binder, K. 1996. *Entropy of glassy polymer melts: Comparison between Gibbs-DiMarzio theory and simulation.* Phys. Rev. E, **54**, 1535.

Workshop, SPHINX. 2002. *Workshop on Glassy Behaviour of Kinetically Constrained Models. In:* Ritort, F., & Sollich, P. (eds), J. Phys.: Cond. Matt. **14**. IOP Publishing.

Wu, F. Y. 1982. *The Potts model.* Rev. Mod. Phys., **54**, 235.

Xia, X., & Wolynes, P.G. 2000. *Fragility of liquids predicted from the random first order transition theory of glasses.* PNAS, **97**, 2990.

Xia, X., & Wolynes, P.G. 2001. Microscopic theory of heterogeneity and nonexponential relaxation in supercooled liquids. Phys. Rev. Lett., **86**, 5526.

Zaccarelli, E., Foffi, G., Dawson, K.A., Sciortino, F., & Tartaglia, P. 2001. *Mechanical properties of a model of attractive colloidal solutions.* Phys. Rev. E, **63**, 031501.

Zachariasen, W.H. 1932. *The atomic arrangement in glass.* J. Amer. Chem. Soc., **54**, 384.

Zanotto, E.D. 1998. *Do cathedral glasses flow?* Am. J. Phys., **66**, 392.

Zanotto, E.D., & Gupta, P.K. 1999. *Do Cathedral Glasses Flow? - Additional Remarks.* Am. J. Phys., **67**, 260–262.

Index

α mode, 56, 84, 120
α particles, 248
α peak, 20
α process, 7, 17, 19, 23, 79, 90, 100, 134, 181, 232, 256, 273, 299
α relaxation, 164, 184, 269
α timescale, 273
β process, 17–19, 23, 55, 90, 145, 181, 232, 256, 299
p-spin model, 32, 75, 218, 239, 240, 276, 280
    Ising, 276, 288
    spherical, 29, 75, 167, 274, 276, 277, 280, 288

acceleration, 51
activation energy, 8, 165, 232
    effective equilibrium, 51
activation free energy, 293, 301, 306, 312, 313, 317
Adam-Gibbs (AG), 301
    relation, 32, 36, 37, 39, 89, 115, 116, 252, 253
    theory, 37
Adam-Gibbs-Di Marzio theory, 227
adiabatic Ansatz, 196, 200, 201, 208
adiabatic approximation, 90, 107, 114, 126, 127, 129, 147, 150, 172, 174–177, 179–181, 185, 187, 192, 194, 196, 205, 206
aging, 23, 27, 59, 70, 71, 75–79, 89, 96, 97, 112, 127, 131, 135, 142, 164, 166, 183, 211, 245, 255, 259, 265, 315
    covariance, 29
    dynamics, 23, 29, 89, 99, 115, 123, 247, 266, 273
    experiment, 29, 57, 71, 141, 211
    full-, 29
    regime, 27–29, 52, 67, 71, 76, 77, 80, 84, 85, 90, 107, 109, 111, 112, 114, 118, 125, 126, 129–131, 134, 135, 139, 143–146, 181, 183, 184, 193, 226, 245, 248, 263
    setup, 110
    sub-, 29

super-, 29
system, 44, 45, 75, 76, 81, 82, 86, 118, 142, 166, 245, 247
time sector, 81
timescale, 146
amorphous ice, 59
amorphous lattice, 306
amorphous magnets, 274
amorphous material, 25, 227
amorphous packing, 9, 59, 235, 266, 300, 305, 307
amorphous phase, 19, 232, 244, 285, 299
amorphous solid, 21, 268, 298, 299, 306, 311
amorphous state, 232, 284, 298, 310
amorphous structure, 24, 43
amorphous system, 30, 226, 239, 247, 256, 265
amorphous-amorphous transition, 59
annealing, 23
antiferromagnetic bound, 274
antiferromagnetic clustering, 296
antiferromagnetic interaction, 294, 295
Arrhenius, 8, 9, 24, 38, 39, 68, 89, 97, 197, 232, 259, 290, 293
    enhanced, 290
    free energy, 290
    glass formers, 38
    law, 8, 24, 26, 31, 65, 68, 70, 97, 103, 109, 111, 177, 197, 289, 293
    relaxation, 106, 166, 177, 261
    sub-, 197
    super-, 291
atactic polystyrene, 61
atatctic-vinyl polymers, 16
avoided critical point, 291
    theory, 13, 269, 297

backgammon model, 29, 147, 164–167, 171, 179, 181, 184–188, 190, 192–194, 197–199, 201
barrier, 19, 24, 25, 31, 34, 120, 165, 195, 213, 217, 219–221, 229, 276, 280, 281, 298, 300–302, 307, 311, 316, 317
    activation, 317